AF389471

Fatigue and Fracture of Traditional and Advanced Structural Alloys

Fatigue and Fracture of Traditional and Advanced Structural Alloys

Editor

Filippo Berto

MDPI • Basel • Beijing • Wuhan • Barcelona • Belgrade • Manchester • Tokyo • Cluj • Tianjin

Editor
Filippo Berto
Norwegian University of Science and Technology
Norway

Editorial Office
MDPI
St. Alban-Anlage 66
4052 Basel, Switzerland

This is a reprint of articles from the Special Issue published online in the open access journal *Metals* (ISSN 2075-4701) (available at: https://www.mdpi.com/journal/metals/special_issues/fatigue_fracture_structural_alloys).

For citation purposes, cite each article independently as indicated on the article page online and as indicated below:

LastName, A.A.; LastName, B.B.; LastName, C.C. Article Title. *Journal Name* **Year**, *Volume Number*, Page Range.

ISBN 978-3-0365-0366-0 (Hbk)
ISBN 978-3-0365-0367-7 (PDF)

© 2021 by the authors. Articles in this book are Open Access and distributed under the Creative Commons Attribution (CC BY) license, which allows users to download, copy and build upon published articles, as long as the author and publisher are properly credited, which ensures maximum dissemination and a wider impact of our publications.

The book as a whole is distributed by MDPI under the terms and conditions of the Creative Commons license CC BY-NC-ND.

Contents

About the Editor

Filippo Berto has been the international chair in fracture mechanics, fatigue, and structural integrity at the Norwegian University of Science and Technology of Trondheim, Norway, since 2016. He was a professor of machine design at the University of Padua, Italy, between 2006 and 2015. He is chairman of the technical committee ESIS TC15 on Structural Integrity of additive manufactured components of European Structural Integrity Society. Filippo Berto has received several awards and distinctions, such as an Award of Merit (2018) and Wohler medal (2020) from the European Structural Integrity Society. He has received the Stephen Timoshenko Fellow (2018) given by University of Stanford (USA). Filippo Berto is part of the Top Italian Scientists Engineering (Italy). He is editor-in-chief or Editor of several scientific journals, such as, Fatigue and Fracture of Engineering Materials and Structures, Material Design and Processing Communications and Materials. In addition, he belongs to the editorial board of several leading scientific journals in the area of fatigue, fracture and structural integrity, such as Materials and Design, International Journal of Fatigue, Safety Science, Theoretical and Applied Fracture Mechanics, Materials Science and Engineering: A, Advanced Engineering Materials, Polymer Testing, Strength of Materials, among others. He is also collaborating with the Centre of Excellence of additive manufacturing established in Auburn and in particular with Prof. Nima Shamsaei and Dr. Mohsen Seifi who are chairs of additive manufacturing programs in ASTM. Filippo Berto is also a voting member of ASTM F42 on Additive Manufacturing Technologies and ASTM E08 on Metal Fatigue and Fracture. The technical chair of TC15 is also the founder of the series of European conferences on additive manufacturing ESIAM.

Editorial

Fatigue and Fracture of Traditional and Advanced Structural Alloys

Filippo Berto

Department of Mechanical and Industrial Engineering, Norwegian University of Science and Technology, Richard Birkelands vei 2b, 7491 Trondheim, Norway; filippo.berto@ntnu.no; Tel.: +47-73593831

Received: 3 December 2020; Accepted: 4 December 2020; Published: 6 December 2020

Prevention and prediction of unexpected fracture and fatigue failures constitute a key objective in any engineering application. Nevertheless, the increase in complexity in modern structures and components makes this key objective very challenging and sometimes only partially reachable. Catastrophic failures often involve the loss of human lives. Within the last few years, a drastic improvement has been achieved in many of the necessary tools and methods to either avoid or mitigate the consequences of failure. This recent development is strongly associated with the ever-increasing performance of traditional and innovative structural alloys as well as with the capacity of controlling process parameters during the realization and manufacturing of the final structure. There is still an impressive room for improvement in this field. This requires a multidisciplinary pool including materials science, structural analysis, manufacturing technologies, quality control and evaluation, mathematics, physics, and probability and reliability. Furthermore, from a scientific point of view, it is also fundamental to increase our knowledge of concepts and approaches that can account for size and time-scaling effects. The Special Issue scope embraces interdisciplinary work aimed at understanding and deploying physics of fatigue and failure techniques of traditional and advanced structural alloys, advancing experimental and theoretical failure analysis, modelling of the structural response with respect to both local and global failures, and structural design that accounts for scale and time effects in preventing engineering failures. This state of the art will help engineers, designers and people from the academy to have an updated state of the art on this very challenging topic which is nowadays very important due to the advances in manufacturing technologies that allow complex new materials to be fabricated.

Thirteen articles have been published in the present Special Issue of *Metals* encompassing the fields of fracture and fatigue damage, high cycle fatigue, fatigue and creep interaction and fatigue in aggressive corrosive media. This grouping is not hard-bound, and thematic links can be established beyond them, which shall be emphasized in the following presentation.

High cycle fatigue is a traditional topic but still very challenging, as is well described in some of the present contributions. In particular, it becomes more and more critical if large scale effect occurs in the full-scale component also in civil construction and megastructures [1]. Aspects related to microstructure and interaction between fatigue properties and metallurgical properties are also very important for the fatigue design.

Interest in fatigue assessment of steels and different alloys under multiaxial loading has increased continuously in recent years. Likewise, the applications of multiaxial fatigue phenomenon of interest have increased substantially in recent years and our capacity to assess fatigue life under complex loading has substantially improved [2]. To provide as optimum a performance as possible in these high demanding conditions, it is necessary to be aware of the application and proper tools required to perform the fatigue assessment under complex loadings.

An aggressive environment due to corrosion and temperature can be extremely critical for the fatigue life of a structure in their service conditions [3]. Designers must consider corrosion in service and high temperature for a proper design against fatigue loadings. Corrosion is also undesirable for reasons related to a safe and economic use of a structure during its service life. Some recent advances of the topic are present in [3] providing a useful and update overview of the problem connected with the fatigue damage of structural materials.

A variety of connected topics have been discussed in the present Special Issue of *Metals*, providing a wide overview of recent developments on different aspects of fracture and fatigue tools available for fracture and fatigue assessment of traditional and innovative materials. Hopefully this Special Issue will be the starting point for future discussions and scientific debate on challenging topics related to fracture and fatigue design. This topic, in fact, remains very actual and has a high and relevant impact for many applications. The selected papers touch different and important topics for fatigue and fracture of structural materials. Scale effect and multiscaling approaches are a fundamental part of these topics that allow a better understanding of the fatigue damage at different scale levels.

As Guest Editor of this Special Issue, I am very satisfied with the final result and hope that the present papers will be useful to researchers and designers working on the demanding objective of failure prevention in present cyclic loadings. I would like to warmly thank all the authors for their contributions and all reviewers for their efforts to ensure high quality publications. At the same time, I would like to thank the many anonymous reviewers who assisted me in the reviewing process. Sincere thanks also to the Editors of *Metals* for their continuous help, and to the *Metals* Editorial Assistants for valuable and inexhaustible engagement and support during the preparation of this volume. In particular, my sincere thanks to Toliver Guo for his help and support.

Conflicts of Interest: The author declares no conflict of interest.

References

1. Li, D.; Zhang, C.; Lu, P. Fatigue Property and Improvement of a Rounded Welding Region between the Diaphragm Plate and Closed Rib of an Orthotropic Steel Bridge Deck. *Metals* **2020**, *10*, 161. [CrossRef]
2. Cruces, A.S.; Lopez-Crespo, P.; Bressan, S.; Itoh, T.; Moreno, B. On the Behaviour of 316 and 304 Stainless Steel under Multiaxial Fatigue Loading: Application of the Critical Plane Approach. *Metals* **2019**, *9*, 978. [CrossRef]
3. He, J.; Bao, J.; Long, K.; Li, C.; Wang, G. Study on the Corrosion Fatigue Properties of 12Cr1MoV Steel at High Temperature in Different Salt Environments. *Metals* **2019**, *9*, 774. [CrossRef]

Publisher's Note: MDPI stays neutral with regard to jurisdictional claims in published maps and institutional affiliations.

© 2020 by the author. Licensee MDPI, Basel, Switzerland. This article is an open access article distributed under the terms and conditions of the Creative Commons Attribution (CC BY) license (http://creativecommons.org/licenses/by/4.0/).

 *metals*

Article

On the Influence of Control Type and Strain Rate on the Lifetime of 50CrMo4

Max Benedikt Geilen [1,*], **Josef Arthur Schönherr** [1], **Marcus Klein** [1], **Dominik Sebastian Leininger** [1], **Alexander Giertler** [2], **Ulrich Krupp** [3] and **Matthias Oechsner** [1]

[1]   Centre for Engineering Materials (MPA-IfW), Technical University of Darmstadt, Grafenstraße 2, 64283 Darmstadt, Germany; schoenherr@mpa-ifw.tu-darmstadt.de (J.A.S.); m.klein@mpa-ifw.tu-darmstadt.de (M.K.); dominik.leininger@t-online.de (D.S.L.); oechsner@mpa-ifw.tu-darmstadt.de (M.O.)

[2]   Institute of Materials Design and Structural Integrity, Osnabrück University of Applied Sciences, Albrechtstraße 30, 49009 Osnabrück, Germany; a.giertler@hs-osnabrueck.de

[3]   IEHK Steel Institute, RWTH Aachen University, Intzestraße 1, 52072 Aachen, Germany; krupp@iehk.rwth-aachen.de

*   Correspondence: geilen@mpa-ifw.tu-darmstadt.de; Tel.: +49-6151-16-25867

Received: 5 October 2020; Accepted: 27 October 2020; Published: 30 October 2020

Abstract: In this study, we investigate the influence of control type and strain rate on the lifetime of specimens manufactured from 50CrMo4. This influence is described by a strain rate dependent method that uses cyclic stress strain curves to correct displacement-controlled cyclic test results. The objective of this correction is to eliminate the stress related differences between displacement-controlled cyclic test results and force-controlled cyclic test results. The method is applied to the results of ultrasonic fatigue tests of six different combinations of heat treatment, specimen geometry (notch factor) and atmosphere. In a statistical analysis, the corrected results show an improved agreement with test results obtained on conventional fatigue testing equipment with similar specimens: the standard deviation in combined data sets is significantly reduced ($p = 4.1\%$). We discuss the literature on intrinsic and extrinsic strain rate effects in carbon steels.

Keywords: ultrasonic cyclic testing; frequency effect; control type effect; strain rate effect; 50CrMo4; SAE 4150; high cycle fatigue; very high cycle fatigue; statistical analyses

1. Introduction

Since the early days of fatigue research, stress based approaches have been used for the fatigue design of components. This approach prevails in the high cycle fatigue (HCF) region to this day. Models of this kind are best calibrated using stress-controlled cyclic tests. In the 1950s and 1960s, strain-based approaches, which differentiate between elastic and plastic strain, were developed to improve predictive power in the low cycle fatigue (LCF) region [1]. Strain-based models are best calibrated using strain-controlled fatigue tests. In the 1980s and 1990s, component failures in the gigacycle range kick-started very high cycle fatigue (VHCF) research [2–12]. In VHCF fatigue initiation, no macroscopically plastic material behavior is observed, therefore stress based models are employed. Ideally, these models should be calibrated using stress-controlled cyclic tests.

Materials testing in the VHCF region is conducted on two kinds of fundamentally different testing setups. The first kind are conventional testing machines—e.g., resonance or servo-hydraulic testing machines—which are also used in the HCF region. These machines are able to operate under stress control but rarely exceed testing frequencies of 200 Hz. While some state of the art testing machines like Rumul Gigaforte 50 reach up to 1000 Hz, typical testing frequencies range from 20 to 100 Hz. The second kind is ultrasonic fatigue testing machines. Ultrasonic fatigue testing machines run at 20 kHz but must be run in pulse pause mode because of a material-dependent specimen's self-heating. They usually reach effective testing frequencies of 2 kHz. Depending on geometry, tolerated temperature increase, material under investigation, load level and cooling equipment, the effective testing frequency varies. Ultrasonic testing machines operate under displacement control [13]. In cyclically softening steel specimens under displacement control, notch stress amplitudes may decrease over the course of the cyclic test.

In research on low strength (ultimate tensile strength 919 MPa) cyclically softening 50CrMo4 [10], differences between the results of displacement-controlled cyclic tests conducted at about 20 kHz and the results of force-controlled cyclic tests conducted between 20 Hz and 700 Hz were observed. Such differences have largely been attributed to a frequency or strain rate effect [6,14–19]. In research on high strength (ultimate tensile strength 1726 MPa) 50CrMo4, the observed differences between such tests were remarkably smaller, which has been attributed to a lower strain rate influence for high strength steels [6]. The understanding of these test results without further processing is limited by the coupling between strain rate and control type.

In components testing, displacement-controlled cyclic tests are conducted to simultaneously test dozens of specimens with a single testing machine [20], which allows testing much larger numbers of specimens at a given budget. These results should be used to design force-loaded components only after correction for control type.

In this study, an algorithm [21] implementing a method for the strain rate dependent correction for control type [10] is utilized to create control type corrected cyclic test results from displacement-controlled cyclic test results for different batches of 50CrMo4. We show that for each batch, these corrected results show an equal or improved agreement with results of force-controlled cyclic tests in the HCF range.

In Section 2, an overview on strain rate effects in fatigue is given. In Section 3, the data used in this investigation are described. In Section 4, the numerical methods of the investigation are described. In Section 5 and 6 , the results of the investigation are presented and analyzed. In Section 7, the investigation is summarized and in Section 8, an outlook is given.

2. Effects of Strain Rate on Deformation and Fatigue

A multitude of studies on fatigue life were conducted at different strain rates on steels [6,16,19,22–32] as well as other materials [22,33–43]. The results of these studies are mixed, some showing momentous differences in fatigue life, some showing ambiguous results, implying that there is no significant strain rate influence in these investigations. In this section, we describe different effects suspected to cause strain rate effects.

Frequency and strain rate effects may be divided into intrinsic effects and extrinsic effects [14]. Effects are called intrinsic if they are caused directly by strain rate dependent material behavior. An intrinsic effect at microscopic scale is a change in the formation process of slip bands or an activation of dynamic strain aging effects. These effects at microscopic scale cause effects at macroscopic scale, for example an increase in monotonic and cyclic yield strength. These effects are well studied for monotonic loading [44–52]. For cyclic loading, fewer studies are available [41,42,45,53]. Extrinsic effects are effects that sometimes or always accompany strain rate effects, but are at least in theory separable from intrinsic strain rate effects. Important extrinsic effects are temperature, hydrogen and chlorine concentration in the crack tip due to diffusion, control type, testing volume and oxidation-induced crack closure.

2.1. Intrinsic Effects

Since cyclic deformation is the sum of a succession of monotonic deformations, it is natural to assume that qualitatively, the same effects apply for monotonic and cyclic deformations. Therefore, we first explain the much better studied monotonic effects on deformation behavior. While the strain rate dependency is fundamentally different for some materials [54], we will focus on the description of the state of research on carbon steels.

At microscopic scale, strain rate is closely linked to the activation energy of dislocations. For dislocations to form or move, a certain energy must be supplied locally [50]. In a given material at a given temperature and strain state, this is a stochastic process with an average rate of dislocations forming and moving per (unit) volume. At higher strain rates, the finite dislocation velocity is not sufficient to accommodate the fixed displacement rate, leading to a higher macroscopic yield stress. Higher strain rates influence the microscopic behavior very similarly to lower temperatures: at a given Peierls stress and a given stress state, fewer dislocations form or move. With increasing temperature, activation energies increase, helping dislocations to move, which explains the observation of lower strain rate sensitivities of flow stresses at 800 °C compared to 25 °C [50]. Between these temperature levels, a maximum in strain rate sensitivity of flow stress is expected due to strain aging. Strain aging is caused by Cottrell atmospheres—carbon atoms pinned in dislocations, necessitating high (localized) forces to unpin and thereby raising the (localized) force necessary to move the dislocation.

At high strain rates, adiabatic conditions promote locally high temperatures. These locally high temperatures promote the strongly localized formation of shear bands: in the formation of dislocations, only a portion of the expended energy is introduced into the crystal structure, another portion is converted into heat [55], which leads to a locally higher temperature because at very high strain rates, heat conduction is slow compared to heat generation. The higher temperature again facilitates the generation of new dislocations, in which new heat is generated, leading to a cycle only broken by the removal of the source of increasing strain. This may be a fracture event in monotonic loading or the onset of unloading in cyclic loading. In the adiabatic regions, recrystallization may occur [56]. In cyclic loading, strongly localized temperature increase depends on load level and pulse length. While the global increase of temperature is well controlled by pulse pause methods, true localized temperatures remain unknown to researchers. Localized temperatures in adiabatic conditions may be computed based on a factor determining the share of energy converted into heat [50,57,58]. This factor has been determined to be close to 100% [57].

On the macroscopic scale, yield stress and ultimate tensile strength increase with increasing strain rate [44,50,59,60]. A widespread model describing this relationship is the Johnson Cook model [61].

From a fracture mechanics point of view, fatigue is a crack growing from a pre-existing defect. This defect may be e.g., an inclusion in the material, a scratch or in- and extrusions caused by persistent slip bands. Crack growth rate is determined by an intrinsic crack growth curve and crack closure effects [62]. Crack growth experiments on 32CrMo4 showed no overall influence of frequency (varied between 0.02 Hz and 20 Hz, measured in the Paris range) on the overall crack growth rate [63]. This has been interpreted as absence of an influence of strain rate on the intrinsic crack growth curve [64]. To draw a generally applicable conclusion, more evidence is necessary, firstly regarding higher strain rates and secondly regarding a wider range of carbon steels.

Plasticity-induced crack closure [62,65–70] depends on strain rate because plasticity depends on strain rate. Kindly note that although the change in plasticity-induced crack closure is an intrinsic effect with regard to intrinsic or extrinsic strain rate effects, the intrinsic crack growth curve is the one measured without any plasticity-induced (or other) crack closure. For small polished specimens, the influence of plasticity-induced crack closure is probably small since loads at the crack tip are below the crack growth threshold until shortly before failure.

No fundamental difference has been observed between the strain rate dependencies of deformations under cyclic compressive and cyclic tensile loads [45].

2.2. Extrinsic Effects

Specimens with larger critically loaded surface areas and volumes show decreased lifetimes compared to specimens with smaller critically loaded surfaces and volumes. This phenomenon is well-studied [71–84] and understood well enough to be included in design standards [85–87].

Ultrasonic fatigue specimens, i.e., all specimens used for very high strain rate fatigue tests, are generally small-sized. This introduces a bias if the size effect is not accounted for. Jeddi and Palin-Luc [15] show that size effects explain the difference in results with different strain rates for several studies [23,24,27], which implies that there is no relevant strain rate effect in these studies. In their analysis, a model assuming that fatigue strength is inversely proportional to specimen diameter is utilized. This model is meant to consider the effect of the critically loaded surface and volume, a consideration of the crack growth phase is not explicitly intended. They [15] conclude that for high strength steels in general, there is no relevant strain rate effect on the fatigue behavior.

Another extrinsic effect is oxide-induced crack closure [62,65,67,69,88–91], which is caused by the following process: the outermost layer of the steel at the flanks of the crack is oxidized over time due to exposure to oxygen. Due to the oxidation, the volume of the outermost layer of the crack is increased. Due to the increased volume, the crack closes at higher loads. This leads to the crack being open for a lower portion of the load cycle, which leads to the generation of fewer dislocations and therefore a lower amount of crack growth per cycle [92]. Although the usual thickness of oxide layers is only a few nanometers, at the crack flanks, the oxide layer may grow to a significant size due to continuous breaking and reforming due to fretting contact between both crack flanks [69]. Since at lower frequencies, more time passes between two cycles, the oxidation effect is stronger and the crack grows slower (less per cycle, not per time) than at higher frequencies. This implies a longer lifetime for low frequencies compared to high frequencies in the absence of other effects. This effect does not occur at high load ratios since the oxide layer is not sufficient to close the crack at high stresses.

Note that the oxide-induced crack closure effect does not dominate the influence of the atmosphere: the average lifetime of specimens tested in argon is longer than that of specimens tested in lab air all else being equal, Section 5. A momentous difference has also been registered between dry air and humid air [93]. This effect has been reported to be pronounced in quenched 42CrMo4 steels [93]. The effect has been shown to decrease with tempering, ceasing to exist in 42CrMo4 tempered at 550 °C for 1 h [93]. For a high hardness batch of 42CrMo4, no obvious difference has been observed in the fatigue life of specimens in high humidity air and specimens in high humidity argon, however it remains unclear if residual adsorbed oxygen in the specimens in argon may be responsible for the behavior [93]. For a 1076 MPa ultimate tensile strength batch of 41NiCrMo7-3-2 steel, slightly higher lifetimes were observed for specimens tested in dry air compared to specimens tested in humid air [94]. For 0.5% carbon steel, momentously longer lifetimes and a higher fatigue limit were observed in vacuum [95]. For 41NiCrMo7-3-2 [96] and 100Cr6 at a high load level [97], slightly longer lifetimes were observed in vacuum compared to lab air [96]. Specimens manufactured from 100Cr6 with artificial surface defects showed higher survival probabilities in lab air than in vacuum when tested at the same load level up to 2×10^9 cycles. Crack growth was observed to accelerate with increasing humidity in argon [98] and air [99,100]. In vacuum, reduced crack growth rates were observed in comparison to lab air [101,102].

The reduced fatigue performance in humid air and argon is largely blamed on the prevention of slip reversal—a mechanism that increases lifetime—in oxidizing environments in general [103–105]. Slip reversal occurs when dislocations are moved along a slip band onto the surface of the specimen during loading and are subsequentially moved back into their old position during unloading. This way, no damage accumulates and therefore, no contribution to failure is made. Adsorption of oxygen may prevent the reverse movement, which makes the slip irreversible [106]. Irreversible slip bands lead to intrusions and extrusions, which may nucleate a crack.

The oxide layers on the surface may have a higher Young's modulus than the base steel. If that's the case, for dislocations to pass the interface, an additional force must be expended. This leads to pileups of dislocations just under the oxide layer, which promote crack initiation [105]. Aluminum shows a momentous influence of environment, however, it has been shown that the surface oxide has a lower Young's modulus than the base metal [107,108].

The environment of internal cracks may be viewed as a vacuum. This view is supported by similar fracture surfaces [96,97,109]. The sometimes distinct segregation of cracks initiating from the surface and cracks initiation from sub-surface with respect to the number of cycles to failure has been attributed to the difference in crack initiation times in vacuum in comparison to lab air [110]. For a batch of JIS-SNCM439, segregation was observed to be present if tests were conducted in lab air but not if tests were conducted in vacuum [96].

In steels, hydrogen is present to some degree, although this degree is very low in high quality steel. Some research [111,112] suggests that VHCF failure from inclusions is connected to hydrogen, even if hydrogen concentrations are not measurable. This may be caused by an inhomogeneous distribution of hydrogen in the specimen: hydrogen does not really follow Fick's law in its diffusion behavior but concentrates at points of high tensile stress. At low strain rates hydrogen has a lot of time to diffuse to the crack tip, at higher strain rates it has less time because the crack tip moves faster. The increased hydrogen concentration implies shorter lifetimes for lower strain rates. In theory, all the hydrogen in the whole specimen diffuses to an infinitesimal small volume around the crack tip because of the stress singularity. The reality is of course less extreme. More on the interaction between hydrogen diffusion and plasticity may be found in [113,114]. If hydrogen is introduced into or removed from the specimen in any way during cyclic loading, this is a time dependent and therefore indirectly strain rate dependent process. In specimens charged with hydrogen, crack growth rates increase momentously with decreasing frequencies [64]. In uncharged austenitic specimens, crack growth rates increase with decreasing frequencies, which has been attributed to non-diffusible hydrogen [64]. In uncharged 34CrMo4 specimens, an increase of crack growth rate with decreasing frequency has not been observed, which may be due to the much higher diffusivity of hydrogen [63,64]. These studies were conducted at low strain rates. Due to the higher diffusivity, hydrogen diffusion may become more influential in carbon steels at higher strain rates, however to our knowledge, there is no experimental evidence available on this issue.

3. Experimental Foundation

In this work, the cyclic properties of the quenched and tempered steel 50CrMo4 are investigated. The material was provided in five batches, having different tempering states with ultimate tensile strengths σ_u of 919 MPa, 1096 MPa, 1170 MPa, 1475 MPa and 1726 MPa. Its chemical compositions are given in Table 1 and tensile test results are given in Table 2.

Most data were obtained in two research projects on the influence of variable amplitude loading in the VHCF range, conducted at the Centre for Engineering Materials (MPA-IfW) and Fraunhofer LBF Darmstadt [5,6,10]. These are all data belonging to the material batches with tensile strengths of 919 MPa, 1170 MPa, 1475 MPa and 1726 MPa. The testing machines used therein for cyclic testing are a 20 kHz ultrasonic testing system (USP), type BOKU Vienna UFTE, a 700 Hz electromechanic test system (EMF) and a 400 Hz servo-hydraulic testing machine (VHF), type Instron Schenck VHF 50D. The results were obtained at testing frequencies of about 19 kHz for the USP, 700 Hz for the EMF and 20–200 Hz for the VHF.

Table 1. Chemical composition (in weight percent).

Material Batch	C	Cr	Mo	Mn	P	S
919 MPa [5]	0.53	1.06	0.19	0.69	<0.01	<0.01
1096 MPa [8]	0.48	1.00	0.18	0.71	0.013	0.010
1170 MPa [10]	0.49	1.01	0.19	0.69	<0.01	0.011
1475 MPa [10]	0.48	1.06	0.19	0.68	0.01	0.012
1726 MPa [6]	0.51	1.09	0.19	0.73	0.016	<0.01

Table 2. Tensile test results.

Material Batch	Young's Modulus E in GPa	0.2% Offset Yield Strength $\sigma_{0.2}$ in MPa	Ultimate Tensile Strength σ_u in MPa
919 MPa [5]	206	842	919
1096 MPa [8]	202	1000	1096
1170 MPa [10]	208	1025	1170
1475 MPa [10]	204	1345	1475
1726 MPa [6]	215	1544	1726

For further validation, test data were provided by Osnabrück University of Applied Sciences [8,115], obtained at 20 kHz with an ultrasonic testing system (USPO) type BOKU Vienna and at about 95 Hz, obtained with a electromechanic resonance testing machine, type Rumul Testronic (RTTO). The investigated material was quenched and tempered to an ultimate tensile strength of 1096 MPa.

The ultrasonic testing equipment was operated in pulse pause mode. All tests were conducted at ambient temperature. The load ratio was fixed to $R = -1$. At load ratios of -1, no cyclic creep is observed in quenched and tempered carbon steel [116]. Therefore, cyclic creep is ignored in this publication.

The fatigue specimen geometries are depicted in Figures 1–4. The metallographic characterization of these specimens showed no irregularities relevant for this investigation; for more details see [5,6,10,115].

To assess the material behavior at high strain rates, fast tensile tests were conducted on a high speed tensile testing machine. On the left hand side of Figure 5, the ultimate tensile strength results for the 919 MPa batch as well as a bi-linear approximation thereof are shown. On the right hand side, these data are combined with data from the 1170 MPa batch and from the 1475 MPa batch. For comparison, the stress is normalized to the particular quasi-static ultimate tensile strength, see Table 2. All three materials show a good agreement in their relative 0.2% offset yield strength $\sigma_{0.2}$. In the tests on the 1170 MPa batch at strain rates of about 15/s, two outliers were observed. In a comparative metallographic examination of an outlier specimen and a non-outlying specimen, the outlier showed a significantly different microstructure. In this investigation, the outliers are considered in the process of fitting the bi-linear curve to uphold consistency, since not all specimens were investigated with regard to their microstructure. The fatigue behaviors of the batches 1170 MPa and 1475 MPa are not assessed in this study because the inhomogenities in the microstructure state were also observed in some cyclic test specimens, and therefore, the comparability of the fatigue behavior is limited.

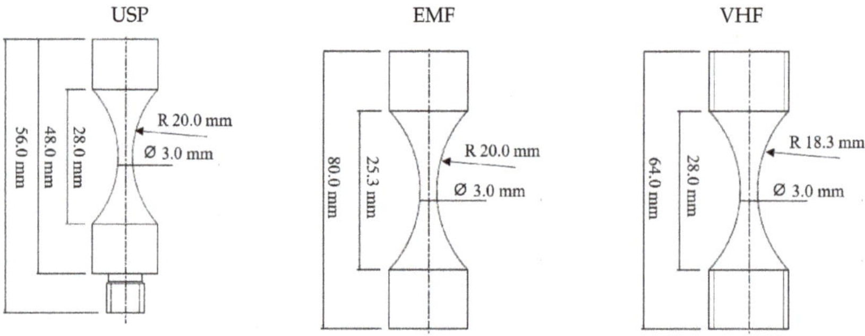

Figure 1. Geometry of specimens with ultimate tensile strength 919 MPa and K = 1 [9,10,21].

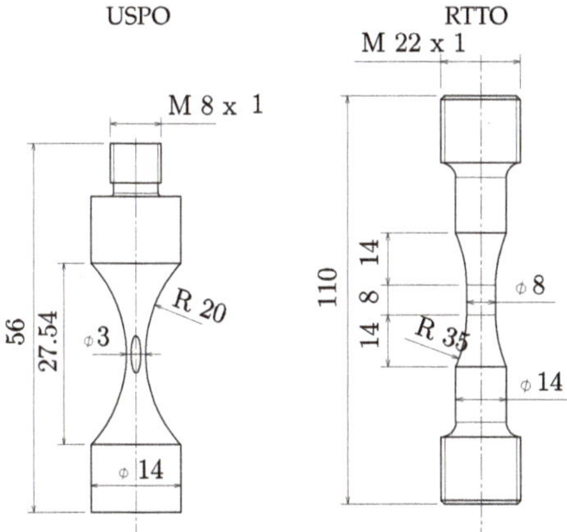

Figure 2. Geometry of specimens with ultimate tensile strength 1096 MPa and K = 1.2 [8,115].

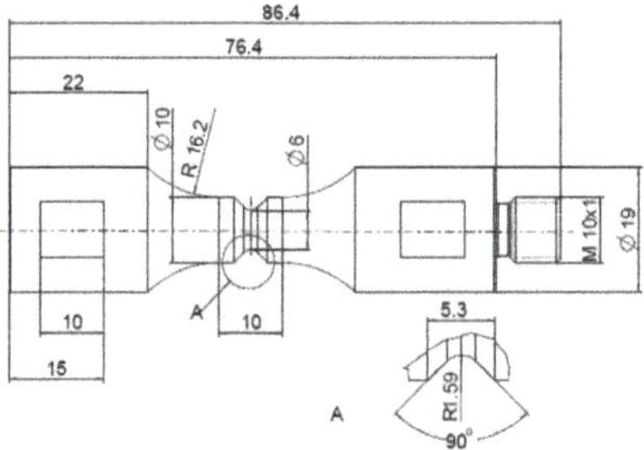

Figure 3. Geometry of ultrasonic testing system (USP) specimens with ultimate tensile strength 919 MPa and K = 1.75 [6], VHF specimens were finalized subsequently [9].

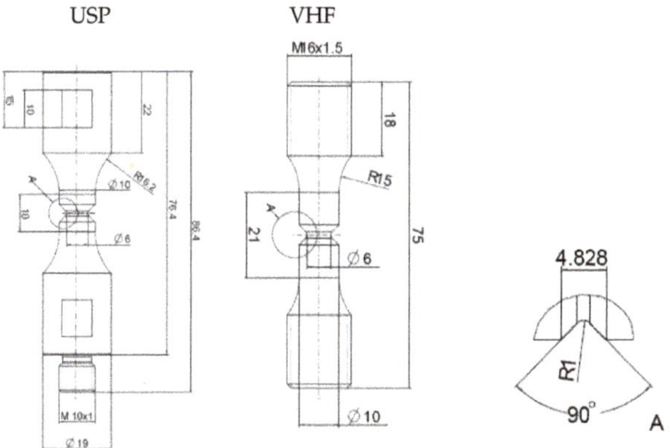

Figure 4. Geometry of specimens with ultimate tensile strength 919 as well as 1726 MPa and K = 2.06 [6].

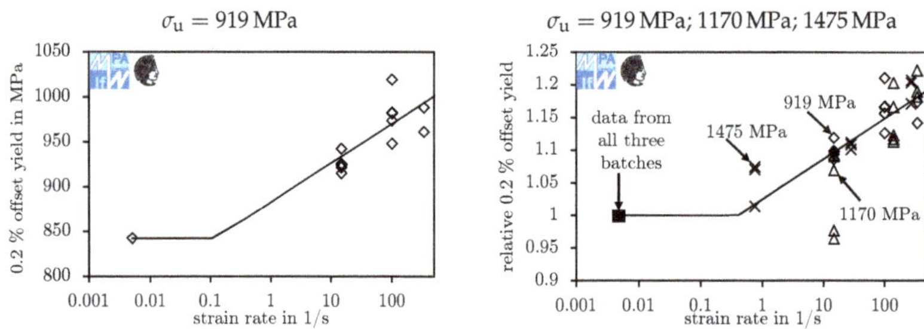

Figure 5. This figure shows 0.2% offset yield depending on the strain rate, for the 919 MPa ultimate tensile strength (**left**) and for the batches 919 MPa, 1170 MPa and 1475 MPa (**right**).

In Figure 6, the stabilized cyclic stress strain test results are shown. The left diagrams show test results for the 919 MPa and 1726 MPa batch. These results were obtained in five constant amplitude tests each. On the right-hand side, the results for the 1096 MPa batch are given in the first row. These were obtained in an Incremental Step Test. As shown in the bottom right diagram, the test results fit together well, since the stress amplitude over strain amplitude values increase with the quasi-static ultimate tensile strength of the material, regardless of the test setting.

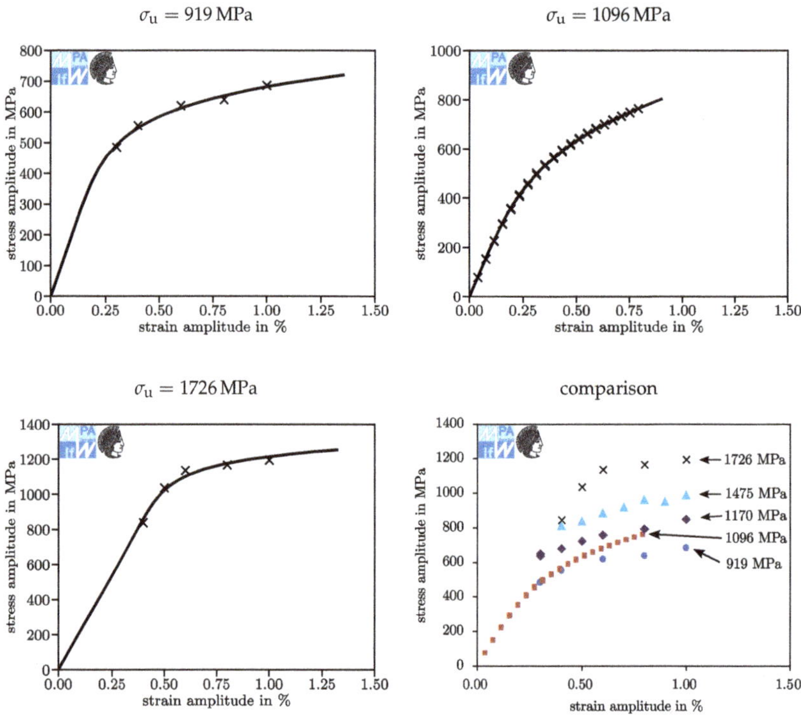

Figure 6. Cyclic stress strain curves. (**First row**): 919 MPa and 1096 MPa ultimate tensile strength. (**Second row**): 1726 MPa ultimate tensile strength (left) and comparison of batches with 919 MPa, 1096 MPa, 1170 MPa, 1475 MPa and 1726 MPa ultimate tensile strength.

The cyclic stress strain curve was generated in a least squares fit in stress direction using the Ramberg–Osgood formulation [117],

$$\varepsilon_a = \epsilon_{a,el} + \epsilon_{a,pl} = \frac{\sigma_a}{E} + \left(\frac{\sigma_a}{K'}\right)^{1/n'},\tag{1}$$

where ε_a is the strain amplitude, σ_a is the stress amplitude, K' is the cyclic hardening coefficient and n' is the cyclic hardening exponent. The parameters obtained for the cyclic stress strain curves provided in the previous figures are summarized in Table 3.

Table 3. Cyclic material constants.

Material Batch	Young's Modulus E in GPa	Hardening Coefficient K' in MPa	Hardening Exponent n'
919 MPa	206	1341	0.135
1096 MPa	202	2816	0.237
1170 MPa	204	1162	0.068
1475 MPa	204	1436	0.073
1726 MPa	215	1687	0.060

4. Method

4.1. Correction for Control Type

The stress levels of displacement-controlled tests are corrected using the Neuber rule [118,119]. Therein, cyclic stress strain curves are used. These curves were obtained either from constant amplitude tests or from Incremental Step Tests by fitting a Ramberg–Osgood curve [117], Figure 6. These Ramberg–Osgood curves are corrected for the specific strain rate, assuming that the cyclic offset yield strength increases proportionally to the monotonous offset yield strength with increasing strain rate. The monotonous offset yield strength was modeled by a bi-linear function in log-lin space, Figure 5. A constant strain rate is assumed, while the strain rate in fact is sinusoidal; the computed strain rate is equal to the real average strain rate.

The strain amplitude ϵ_a according to Ramberg–Osgood is computed as the sum of the elastic strain amplitude $\epsilon_{a,el}$ and the plastic strain amplitude $\epsilon_{a,pl}$, which may be computed from the stress amplitude σ_a, the 0.2% offset yield strength $\sigma_{0.2}$, the yield offset α, Young's modulus E and the cyclic hardening exponent n' :

$$\epsilon_a = \epsilon_{a,el} + \epsilon_{a,pl} = \frac{\sigma_a}{E} + \alpha\,\frac{\sigma_a}{E}\left(\frac{\sigma_a}{\sigma_{0.2}}\right)^{1/n'-1}\tag{2}$$

This formulation differs from the formulation in Equation (1) only in the variables used, the relationship between the variables is:

$$\alpha = \left(\frac{E}{K'}\right)^{1/n'}\left(\frac{\sigma_{0.2}}{E}\right)^{1/n'-1}\tag{3}$$

Here, the less customary version of the formula is used because the offset yield strength $\sigma_{0.2}$ is explicit, which is advantageous for the algorithm. The yield offset α is computed as:

$$\alpha = \frac{0.002\,E}{\sigma_{0.2}}\tag{4}$$

The cyclic offset yield strength $\sigma_{0.2}$ is strain rate dependent and therefore computed based on the ratio of the strain rate $\dot{\epsilon}$ and the reference strain rate $\dot{\epsilon}_{ref}$. To model the behavior, the bi-linear function x is used:

$$\sigma_{0.2}(\dot{\epsilon}) = \sigma_{0.2}(\dot{\epsilon}_{ref}) \cdot x\left(\frac{\dot{\epsilon}}{\dot{\epsilon}_{ref}}\right) \tag{5}$$

The strain rate $\dot{\epsilon}$ is computed using the strain amplitude ϵ_a and the frequency f, assuming that the strain rate was constant during a single fatigue test:

$$\dot{\epsilon} = 4 \cdot \epsilon_a \cdot f \tag{6}$$

Neuber's rule is applied by finding the point in the Ramberg–Osgood formulation of the cyclic material behavior (Equation (2)) for which the strain energy density is equal to the strain energy density in the elastic computation, which is derived from the elastically computed stress amplitude and the elastically computed strain amplitude :

$$\min_{\sigma_a}(\sigma_a \cdot \epsilon_a(\sigma_a) - \sigma_{a,elastic} \cdot \epsilon_{a,elastic})^2 \tag{7}$$

Since the Ramberg–Osgood formulation is not static but depends on the strain rate, the optimization problem is solved iteratively and in each iteration, the strain rate from the last iteration is used. The solving of the minimization problem is repeated until the change in $\dot{\epsilon}$ over an iteration comes below a threshold. For this investigation, the threshold 10^{-15} was used. The computation procedure is displayed in Figure 7. Due to the iterative character of the algorithm, an index i is added to some variables.

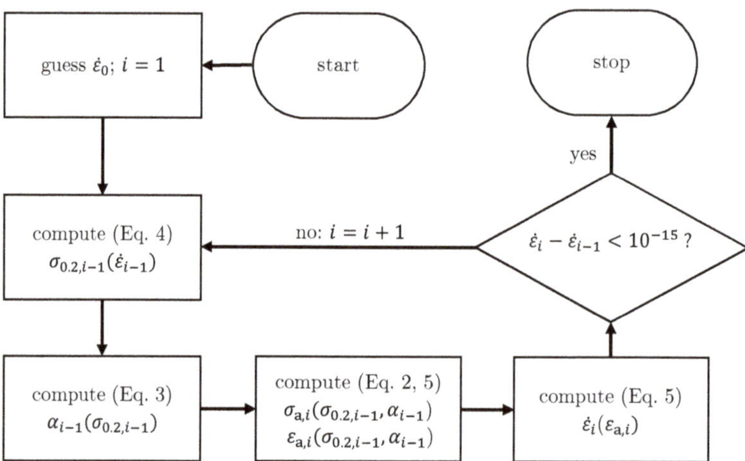

Figure 7. Algorithm used in this study to compute corrected load levels.

An earlier version of the algorithm used herein is described in [21]. The difference between both versions of the algorithm is that the algorithm used herein models the offset yield strength as a bi-linear function, while the algorithm described there utilizes a linear function. The change made for this study leads to slightly smaller correction offsets at high strain rates and slightly higher offsets at low strain rates (as long as these low strain rates are above the lowest strain rate tested). The change is motivated by the fact that for medium to low strain rates, the offset yield strength is not a function of the strain rate.

4.2. Statistical Analysis

The goal of the correction is to create data from displacement-controlled cyclic tests that agree well with data from force-controlled cyclic tests. To evaluate whether the agreement is improved by the correction, a statistical analysis is conducted. Two types of data sets are created, merging results of displacement-controlled cyclic tests before (type 1) and after (type 2) correction with results of force-controlled cyclic tests.

For each data set, a lifetime model assuming a linear course of the fatigue curve in log-log space and a constant log-normal distribution of fracture cycles at given load levels is fitted to all events occurring at up to 3×10^6 cycles, using the maximum likelihood method. For these models, the standard deviation is estimated in the fitting process.

The log-normal distribution is generally accepted as a good approximation of the fatigue behavior of steels in the HCF range [120–123]. Overall, the assumption of log-normality is reasonable, although some of the data sets under investigation deviate significantly from log-normal distributions: in data sets with two populations that differ from each other in their average lifetimes at a given load level (e.g. due to different control types), the assumption of a log-normal distribution is obviously not true. More data on the question whether the data sets under investigations are well described by models using constant log-normal distributions may be found in Appendix A.

The estimated standard deviation provides an intuitive measure of the agreement between two underlying data sets: if the points of both data sets are far away from each other, the estimated standard deviation is high and the intuitively estimated agreement is low. If the points of each data set are close to the point of the other data set, the estimated standard deviation is low and the intuitively estimated agreement is high.

To further reduce subjectivity, the agreement is also evaluated using the mathematically motivated measure of the log-likelihood function. A higher log-likelihood value implies a better agreement of all data in the data set and therefore also implies a better agreement between both merged data sets. Therefore, the log-likelihood value is used as a more objective comparative measure of agreement.

5. Results

All tests performed with electromagnetic and servo-hydraulic testing machines were conducted force-controlled, whereas the ultrasonic testing system operates displacement-controlled. In Figure 8, the fatigue behavior, namely the stress amplitudes plotted against the number of cycles for specimens quenched and tempered to ultimate tensile strength levels of 919 MPa are given. The tests were conducted with unnotched round specimens (stress concentration factor $K_t = 1$) in laboratory air. On the left hand side, a purely elastic material law was used for evaluation. The data presented there have not undergone any correction. On the right hand side, the displacement-controlled test results were corrected using the elastic-plastic material law given in Section 3 and the method described in Section 4.1. The continuous lines depict the fatigue curves for 50% survival probability, the dashed ones for 10% and 90%, respectively. The curves were generated using the maximum likelihood method, fitting the slope exponent k_1, the standard deviation SD and a reference point of the fatigue curve. The fatigue curve describes a field of failure probability densities f with the reference logarithmic number of cycles $N_{ref,log}$, the reference logarithmic normalized stress amplitude $\sigma_{ref,log}$ and the standard deviation SD that depends on the logarithm of the number of cycles N_{log} and the logarithm of the normalized stress amplitude $\sigma_{amplitude,log}$:

$$f(N_{log}, \sigma_{amplitude,log}) = \frac{1}{\sqrt{2\pi SD^2}} \exp\left(-\frac{\left(N_{log} - N_{ref,log} - \left(\sigma_{amplitude,log} - \sigma_{ref,log}\right) \cdot k_1\right)^2}{2\,SD^2}\right) \tag{8}$$

For the generation of these curves, only events occurring at up to 3×10^6 cycles were taken into account.

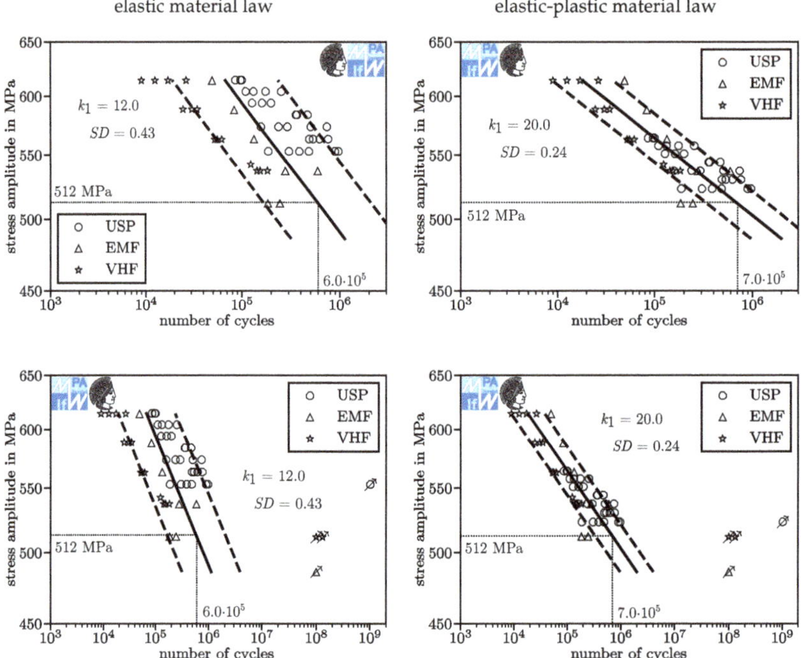

Figure 8. Fatigue behavior of 50CrMo4 in laboratory air; 919 MPa ultimate tensile strength; $K_t = 1$; Curves of 10, 50 and 90% probability of survival; run-outs are marked with an arrow; left side: before correction; right side: after correction; top: only HCF region; bottom: including VHCF region.

In the HCF range (top row of Figure 8), specimens tested with the USP machine last longer at a certain load level compared to the force-controlled machines EMF and VHF, if the stress was calculated using the linear elastic material law (left column). After applying the correction by using the elastic-plastic material law (right column), the USP data points were all moved to a lower load level and agree better with the other results. The EMF and VHF results were not changed. Thereby, the slope exponent increases from $k_1 = 12.0$ to $k_1 = 20.0$ after correction and the standard deviation of the fatigue curve decreases from 0.43 to 0.24. In the bottom row diagram, the VHCF region may be observed. The run-out obtained on the USP is moved closer to the run-outs of the EMF and the VHF. In the corrected data, the results of the VHF occur earlier than results obtained on the other testing setups, except the lowest load level of the EMF.

To objectively evaluate whether the improvement of the fit is significant, a *t*-test is conducted against the null-hypothesis that the average logarithmic deviation of the fracture events from the 50% lifetime prediction is the same for both the evaluation with correction and the evaluation without correction. The logarithmic deviation is measured horizontally. Since both curves are obtained using the same specimens, a paired-sample test is conducted. With predicted logarithmic lifetimes according to the model with correction $N_{\log,\text{predicted,corr},i}$ and according to the model without correction $N_{\log,\text{predicted,no\_corr},i}$ as well as the measured lifetimes $N_{\log,\text{fracture},i}$, differences in logarithmic prediction error $\Delta N_{\log,i}$ are computed and used to compute the *t*-statistic and the *p*-value according to statistical standard theory utilizing the Python module SciPy [124].

$$\Delta N_{\log,i} = \left| N_{\log,\text{fracture},i} - N_{\log,\text{predicted,corr},i} \right| - \left| N_{\log,\text{fracture},i} - N_{\log,\text{predicted,no\_corr},i} \right| \tag{9}$$

The null hypothesis H_0 and the hypothesis H_1 are mathematically defined as:

$$H_0 : \sum_i \Delta N_{\log,i} = 0 \quad \text{vs.} \quad H_1 : \sum_i \Delta N_{\log,i} \neq 0 \tag{10}$$

Computation yields a t-statistic of 6.3 at 49 degrees of freedom and a p-value $7.5 \times 10^{-6}\%$. Simplified, a p-value of $7.5 \times 10^{-6}\%$ means: If the true average of $\Delta N_{\log,i}$ was zero, there would be a $7.5 \times 10^{-6}\%$ chance to obtain a mean with a magnitude as great or greater than we actually obtained. Less simplified, this includes the assumption that our estimated standard deviation is true. Overly simplified, we are $100\% - 7.5 \times 10^{-6}\% = 99.99\ldots\%$ certain that the correction for control type leads to models with lower prediction error, which is, strictly speaking, wrong. At a 5% threshold for significance, the difference in prediction quality is statistically significant and the null hypothesis is rejected. Kindly note that this only applies to the batch under investigation, not for 50CrMo4 in general. The average magnitudes of the logarithmic prediction error are 0.376 without and 0.203 with correction.

The observations made about the results presented in Figure 8 also hold true for test results of notched specimens with $K_t = 1.75$ that were quenched and tempered to ultimate tensile strength levels of 919 MPa, tested in laboratory air; see Figure 9. If the stress was calculated using the elastic material law, the USP specimens also last longer at a given stress level. The correction routine decreases the standard deviation from 0.20 to 0.16. The run-outs are moved further apart. Before correction, the fracture events obtained on the VHF occur earlier (tend to be on the left side of the fatigue curve for 50% survival probability) than fracture events obtained on the USP (which tend to be on the right side of the fatigue curve for 50% survival probability). After correction, this relationship is inverted.

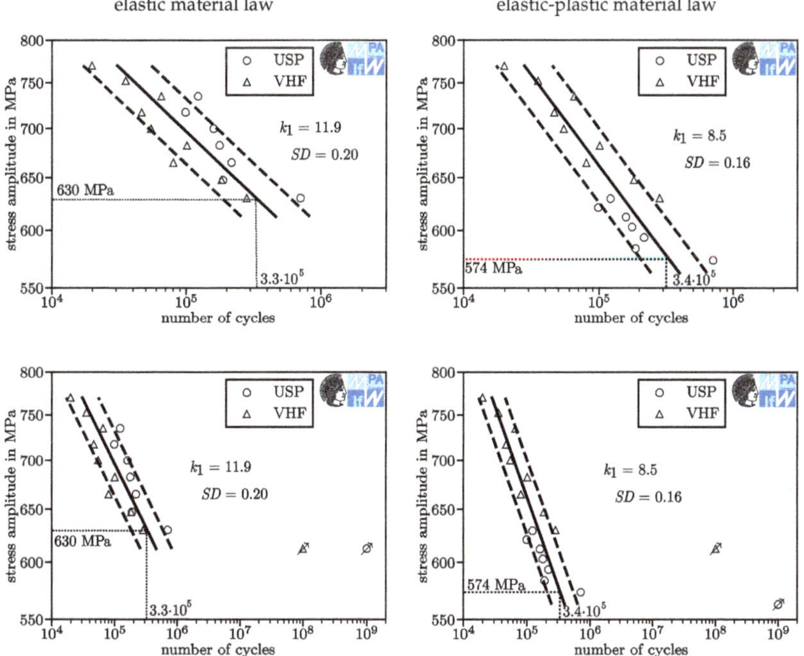

Figure 9. Fatigue behavior of 50CrMo4 in laboratory air; 919 MPa ultimate tensile strength; $K_t = 1.75$; Curves of 10, 50 and 90% probability of survival; run-outs are marked with an arrow; left side: before correction; right side: after correction; top: only HCF region; bottom: including VHCF region.

Testing the difference in logarithmic prediction error $\Delta N_{\log,i}$ according to Equation (9) yields a t-statistic of 1.1 at 15 degrees of freedom and a p-value of 28.8%, which implies that the null hypothesis cannot be rejected based on the data available. Absence of evidence is not evidence of absence. The low p-value may be caused by unfavorable randomness or low number of specimens. Additional tests may yield significant results. The average magnitudes of the logarithmic prediction error are 0.174 without and 0.137 with correction.

Figure 10 shows the results for notched specimens with $K_t = 1.75$ quenched and tempered to ultimate tensile strength levels of 919 MPa, tested in an argon inert gas atmosphere. The diagrams before and after correction show a very similar behavior compared to the previous tests. This data set includes an outlier failure on the VHF. This outlier cannot be explained by the correction method. Kindly note the extremely low standard deviation of 0.11 after correction (ignoring the outlier since it occurred at over 3×10^6 cycles). While we would expect tests in argon to yield results with lower scatter due to a controlled atmosphere, the reduced standard deviation may also be caused by the elimination of extrinsic effects, e.g., oxide-induced crack closure.

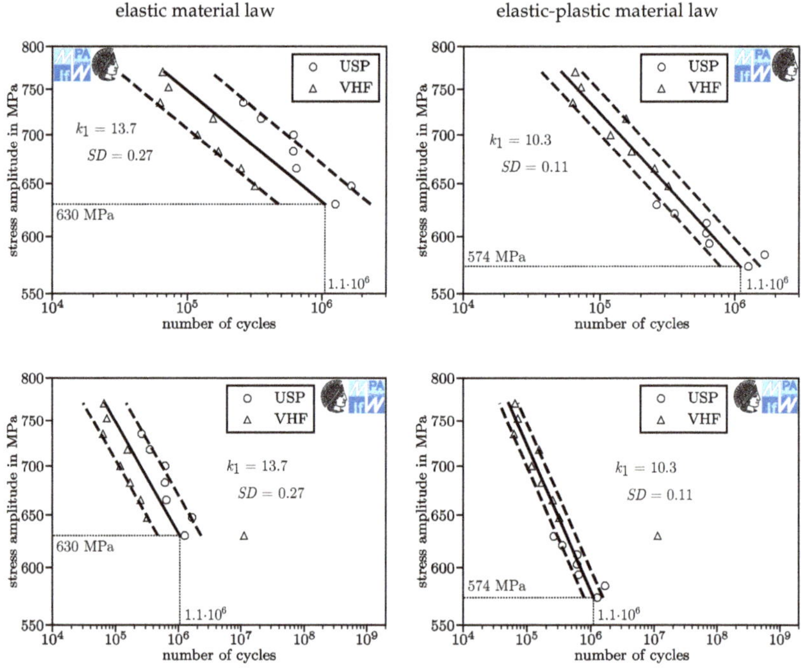

Figure 10. Fatigue behavior of 50CrMo4 in argon; 919 MPa ultimate tensile strength; $K_t = 1.75$; Curves of 10, 50 and 90% probability of survival; run-outs are marked with an arrow; left side: before correction; right side: after correction; top: only HCF region; bottom: including VHCF region.

In fact, if there was a momentous bias between the data obtained on the USP compared to the data obtained on the VHF, the standard deviation would be greater. Therefore, the bias must be small. Consequently, the sum of all intrinsic and extrinsic effects not accounted for is small. The natural conclusion is that the individual effects are small, however more research is necessary to eliminate the possibility that the effects are large and cancel each other out.

Some atmospheric influences, e.g., oxide-induced crack closure, lead to a longer lifetime for low frequencies compared to high frequencies, see Section 2.2. Longer lifetimes at low frequencies can be observed in Figure 9 (laboratory air, 6 out of 7 USP results on the left side of the 50% fatigue curve), but not in Figure 10 (argon, 4 out of 7 USP results on the left side of the 50% fatigue curve) on the right. The underlying experiments only differ in the atmosphere and the load levels; the specimens were from the same batch. At the current state of research, the atmospheric influences are the best explanation for the difference in lifetime since intrinsic effects can be ruled out, since they would also occur in argon.

The average lifetime of specimens tested in argon is longer than that of specimens tested in lab air all else being equal, Figures 9 and 10. At first, this seems to be in dissent with the observation of different lifetime prolongations due to atmospheric influences. This ostensible contradiction can be explained by the dominance of the effect of the prevention of slip-reversal, which only requires small amounts of oxygen in the atmosphere and does not depend on frequency in lab air because the oxygen concentration there exceeds the necessary concentration by orders of magnitude (for more on this effect, see Section 2.2).

Testing the difference in logarithmic prediction error $\Delta N_{\log,i}$ yields a t-statistic of 2.3 at 15 degrees of freedom and a p-value of 3.5%, which implies significance. The average magnitudes of the logarithmic prediction error are 0.288 without and 0.174 with correction.

Regarding the test results for notched specimens with a stress concentration factor of $K_t = 2.06$, the data obtained in the HCF range show a better agreement, see first row of Figure 11. The fracture events in the VHCF range (second row) also show a higher concordance after correction, but are obviously not comprised within the linear fatigue curve approximation. Hence, a bi-linear approximation of the fatigue curve is necessary to comply with the data, which is out of scope for this paper.

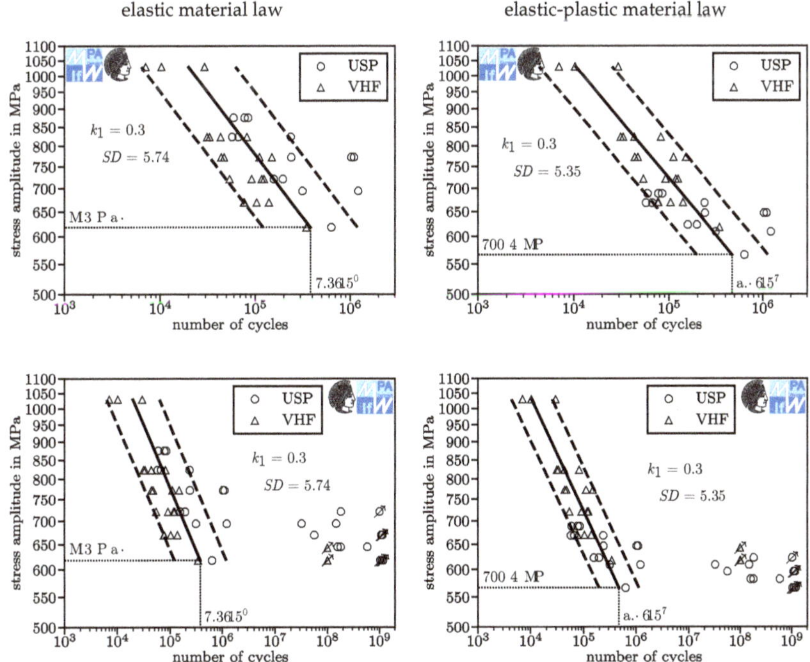

Figure 11. Fatigue behavior of 50CrMo4 in laboratory air; 919 MPa ultimate tensile strength; $K_t = 2.06$; run-outs are marked with an arrow; left side: before correction; right side: after correction; top: only HCF region; bottom: including VHCF region.

Testing the difference in logarithmic prediction error $\Delta N_{\log,i}$ yields a *t*-statistic of 2.6 at 40 degrees of freedom and a *p*-value of 1.3%, which implies significance. The average magnitudes of the logarithmic prediction error are 0.719 without and 0.640 with correction.

The results for specimens with a slight stress concentration of $K_t = 1.2$, quenched and tempered to ultimate tensile strength levels of 1096 MPa, tested in laboratory air, are given in Figure 12. In the HCF range, the non-corrected data show a higher lifetime at similar stress levels (around 625 MPa stress amplitude) for tests conducted with the ultrasonic testing machine USPO compared to those obtained with an electromagnetic resonance testing machine (RTTO). After correction, the stress amplitudes match better and the standard deviation reduces—even though there still is a distinct scatter range. Regarding the VHCF range (second row), the stress level of the USPO run-outs is reduced, even though it is still quite high, compared to the results obtained with the RTTO. The observed differences in fatigue behavior are partially or completely caused by differences in specimen geometry. Since the interaction between correction for control type and correction for critically loaded surface area is a nontrivial problem that requires an evaluation of its own, we do not conduct the correction for critically loaded surface area here.

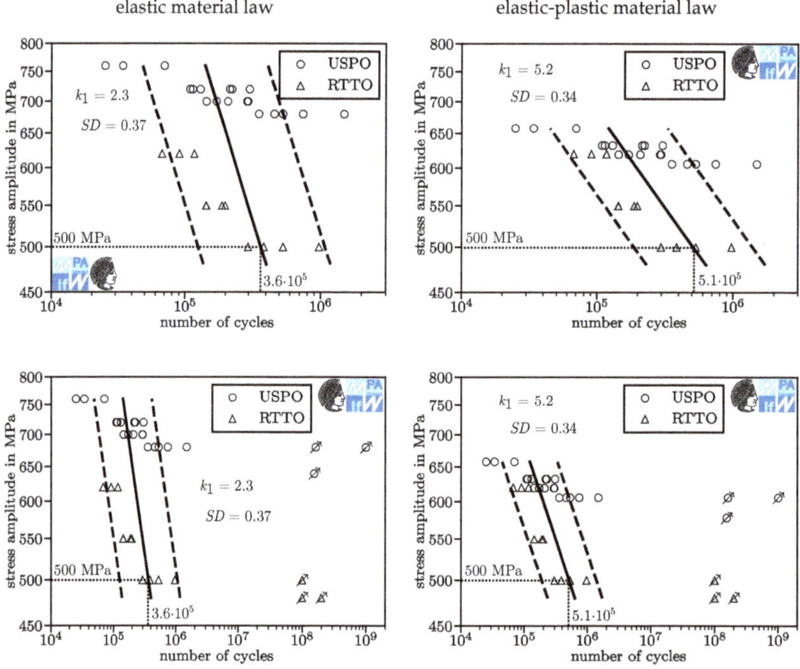

Figure 12. Fatigue behavior of 50CrMo4 in laboratory air; 1096 MPa ultimate tensile strength; $K_t = 1.2$; run-outs are marked with an arrow; left side: before correction; right side: after correction; top: only HCF region; bottom: including VHCF region.

Testing the difference in logarithmic prediction error $\Delta N_{\log,i}$ yields a *t*-statistic of 1.6 at 29 degrees of freedom and a *p*-value of 12.2%, which implies rejection of significance. The average magnitudes of the logarithmic prediction error are 0.295 without and 0.276 with correction.

Figure 13 depicts the test results for specimens quenched and tempered to ultimate tensile strength levels of 1726 MPa. Notched specimens with $K_t = 2.06$ were tested in laboratory air. Here, the proposed correction does not show a relevant change because of the comparably high cyclic 0.2% offset yield strength stress and especially high strain rates.

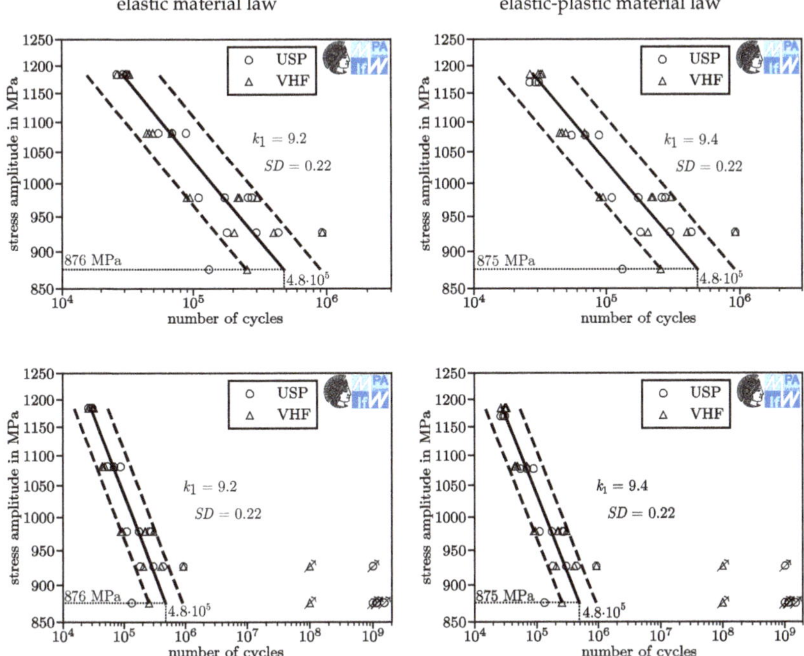

Figure 13. Fatigue behavior of 50CrMo4 in laboratory air; 1726 MPa ultimate tensile strength; $K_t = 2.06$; run-outs are marked with an arrow; left side: before correction; right side: after correction; top: only HCF region; bottom: including VHCF region.

Testing the difference in logarithmic prediction error $\Delta N_{\log,i}$ yields a t-statistic of 1.04 at 31 degrees of freedom and a p-value of 30.5%, which implies rejection of significance. The very high p-value is not surprising because the intersubjective impression is that there is no relevant difference between both data sets. The average magnitudes of the logarithmic prediction error are 0.159 without and 0.162 with correction, meaning that the method for correction worsened prediction quality (insignificantly).

To determine whether there is an overall positive influence on the logarithmic prediction error $\Delta N_{\log,i}$, a t-test according to (10) is conducted using data from all cyclic tests presented above. The test yields a t-statistic of 6.5 at 184 degrees of freedom and a p-value of 7.1×10^{-8}%.

6. Analysis

In Table 4, the parameters of the fatigue curves in Figures 8–13 are summarized. Regarding the slope exponent k_1, there is no trend whether the slope increases or decreases after correction. In all tests conducted, the standard deviation SD decreases or remains unchanged after correction. Furthermore, for all test results, the logarithmic likelihood value $L_{\log}$ increases or remains unchanged, which implies that the corrected regression lines fit the test results better than the non-corrected ones.

Table 4. Parameters, log-likelihood values of fitted fatigue curves as well as number of specimens and p-values obtained in t-test.

	σ_u	K_t	Atmosphere	Corrected?	σ_{ref}	N_{ref}	k_1	SD	L_{log}	# Specimens p-Value
Figure 8	919 MPa	1	lab air	no	512 MPa	6.0×10^5	12.0	0.43	−28.4	50
Figure 8	919 MPa	1	lab air	yes	512 MPa	7.0×10^5	20.0	0.24	−0.4	$7.5 \times 10^{-6}\%$
Figure 9	919 MPa	1.75	lab air	no	630 MPa	3.3×10^5	11.9	0.20	3.2	16
Figure 9	919 MPa	1.75	lab air	yes	574 MPa	3.4×10^5	8.5	0.16	6.5	28.8%
Figure 10	919 MPa	1.75	argon	no	630 MPa	1.1×10^6	13.7	0.27	−1.5	16
Figure 10	919 MPa	1.75	argon	yes	574 MPa	1.1×10^6	10.3	0.11	11.2	3.5%
Figure 11	919 MPa	2.06	lab air	no	618 MPa	3.8×10^5	5.8	0.39	−16.1	41
Figure 11	919 MPa	2.06	lab air	yes	566 MPa	4.7×10^5	6.3	0.30	−7.2	1.3%
Figure 12	1096 MPa	1.2	lab air	no	500 MPa	3.6×10^5	2.3	0.37	−12.34	30
Figure 12	1096 MPa	1.2	lab air	yes	500 MPa	5.1×10^5	5.2	0.34	−10.2	12.2%
Figure 13	1726 MPa	2.06	lab air	no	876 MPa	4.8×10^5	9.2	0.22	3.6	32
Figure 13	1726 MPa	2.06	lab air	yes	875 MPa	4.8×10^5	9.4	0.22	3.6	30.6%

Complementing the intuitive interpretation of Table 4, statistical tests were conducted. A paired-sample t-test was conducted for the log-likelihoods. The differences in log-likelihood, 28.0, 3.3, 12.7, 8.9, 2.14 and 0.0, yield a t-statistic of 2.17, which translates to a p-value of 8.2%. The average difference of the log-likelihoods is 9.2. At a 5% threshold for acceptance of significance, the reduction of log-likelihood is not significant. As per usual in science and especially in statistical testing, absence of evidence is not evidence for absence. In the case of the log-likelihood, this means that the reason for not rejecting the null hypothesis may well be the low sample number of 6 and the high standard deviation which was estimated to be 10.3.

Another paired-sample t-test was conducted for the standard deviations. The differences in standard deviation, 0.19, 0.04, 0.16, 0.09, 0.03 and 0.00, yield a t-statistic of 2.74, which translates to a p-value of 4.1%. Therefore, the reduction of standard deviation is significant considering the threshold of 5%. The average difference in standard deviation is 0.085. This implies that the method for correction for control type significantly decreases standard deviation, not only of the data sets under investigation, but also of all data sets generated in similar tests on batches of 50CrMo4 that are comparable to the ones under investigation.

Testing the average magnitudes of logarithmic prediction error of the six different tests in a paired-sample t-test yields a t-statistic of 2.68, which translates to a p-value of 4.4% at average magnitudes of logarithmic prediction errors of 0.335 without and 0.265 with correction for control type. This also implies significance.

7. Summary

In this study, cyclic test results obtained in displacement-controlled cyclic tests at about 20 kHz were corrected for control type. Two kinds of combined sets were created. The first kind of set contains cyclic test results from force-controlled tests and uncorrected results from displacement-controlled tests. The second kind of set contains cyclic test results from force-controlled tests and control type corrected results from displacement-controlled tests.

The agreement of the cyclic test results with each other inside each data set created this way was quantified through the standard deviation and the likelihood value of a model fitting the data set in the HCF range. Throughout the investigation, the agreements of the data sets with control type corrected data from displacement-controlled tests were equal or superior to the agreements of the data sets with uncorrected data from displacement-controlled tests. This means that the correction method offers significantly improved comparability of displacement-controlled cyclic tests with force-controlled cyclic tests in the HCF range. In the VHCF range, additional investigations are necessary.

8. Outlook

For 50CrMo4, the state of research regarding the influence of the strain rate on the fatigue life may be advanced by incorporating the method presented in this paper. A higher strain rate effect has been reported in low strength 50CrMo4 than in high strength 50CrMo4. However, the difference in strain rate effects may have been overestimated due to an accidental bias in experimental design: the state of research is based on experiments in which low strength 50CrMo4 was tested at load levels momentously exceeding its cyclic yield strength, but high strength 50CrMo4 was tested only at load levels under or just over its cyclic yield strength. In these experiments, strain rate effects cannot be separated from control type effects statistically. An analysis of the experimental data considering and correcting for control type effects would reduce and may practically eliminate the observed difference in strain rate effects. The current state of research is consistent with theoretical considerations: in low strength 50CrMo4, the strain rate dependent Peierls stress makes up a greater share of the total load stress necessary to move a dislocation than in high strength 50CrMo4. Stresses caused by nearby dislocations do not depend on strain rate. However, this does not include any prediction of the size of the strain rate effect, therefore a strain rate effect close to zero in low strength 50CrMo4 would also be consistent with theoretical considerations. Especially the extremely low standard deviation for the specimens tested in argon implies only a small intrinsic strain rate effect for low strength 50CrMo4.

Author Contributions: Conceptualization, M.B.G.; methodology, M.B.G.; software, M.B.G., D.S.L. and J.A.S.; validation, M.B.G. and J.A.S.; formal analysis, M.B.G.; investigation, M.B.G. and M.K.; resources, M.O., M.K. and U.K.; data curation, M.B.G. and A.G.; writing—original draft preparation, M.B.G., M.K. and J.A.S.; writing—review and editing, M.B.G., J.A.S., M.K., D.S.L., A.G., U.K. and M.O.; visualization, M.B.G.; supervision, M.O.; project administration, M.K.; funding acquisition, M.O. All authors have read and agreed to the published version of the manuscript.

Funding: The authors thank the Federal Ministry of Economics and Technology, the German Federation of Industrial Research Associations 'Otto von Guericke' e.V. (BMWi/AiF no. 324ZN and AiF no. 18198 N) and the Forschungskuratorium Maschinenbau e.V. (FKM) for the financial support. We acknowledge support by the German Research Foundation and the Open Access Publishing Fund of Technical University Darmstadt.

Acknowledgments: Some of the data used in this paper was generated by Fraunhofer LBF in BMWi/AiF no. 324ZN and AiF no. 18198. We thank Klaus Störzel and Tobias Melz for the fruitful cooperation.

Conflicts of Interest: The authors declare no conflict of interest.

Appendix A. Do Models Assuming Log-Normal Distributions Describe the Data Sets Well?

In this appendix, we investigate whether models assuming normal distributions describe the data sets under investigation well. To this means, all fatigue events with numbers of cycles to failure 3×10^6 of each test set are shifted to a single virtual load level and compared graphically to the cumulative probability function of the fitted log-normal distribution. The relative numbers of cycles until failure are standardized over the estimated standard deviations, Figures A1–A6.

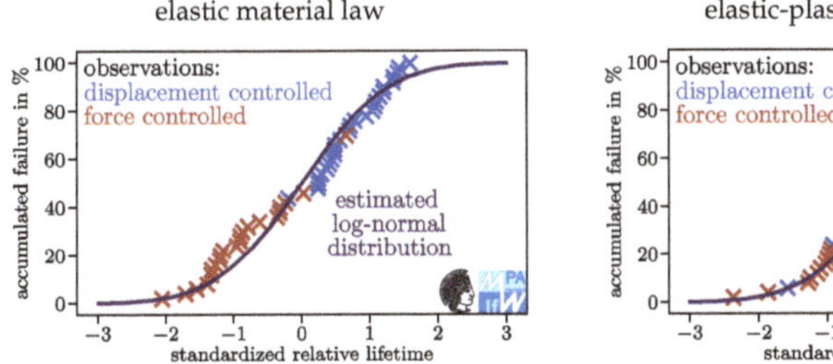

Figure A1. Estimated and observed accumulated failure rates for 50CrMo4 specimens in laboratory air; 919 MPa ultimate tensile strength; $K_t = 1$.

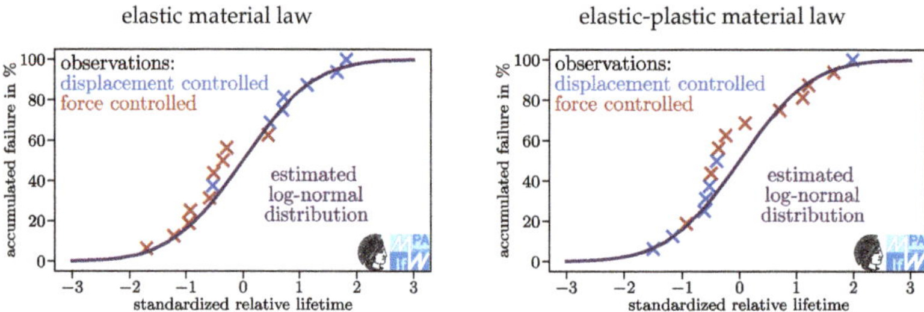

Figure A2. Estimated and observed accumulated failure rates for 50CrMo4 specimens in laboratory air; 919 MPa ultimate tensile strength; $K_t = 1.75$.

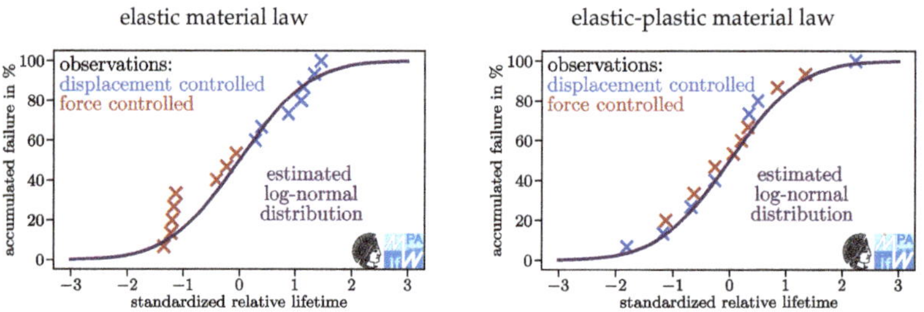

Figure A3. Estimated and observed accumulated failure rates for 50CrMo4 specimens in argon; 919 MPa ultimate tensile strength; $K_t = 1.75$.

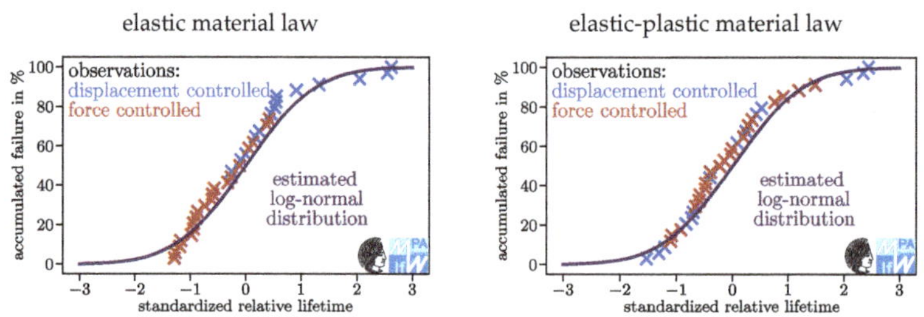

Figure A4. Estimated and observed accumulated failure rates for 50CrMo4 specimens in laboratory air; 919 MPa ultimate tensile strength; $K_t = 2.06$.

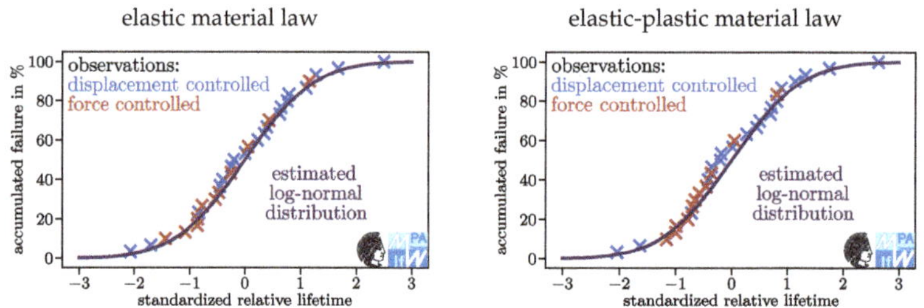

Figure A5. Estimated and observed accumulated failure rates for 50CrMo4 specimens in laboratory air; 1096 MPa ultimate tensile strength; $K_t = 1.2$.

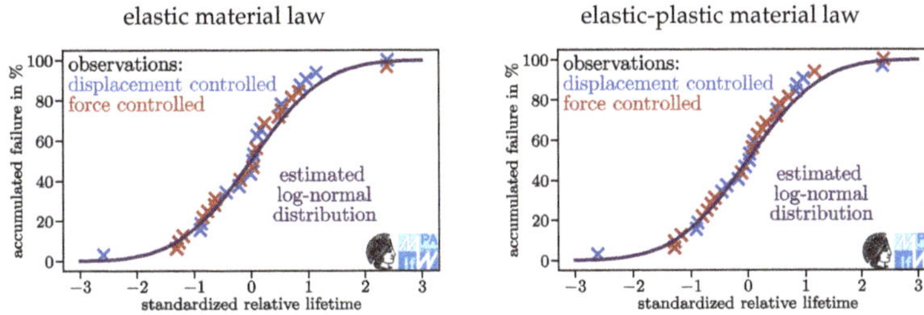

Figure A6. Estimated and observed accumulated failure rates for 50CrMo4 specimens in laboratory air; 1726 MPa ultimate tensile strength; $K_t = 2.06$.

Curves describing the observed accumulated failure rate would run horizontally from each observed data point and jump to the next level just before the next observation. This explains the impression that the estimated log-normal distributions are biased towards higher standardized relative failure rates.

One-sided tendencies of failure events under displacement and force control towards earlier or later lifetimes reflect systematic differences between the results obtained on both experimental setups. Such tendencies are eminent in all non-corrected test results obtained on specimens that were quenched and tempered to ultimate tensile strength levels of 919 MPa.

Quantile-quantile plots are given in Figures A7–A12. These figures show the same as Figures A1–A6, just in a differently scaled coordinate system with switched axes in which the cumulative function of the log-normal distribution is the 45° line. Good compliance with this line means that the data is well described by a normal distribution. In each figure, the p-value of the corresponding Shapiro-Wilk test is given. Low values imply a rejection of the null hypothesis. The null hypothesis is that the data set under investigation follows a log-normal distribution. The power of Shapiro-Wilk tests is lower in data sets with fewer specimens than in data sets with more specimens.

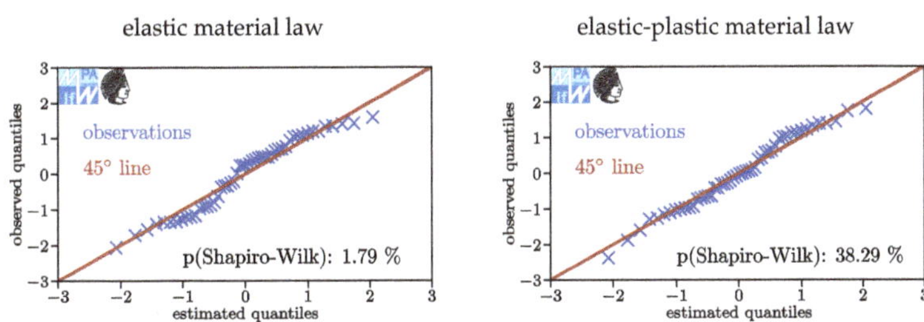

Figure A7. Quantile-quantile plots for 50CrMo4 specimens in laboratory air; 919 MPa ultimate tensile strength; $K_t = 1$.

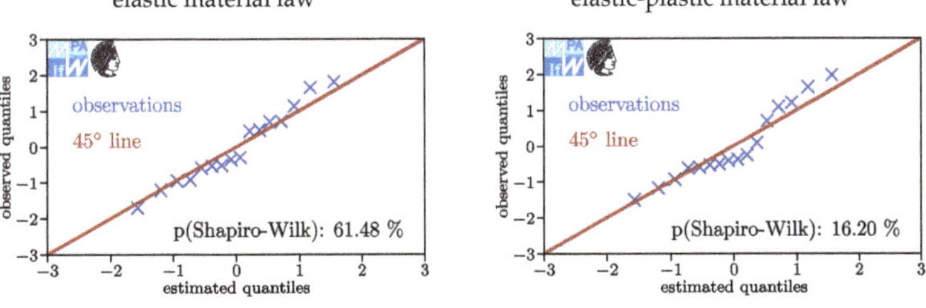

Figure A8. Quantile-quantile plots for 50CrMo4 specimens in laboratory air; 919 MPa ultimate tensile strength; $K_t = 1.75$.

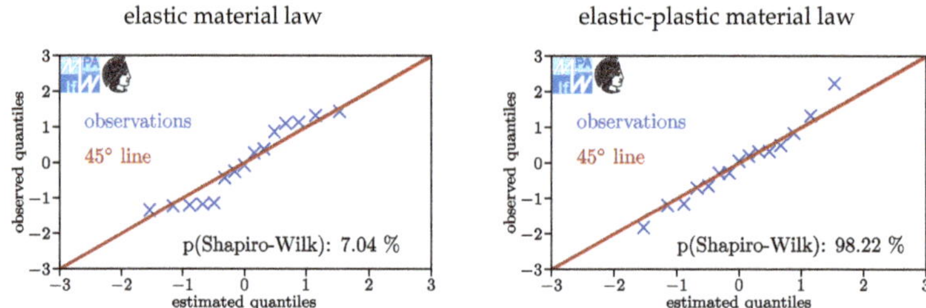

Figure A9. Quantile-quantile plots for 50CrMo4 specimens in argon; 919 MPa ultimate tensile strength; $K_t = 1.75$.

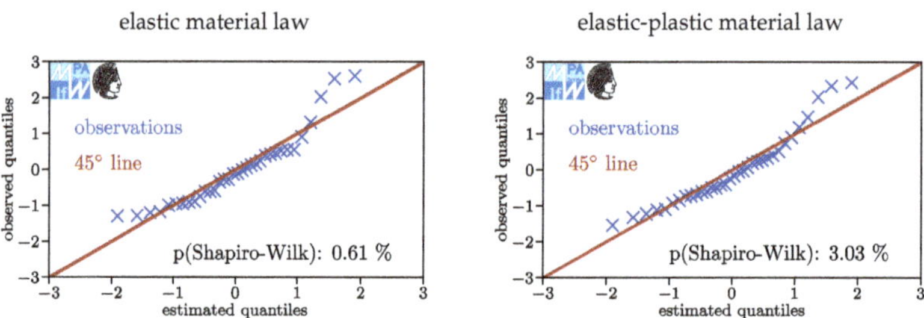

Figure A10. Quantile-quantile plots for 50CrMo4 specimens in laboratory air; 919 MPa ultimate tensile strength; $K_t = 2.06$.

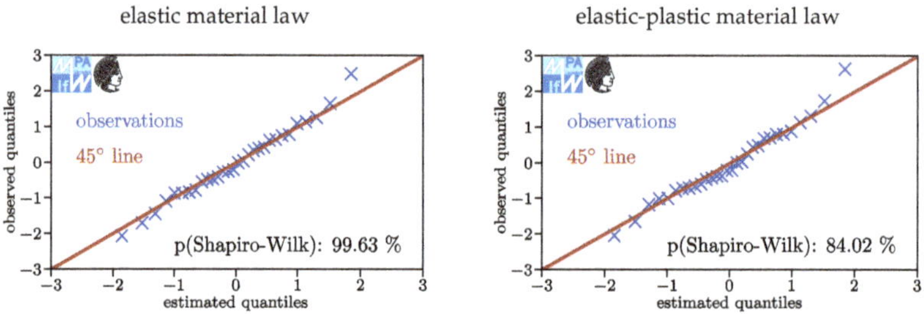

Figure A11. Quantile-quantile plots for 50CrMo4 specimens in laboratory air; 1096 MPa ultimate tensile strength; $K_t = 1.2$.

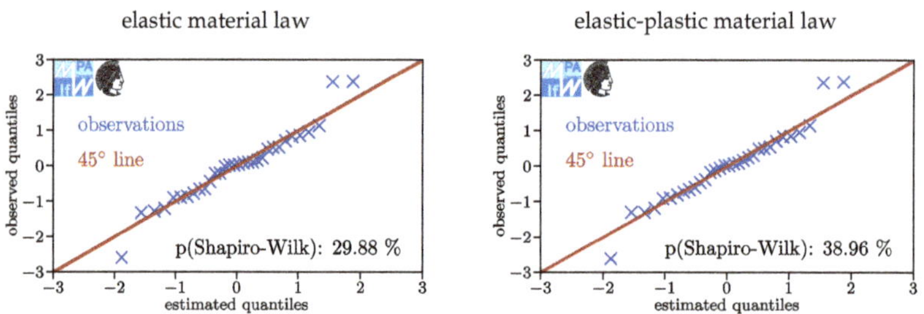

Figure A12. Quantile-quantile plots for 50CrMo4 specimens in laboratory air; 1726 MPa ultimate tensile strength; $K_t = 2.06$.

Although some systematic deviations are present, overall, the data sets agree with the models based on constant log-normal distributions, which implies that the assumption of a log-normal distribution is reasonable. This does not mean that the homoscedastic log-normal distribution is a good model overall. Actually, problems about model consistency and accordance to reality have been brought up in literature [81,125–131]. At a significance level of 5%, the Shapiro-Wilk test rejects normality for three out of twelve data sets.

References

1. Dowling, N.E. *Mechanical Behavior of Materials: Engineering Methods for Deformation, Fracture, and Fatigue*; Pearson: Boston, MA, USA, 2012.
2. Bathias, C. There is no infinite fatigue life in metallic materials. *Fatigue Fract. Eng. Mater. Struct.* **1999**, *22*, 559–565. [CrossRef]
3. Bathias, C.; Drouillac, L.; Le Francois, P. How and why the fatigue S-N curve does not approach a horizontal asymptote. *Int. J. Fatigue* **2001**, *23*, 143–151. [CrossRef]
4. Schwerdt, D. *Schwingfestigkeit und Schädigungsmechanismen der Aluminiumlegierungen EN AW-6056 und EN AW-6082 sowie des Vergütungsstahls 42CrMo4 bei sehr hohen Schwingspielzahlen*; TU Darmstadt: Darmstadt, Germany, 2011.
5. Oechsner, M.; Hanselka, H.; Schneider, N.; Eufinger, J.; Pyttel, B.; Berger, C. *Schlussbericht zu BMWi 324 ZN: Bauteilauslegung unter Berücksichtigung von Beanspruchungen mit variablen Amplituden und sehr hohen Schwingspielzahlen*; MPA-IfW—Technische Universität Darmstadt: Darmstadt, Germany, 2012.
6. Schneider, N. Ermüdungsfestigkeit bei sehr hohen Schwingspielzahlen unter Berücksichtigung des Einflusses der Prüffrequenz. Ph.D. Thesis, TU Darmstadt, Darmstadt, Germany, 2014.
7. Schneider, N.; Schwerdt, D.; Wuttke, U.; Oechsner, M. Fatigue at high number of cyclic loads: Testing methods and failure mechanism. *Mater. Und Werkst.* **2015**, *46*, 931–941. [CrossRef]
8. Krupp, U.; Giertler, A.; Koschella, K. Microscopic Damage Evolution During Very High Cycle Fatigue (VHCF) of Tempered Martensitic Steel. *Procedia Eng.* **2016**, *160*, 231–238. [CrossRef]
9. Schneider, N.; Bödecker, J.; Berger, C.; Oechsner, M. Frequency effect and influence of testing technique on the fatigue behaviour of quenched and tempered steel and aluminium alloy. *Int. J. Fatigue* **2016**, *93*, 224–231. [CrossRef]
10. Oechsner, M.; Melz, T.; Kaffenberger, M.; Störzel, K. *Schlussbericht zu AiF A 18198 N: Bauteilauslegung unter Berücksichtigung von Beanspruchungen mit variablen Amplituden und sehr hohen Schwingspielzahlen*; MPA-IfW—Technische Universität Darmstadt: Darmstadt, Germany, 2017.
11. Pyttel, B.; Brunner, I.; Kaiser, B.; Berger, C.; Mahendran, M. Fatigue behaviour of helical compression springs at a very high number of cycles—Investigation of various influences. *Int. J. Fatigue* **2014**, *60*, 101–109. [CrossRef]
12. Pyttel, B.; Fernández Canteli, A.; Argente Ripoll, A. Comparison of different statistical models for description of fatigue including very high cycle fatigue. *Int. J. Fatigue* **2016**, *93*, 435–442. [CrossRef]
13. Mayer, H.; Schuller, R.; Karr, U.; Fitzka, M.; Irrasch, D.; Hahn, M.; Bacher-Höchst, M. Mean stress sensitivity and crack initiation mechanisms of spring steel for torsional and axial VHCF loading. *Int. J. Fatigue* **2016**, *93*, 309–317. [CrossRef]
14. Mayer, H. Recent developments in ultrasonic fatigue. *Fatigue Fract. Eng. Mater. Struct.* **2016**, *39*, 3–29. [CrossRef]
15. Jeddi, D.; Palin-Luc, T. A review about the effects of structural and operational factors on the gigacycle fatigue of steels. *Fatigue Fract. Eng. Mater. Struct.* **2018**, *41*, 969–990. [CrossRef]
16. Bach, J. Ermüdungsverhalten von Niedrig Legierten Stählen im HCF-und VHCF-Bereich. Ph.D. Thesis, Friedrich-Alexander-Universität Erlangen-Nürnberg, Erlangen, Germany, 2018.
17. Bach, J.; Göken, M.; Höppel, H.W. Fatigue of low alloyed carbon steels in the HCF/VHCF-regimes. In *Fatigue of Materials at Very High Numbers of Loading Cycles*; Springer Fachmedien Wiesbaden: Wiesbaden, Germany, 2018; pp. 1–23.
18. Cai, Y.; Zhao, Y.; Ma, X.; Yang, Z.; Ding, Y. An extended model for fatigue life prediction and acceleration considering load frequency effect. *IEEE Access* **2018**, *6*, 21064–21074. [CrossRef]

19. Hu, Y.; Sun, C.; Xie, J.; Hong, Y. Effects of loading frequency and loading type on high-cycle and very-high-cycle fatigue of a high-strength steel. *Materials* **2018**, *11*, 1456. [CrossRef] [PubMed]

20. Geilen, M.B.; Klein, M.; Oechsner, M. Very high cycle fatigue behaviour of compression springs under constant and variable amplitude loading. *Mater. Und Werkst.* **2019**, *50*, 1301–1316. [CrossRef]

21. Geilen, M.B.; Klein, M.; Oechsner, M.; Kaffenberger, M.; Störzel, K.; Melz, T. A Method for the Strain Rate Dependent Correction for Control Type of Fatigue Tests. *Int. J. Fatigue* **2020**, *138*, 105726. [CrossRef]

22. Atrens, A.; Hoffelner, W.; Duerig, T.W.; Allison, J.E. Subsurface crack initiation in high cycle fatigue in Ti6A14V and in a typical martensitic stainless steel. *Scr. Metall.* **1983**, *17*, 601–606. [CrossRef]

23. Furuya, Y.; Matsuoka, S.; Abe, T.; Yamaguchi, K. Gigacycle fatigue properties for high-strength low-alloy steel at 100 Hz, 600 Hz, and 20 kHz. *Scr. Mater.* **2002**, *46*, 157–162. [CrossRef]

24. Furuya, Y.; Matsuoka, S. Gigacycle fatigue properties of a modified-ausformed Si-Mn steel and effects of microstructure. *Metall. Mater. Trans. A* **2004**, *35*, 1715–1723. [CrossRef]

25. Tsutsumi, N.; Murakami, Y.; Doquet, V. Effect of test frequency on fatigue strength of low carbon steel. *Fatigue Fract. Eng. Mater. Struct.* **2009**, *32*, 473–483. [CrossRef]

26. Müller-Bollenhagen, C.; Zimmermann, M.; Christ, H.J. Adjusting the very high cycle fatigue properties of a metastable austenitic stainless steel by means of the martensite content. *Procedia Eng.* **2010**, *2*, 1663–1672. [CrossRef]

27. Pyttel, B.; Schwerdt, D.; Berger, C. Fatigue strength and failure mechanisms in the VHCF-region for quenched and tempered steel 42CrMoS4 and consequences to fatigue design. *Procedia Eng.* **2010**, *2*, 1327–1336. [CrossRef]

28. Setowaki, S.; Ichikawa, Y.; Nonaka, I. Effect of frequency on high cycle fatigue strength of railway axle steel. In Proceedings of fifth International Conference on Very High Cycle Fatigue, VHCF5, Berlin, Germany, 28–30 June 2011; pp. 153–158.

29. Nonaka, I.; Setowaki, S.; Ichikawa, Y. Effect of load frequency on high cycle fatigue strength of bullet train axle steel. *Int. J. Fatigue* **2014**, *60*, 43–47. [CrossRef]

30. Guennec, B.; Ueno, A.; Sakai, T.; Takanashi, M.; Itabashi, Y. Effect of the loading frequency on fatigue properties of JIS S15C low carbon steel and some discussions based on micro-plasticity behavior. *Int. J. Fatigue* **2014**, *66*, 29–38. [CrossRef]

31. Zhu, M.L.; Liu, L.L.; Xuan, F.Z. Effect of frequency on very high cycle fatigue behavior of a low strength Cr–Ni–Mo–V steel welded joint. *Int. J. Fatigue* **2015**, *77*, 166–173. [CrossRef]

32. Torabian, N.; Favier, V.; Dirrenberger, J.; Adamski, F.; Ziaei-Rad, S.; Ranc, N. Correlation of the high and very high cycle fatigue response of ferrite based steels with strain rate-temperature conditions. *Acta Mater.* **2017**, *134*, 40–52. [CrossRef]

33. Mayer, H.; Papakyriacou, M.; Pippan, R.; Stanzl-Tschegg, S. Influence of loading frequency on the high cycle fatigue properties of AlZnMgCu1. 5 aluminium alloy. *Mater. Sci. Eng. A* **2001**, *314*, 48–54. [CrossRef]

34. Papakyriacou, M.; Mayer, H.; Pypen, C.; Plenk, H., Jr.; Stanzl-Tschegg, S. Influence of loading frequency on high cycle fatigue properties of bcc and hcp metals. *Mater. Sci. Eng. A* **2001**, *308*, 143–152. [CrossRef]

35. Mayer, H.; Ede, C.; Allison, J.E. Influence of cyclic loads below endurance limit or threshold stress intensity on fatigue damage in cast aluminium alloy 319-T7. *Int. J. Fatigue* **2005**, *27*, 129–141. [CrossRef]

36. Engler-Pinto, C.C., Jr.; Frisch Sr, R.J.; Lasecki, J.V.; Mayer, H.; Allison, J.E. Effect of frequency and environment on high cycle fatigue of cast aluminum alloys. In Proceedings of the VHCF-Fourth International Conference on Very High Cycle Fatigue, Ann Arbor, MI, USA, 19–22 August 2007; pp. 421–427.

37. Takeuchi, E.; Furuya, Y.; Nagashima, N.; Matsuoka, S. The effect of frequency on the giga–cycle fatigue properties of a Ti–6Al–4V alloy. *Fatigue Fract. Eng. Mater. Struct.* **2008**, *31*, 599–605. [CrossRef]

38. Zhu, X.; Jones, J.W.; Allison, J.E. Effect of frequency, environment, and temperature on fatigue behavior of E319 cast aluminum alloy: Stress-controlled fatigue life response. *Metall. Mater. Trans. A* **2008**, *39*, 2681–2688. [CrossRef]

39. Begum, S.; Chen, D.L.; Xu, S.; Luo, A.A. Effect of strain ratio and strain rate on low cycle fatigue behavior of AZ31 wrought magnesium alloy. *Mater. Sci. Eng. A* **2009**, *517*, 334–343. [CrossRef]

40. Protz, R.; Kosmann, N.; Gude, M.; Hufenbach, W.; Schulte, K.; Fiedler, B. Voids and their effect on the strain rate dependent material properties and fatigue behaviour of non-crimp fabric composites materials. *Compos. Part B Eng.* **2015**, *83*, 346–351. [CrossRef]

41. Jost, B.; Klein, M.; Beck, T.; Eifler, D. Temperature dependent cyclic deformation and fatigue life of EN-GJS-600 (ASTM 80-55-06) ductile cast iron. *Int. J. Fatigue* **2017**, *96*, 102–113. [CrossRef]

42. Jost, B.; Klein, M.; Beck, T.; Eifler, D. Temperature and frequency influence on the cyclic deformation behavior of EN-GJS-600 (ASTM 80-55-06) ductile cast iron at 0.005 and 5 Hz. *Int. J. Fatigue* **2018**, *110*, 225–237. [CrossRef]

43. Wang, M.; Pang, J.C.; Liu, H.Q.; Zou, C.L.; Li, S.X.; Zhang, Z.F. Deformation mechanism and fatigue life of an Al-12Si alloy at different temperatures and strain rates. *Int. J. Fatigue* **2019**, *127*, 268–274. [CrossRef]

44. Campbell, J.D.; Ferguson, W.G. The temperature and strain-rate dependence of the shear strength of mild steel. *Philos. Mag.* **1970**, *21*, 63–82. [CrossRef]

45. Jaske, C.E.; Leis, B.N.; Pugh, C.E. Monotonic and cyclic stress-strain response of annealed 2 1/4 Cr–1 Mo steel; Oak Ridge National Lab, Tenn. USA, 1975.

46. El-Magd, E. Influence of strain rate on ductility of metallic materials. *Steel Res.* **1997**, *68*, 67–71. [CrossRef]

47. Bleck, W.; Schael, I. Determination of crash–relevant material parameters by dynamic tensile tests. *Steel Res.* **2000**, *71*, 173–178. [CrossRef]

48. Børvik, T.; Hopperstad, O.S.; Berstad, T. On the influence of stress triaxiality and strain rate on the behaviour of a structural steel. Part II. Numerical study. *Eur. J. Mech.-A/Solids* **2003**, *22*, 15–32. [CrossRef]

49. Hopperstad, O.S.; Børvik, T.; Langseth, M.; Labibes, K.; Albertini, C. On the influence of stress triaxiality and strain rate on the behaviour of a structural steel. Part I. Experiments. *Eur. J. Mech.-A/Solids* **2003**, *22*, 1–13. [CrossRef]

50. Lee, W.S.; Liu, C.Y. The effects of temperature and strain rate on the dynamic flow behaviour of different steels. *Mater. Sci. Eng. A* **2006**, *426*, 101–113. [CrossRef]

51. Böhme, W.; Luke, M; Blauel, J. G.; Sun, D.- Z. Rohr, I.; Harwick, W *FAT-Richtlinie Dynamische Werkstoffkennwerte für die Crashsimulation*; FAT-Schriftenreihe: Frankfurt am Main, Germany, 2007; Volume 211.

52. Emde, T. *Mechanisches Verhalten metallischer Werkstoffe über weite Bereiche der Dehnung, der Dehnrate und der Temperatur*; Verlagshaus Mainz GmbH: Aachen, Germany, 2009.

53. Benson, D.K.; Hancock, J.R. The effect of strain rate on the cyclic response of metals. *Metall. Trans.* **1974**, *5*, 1711–1715. [CrossRef]

54. Pineau, A.; Benzerga, A.A.; Pardoen, T. Failure of metals I: Brittle and ductile fracture. *Acta Mater.* **2016**, *107*, 424–483. [CrossRef]

55. Hodowany, J.; Ravichandran, G.; Rosakis, A.J.; Rosakis, P. Partition of plastic work into heat and stored energy in metals. *Exp. Mech.* **2000**, *40*, 113–123. [CrossRef]

56. Cho, K.M.; Lee, S.; Nutt, S.R.; Duffy, J. Adiabatic shear band formation during dynamic torsional deformation of an HY-100 steel. *Acta Metall. Mater.* **1993**, *41*, 923–932. [CrossRef]

57. Kapoor, R.; Nemat-Nasser, S. Determination of temperature rise during high strain rate deformation. *Mech. Mater.* **1998**, *27*, 1–12. [CrossRef]

58. Rosakis, P.; Rosakis, A.J.; Ravichandran, G.; Hodowany, J. A thermodynamic internal variable model for the partition of plastic work into heat and stored energy in metals. *J. Mech. Phys. Solids* **2000**, *48*, 581–607. [CrossRef]

59. Holzer, A.J.; Brown, R.H. Mechanical behavior of metals in dynamic compression. *J. Eng. Mater. Technol.* **1979**, *101*, 238–247. [CrossRef]

60. Klepaczko. An experimental technique for shear testing at high and very high strain rates. The case of a mild steel. *Int. J. Impact Eng.* **1994**, *15*, 25–39. [CrossRef]

61. Johnson, G.R.; Cook, W.H. Fracture characteristics of three metals subjected to various strains, strain rates, temperatures and pressures. *Eng. Fract. Mech.* **1985**, *21*, 31–48. [CrossRef]

62. Zerbst, U.; Vormwald, M.; Pippan, R.; Gänser, H.P.; Sarrazin-Baudoux, C.; Madia, M. About the fatigue crack propagation threshold of metals as a design criterion–a review. *Eng. Fract. Mech.* **2016**, *153*, 190–243. [CrossRef]

63. Tanaka, H.; Homma, N.; Matsuoka, S.; Murakami, Y. Effect of hydrogen and frequency on fatigue behavior of SCM435 steel for storage cylinder of hydrogen station. *Nihon Kikai Gakkai Ronbunshu Hen/Trans. Jpn. Soc. Mech. Eng. Part A* **2007**, *73*, 1358–1365.

64. Murakami, Y.; Matsuoka, S. Effect of hydrogen on fatigue crack growth of metals. *Eng. Fract. Mech.* **2010**, *77*, 1926–1940. [CrossRef]

65. Suresh, S.; Ritchie, R.O. Near-threshold fatigue crack propagation: A perspective on the role of crack closure. In *Fatigue Crack Growth Threshold Concepts*; The Metallurgical Society Inc.: Warrendale, PA, USA, 1983.

66. Blom, A.F.; Holm, D.K. An experimental and numerical study of crack closure. *Eng. Fract. Mech.* **1985**, *22*, 997–1011. [CrossRef]

67. Vasudeven, A.K.; Sadananda, K.; Louat, N. A review of crack closure, fatigue crack threshold and related phenomena. *Mater. Sci. Eng. A* **1994**, *188*, 1–22. [CrossRef]

68. Chermahini, R.G.; Shivakumar, K.N.; Newman, J.C., Jr.; Blom, A.F. Three-dimensional aspects of plasticity-induced fatigue crack closure. *Eng. Fract. Mech.* **1989**, *34*, 393–401. [CrossRef]

69. Pippan, R.; Hohenwarter, A. Fatigue crack closure: A review of the physical phenomena. *Fatigue Fract. Eng. Mater. Struct.* **2017**, *40*, 471–495. [CrossRef]

70. Jiang, Y.; Feng, M.; Ding, F. A reexamination of plasticity-induced crack closure in fatigue crack propagation. *Int. J. Plast.* **2005**, *21*, 1720–1740. [CrossRef]

71. Weibull, W. *A Statistical Theory of Strength of Materials*; Generalstabens litografiska anstalts förlag: Stockholm, Sweden, 1939.

72. Kuhn, P. Effect of geometric size on notch fatigue. In *Colloquium on Fatigue*; Springer: Berlin/Heidelberg, Germany, 1956; pp. 131–140.

73. Böhm, J.; Heckel, K. Die Vorhersage der Dauerschwingfestigkeit unter Berücksichtigung des statistischen Größeneinflusses. *Mater. Und Werkst.* **1982**, *13*, 120–128. [CrossRef]

74. Liu, J.; Zenner, H. Berechnung der Dauerschwingfestigkeit unter Berücksichtigung der spannungsmechanischen und statistischen Stützziffer. *Mater. Und Werkst.* **1991**, *22*, 187–196. [CrossRef]

75. Liu, J.; Zenner, H. Berechnung von Bauteilwöhlerlinien unter Berücksichtigung der statistischen und spannungsmechanischen Stützziffer. *Mater. Und Werkst.* **1995**, *26*, 14–21. [CrossRef]

76. Kloos, K.H.; Kaiser, B.; Friederich, H. Einfluß der Probengröße auf das Verformungs–und Versagensverhalten von Vergütungsstahl 42 CrMo 4 unter Schwingbeanspruchung. *Mater. Und Werkst.* **1995**, *26*, 330–336. [CrossRef]

77. Friederich, H.; Kaiser, B.; Kloos, K.H. Anwendung der Fehlstellentheorie nach Weibull zur Berechnung des statistischen Größeneinflusses bei Dauerschwingbeanspruchung. *Mater. Und Werkst.* **1998**, *29*, 178–184. [CrossRef]

78. Bomas, H.; Linkewitz, T.; Mayr, P. Application of a weakest–link concept to the fatigue limit of the bearing steel SAE 52100 in a bainitic condition. *Fatigue Fract. Eng. Mater. Struct.* **1999**, *22*, 733–741. [CrossRef]

79. Makkonen, M. Statistical size effect in the fatigue limit of steel. *Int. J. Fatigue* **2001**, *23*, 395–402. [CrossRef]

80. Flacelière, L.; Morel, F. Probabilistic approach in high–cycle multiaxial fatigue: Volume and surface effects. *Fatigue Fract. Eng. Mater. Struct.* **2004**, *27*, 1123–1135. [CrossRef]

81. Castillo, E.; Fernandez-Canteli, A. *A Unified Statistical Methodology for Modeling Fatigue Damage*; Springer Science & Business Media: Berlin/Heildelberg, Germay, 2009.

82. Hertel, O.; Vormwald, M. Statistical and geometrical size effects in notched members based on weakest-link and short-crack modelling. *Eng. Fract. Mech.* **2012**, *95*, 72–83. [CrossRef]

83. Ai, Y.; Zhu, S.P.; Liao, D.; Correia, J.A.; de Jesus, A.M.P.; Keshtegar, B. Probabilistic modelling of notch fatigue and size effect of components using highly stressed volume approach. *Int. J. Fatigue* **2019**, *127*, 110–119. [CrossRef]

84. Vaara, J.; Väntänen, M.; Kämäräinen, P.; Kemppainen, J.; Frondelius, T. Bayesian analysis of critical fatigue failure sources. *Int. J. Fatigue* **2020**, *130*, 105282. [CrossRef]

85. Rennert, R.; Kullig, E.; Vormwald, M.; Esderts, A.; Siegele, D. *Rechnerischer Festigkeitsnachweis für Maschinenbauteile aus Stahl, Eisenguss- und Aluminiumwerkstoffen*, 6th ed.; FKM-Richtlinie, VDMA Verlag: Frankfurt am Main, Germany, 2012.

86. Fiedler, M.; Wächter, M.; Varfolomeev, I.; Vormwald, M.; Esderts, A. *Rechnerischer Festigkeitsnachweis unter Expliziter Erfassung Nichtlinearen Werkstoffverhaltens für Maschinenbauteile aus Stahl, Stahlguss und Aluminiumknetlöegierungen*, 1st ed.; FKM-Richtlinie, VDMA Verlag: Frankfurt am Main, Germany, 2019.

87. Kletzin, U.; Reich, R.; Oechsner, M.; Spies, A.; Pyttel, B.; Hanning, G.; Rennert, R.; Kullig, E. *Rechnerischer Festigkeitsnachweis für Federn und Federelemente*, 1st ed.; FKM-Richtlinie, VDMA Verlag: Frankfurt am Main, Germany, 2020.

88. Skelton, R.P.; Haigh. Fatigue crack growth rates and thresholds in steels under oxidising conditions. *Mater. Sci. Eng.* **1978**, *36*, 17–25. [CrossRef]

89. Suresh, S.; Zamiski, G.F.; Ritchie, D.R.O. Oxide-induced crack closure: An explanation for near-threshold corrosion fatigue crack growth behavior. *Metall. Mater. Trans. A* **1981**, *12*, 1435–1443. [CrossRef]
90. Liaw, P.K.; Leax, T.R.; Williams, R.S.; Peck, M.G. Influence of oxide-induced crack closure on near-threshold fatigue crack growth behavior. *Acta Metall.* **1982**, *30*, 2071–2078. [CrossRef]
91. Maierhofer, J.; Simunek, D.; Gänser, H.P.; Pippan, R. Oxide induced crack closure in the near threshold regime: The effect of oxide debris release. *Int. J. Fatigue* **2018**, *117*, 21–26. [CrossRef]
92. Riemelmoser, F.O.; Gumbsch, P.; Pippan, R. Dislocation modelling of fatigue cracks: An overview. *Mater. Trans.* **2001**, *42*, 2–13. [CrossRef]
93. Lee, H.H.; Uhlig, H.H. Corrosion fatigue of type 4140 high strength steel. *Metall. Mater. Trans. B* **1972**, *3*, 2949–2957. [CrossRef]
94. Shives, T.R.; Bennett, J.A. *The Effect of Environment on the Fatigue Strengths of Four Selected Alloys*; National Aeronautics and Space Administration: Washington, DC, USA, 1965.
95. Wadsworth, N.J. The influence of atmospheric corrosion on the fatigue limit of iron-0.5% carbon. *Philos. Mag.* **1961**, *6*, 397–401. [CrossRef]
96. Yoshinaka, F.; Nakamura, T. Effect of vacuum environment on fatigue fracture surfaces of high strength steel. *Mech. Eng. Lett.* **2016**, *2*, 15–00730. [CrossRef]
97. Spriestersbach, D. VHCF-Verhalten des hochfesten Stahls 100Cr6: Rissinitiierungsmechanismen und Schwellenwerte. Ph.D. Thesis, TU Kaiserslautern, Kaiserslautern, Germany, 2020.
98. Wei, R.; Talda, P.; Li, C.Y. Fatigue-crack propagation in some ultrahigh-strength steels. In *Fatigue Crack Propagation*; ASTM International: West Conshohocken, PA, USA, 1967.
99. Dahlberg, E.P. Fatigue-crack propagation in high-strength 4340 steel in humid air(Water vapor effects on fatigue crack propagation rates in high strength 4340 steel examined by optical and electron fractography). *ASM Trans. Q.* **1965**, *58*, 46–53.
100. Achter, M. Effect of environment on fatigue cracks. In *Fatigue Crack Propagation*; ASTM International: West Conshohocken, PA, USA, 1967.
101. Henaff, G.; Marchal, K.; Petit, J. On fatigue crack propagation enhancement by a gaseous atmosphere: experimental and theoretical aspects. *Acta Metall. Mater.* **1995**, *43*, 2931–2942. [CrossRef]
102. Spriestersbach, D.; Grad, P.; Brodyanski, A.; Lösch, J.; Kopnarski, M.; Kerscher, E. Very high cycle fatigue crack initiation: Investigation of fatigue mechanisms and threshold values for 100Cr6. In *Fatigue of Materials at Very High Numbers of Loading Cycles*; Springer: Berlin, Germany, 2018; pp. 167–210.
103. Thompson, N.; Wadsworth, N.J.; Louat, N. The origin of fatigue fracture in copper. *Philos. Mag.* **1956**, *1*, 113–126. [CrossRef]
104. Duquette, D.J. *Environmental Effects on General Fatigue Resistance and Crack Nucleation in Metals and Alloys*; Technical Report; Office of Naval Research: Troy, NY, USA, 1978.
105. Hudson, C.M.; Seward, S.K. A literature review and inventory of the effects of environment on the fatigue behavior of metals. *Eng. Fract. Mech.* **1976**, *8*, 315–329. [CrossRef]
106. Krupp, U. Mikrostrukturelle Aspekte der Rissinitiierung und-ausbreitung in metallischen Werkstoffen. Habilitation Thesis, Universität Siegen, Siegen, Germany, 2004.
107. Bradhurst, D.H.; Leach, J.L. The mechanical properties of thin anodic films on aluminum. *J. Electrochem. Soc.* **1966**, *113*, 1245. [CrossRef]
108. Grosskreutz, J.C. The effect of oxide films on dislocation-surface interactions in aluminum. *Surf. Sci.* **1967**, *8*, 173–190. [CrossRef]
109. Nakamura, T.; Oguma, H.; Shinohara, Y. The effect of vacuum-like environment inside sub-surface fatigue crack on the formation of ODA fracture surface in high strength steel. *Procedia Eng.* **2010**, *2*, 2121–2129. [CrossRef]
110. Petit, J.; Sarrazin-Baudoux, C. An overview on the influence of the atmosphere environment on ultra-high-cycle fatigue and ultra-slow fatigue crack propagation. *Int. J. Fatigue* **2006**, *28*, 1471–1478. [CrossRef]
111. Murakami, Y.; Nomoto, T.; Ueda, T.; Murakami, Y. On the mechanism of fatigue failure in the superlong life regime (N> 10000000 cycles). Part 1: Influence of hydrogen trapped by inclusions. *Fatigue Fract. Eng. Mater. Struct.* **2000**, *23*, 893–902. [CrossRef]

112. Murakami, Y.; Nomoto, T.; Ueda, T.; Murakami, Y. On the mechanism of fatigue failure in the superlong life regime (N> 10000000 cycles). Part II: Influence of hydrogen trapped by inclusions. *Fatigue Fract. Eng. Mater. Struct.* **2000**, *23*, 903–910. [CrossRef]

113. Barrera, O.; Tarleton, E.; Tang, H.W.; Cocks, A.C. Modelling the coupling between hydrogen diffusion and the mechanical behaviour of metals. *Comput. Mater. Sci.* **2016**, *122*, 219–228. [CrossRef]

114. Barrera, O.; Bombac, D.; Chen, Y.; Daff, T.D.; Galindo-Nava, E.; Gong, P.; Haley, D.; Horton, R.; Katzarov, I.; Kermode. Understanding and mitigating hydrogen embrittlement of steels: A review of experimental, modelling and design progress from atomistic to continuum. *J. Mater. Sci.* **2018**, *53*, 6251–6290. [CrossRef]

115. Giertler, A. Volume 2020, Band 2, Berichte aus dem Institut für Eisenhüttenkunde. In *Mechanismen der Rissentstehung und -ausbreitung im Vergütungsstahl 50CrMo4 bei sehr hohen Lastspielzahlen*; Shaker Verlag: Düren, Germany, 2020.

116. Glaser, A. Mittelspannungseinfluß auf das Verformungsverhalten von Ck 45 und 42 CrMo 4 bei Spannungs-Und Dehnungskontrollierter Homogen-Einachsiger Schwingbeanspruchung. Ph.D. Thesis, KIT, Karlsruhe, Germany, 1988.

117. Ramberg, W.; Osgood, W.R. *Description of Stress-Strain Curves by Three Parameters*; National Advisory Comitee for Aeronatuics: Washington, DC, USA, 1943.

118. Neuber, H. Über die Berücksichtigung der Spannungskonzenlration bei Festigkeitsproblemen. *Konstr. Berl.* **1968**, *20*, 245-251.

119. Topper, T.; Wetzel, R.M.; Morrow, J. *Neuber's Rule Applied to Fatigue of Notched Specimens*; U.S. Naval Air Engineering Center: Philadelphia, PA, USA, 1967.

120. Wächter, M. *Zur Ermittlung von zyklischen Werkstoffkennwerten und Schädigungsparameterwöhlerlinien*; Technische Universität Clausthal: Clausthal-Zellerfeld, Germany, 2016.

121. Müller, C. *Zur Statistischen Auswertung Experimenteller Wöhlerlinien*; Universitätsbibliothek Clausthal: Clausthal-Zellerfeld, Germany, 2015.

122. Müller, C.; Wächter, M.; Masendorf, R.; Esderts, A. Distribution functions for the linear region of the SN curve. *Mater. Test.* **2017**, *59*, 625–629. [CrossRef]

123. Radaj, D.; Vormwald, M. *Ermüdungsfestigkeit*; Springer: Berlin, Germany, 2007.

124. Virtanen, P.; Gommers, R.; Oliphant, T.E.; Haberland, M.; Reddy, T.; Cournapeau, D.; Burovski, E.; Peterson, P.; Weckesser, W.; Bright, J. SciPy 1.0–Fundamental Algorithms for Scientific Computing in Python. *arXiv* **2019**, arXiv:1907.10121.

125. Geilen, M.B.; Klein, M.; Oechsner, M. On the Influence of Ultimate Number of Cycles on Lifetime Prediction for Compression Springs Manufactured from VDSiCr Class Spring Wire. *Materials* **2020**, *13*, 3222. [CrossRef]

126. Bomas, H.; Burkart, K.; Zoch, H.W. Evaluation of S–N curves with more than one failure mode. *Int. J. Fatigue* **2011**, *33*, 19–22. [CrossRef]

127. Paolino, D.S.; Chiangdussi, G.; Rosetto, M. A unified statistical model for S–N fatigue curves: Probabilistic definition. *Fatigue Fract. Eng. Mater. Struct.* **2013**, *36*, 187–201. [CrossRef]

128. Haydn, W.; Schwabe, F. Ermittlung geschlossener probabilistischer Auslegungskennlinien für einstufige Schwingfestigkeitsbeanspruchungen vom LCF- bis VHCF-Bereich. *Mater. Und Werkst.* **2015**, *46*, 591–602. [CrossRef]

129. Fernández-Canteli, A.; Blasón, S.; Pyttel, B.; Muniz-Calvente, M.; Castillo, E. Considerations about the existence or non-existence of the fatigue limit: Implications on practical design. *Int. J. Fract.* **2020**, *223*, 189–196. [CrossRef]

130. Muniz-Calvente, M.; Fernández-Canteli, A.; Pyttel, B.; Castillo, E. Probabilistic assessment of VHCF data as pertaining to concurrent populations using a Weibull regression model. *Fatigue Fract. Eng. Mater. Struct.* **2017**, *40*, 1772–1782. [CrossRef]

131. Kobelev, V. A proposal for unification of fatigue crack growth. *J. Phys. Conf. Ser.* **2018**, *843*, 012022. [CrossRef]

Publisher's Note: MDPI stays neutral with regard to jurisdictional claims in published maps and institutional affiliations.

© 2020 by the authors. Licensee MDPI, Basel, Switzerland. This article is an open access article distributed under the terms and conditions of the Creative Commons Attribution (CC BY) license (http://creativecommons.org/licenses/by/4.0/).

 *metals*

Article

Fatigue Fracture Mechanism of a Nickel-Based Single Crystal Superalloy with Partially Recrystallized Grains at 550 °C by In Situ SEM Studies

Hao Yang [1], Jishen Jiang [2], Zhuozheng Wang [2], Xianfeng Ma [2,*], Jiajun Tu [2], Hui-ji Shi [1], Hailin Zhai [2] and Wenjie Zhang [2]

[1] AML, Department of Engineering Mechanics, Tsinghua University, Beijing 100084, China; yanghao06@tsinghua.org.cn (H.Y.); shihj@mail.tsinghua.edu.cn (H.-j.S.)

[2] Sino-French Institute of Nuclear Engineering and Technology, Sun Yat-sen University, Zhuhai 519082, China; jiangjsh3@mail.sysu.edu.cn (J.J.); xll0302@foxmail.com (Z.W.); timothee_tujiajun@foxmail.com (J.T.); zhaihlin@mail2.sysu.edu.cn (H.Z.); zhangwj25@mail2.sysu.edu.cn (W.Z.)

* Correspondence: maxf6@mail.sysu.edu.cn; Tel.: +86-756-366-8090

Received: 26 June 2020; Accepted: 21 July 2020; Published: 27 July 2020

Abstract: The fatigue fracture mechanism of a nickel-based single crystal (NBSC) superalloy with recrystallized grains was studied at 550 °C by in situ observation with a scanning electron microscope (SEM) for the first time. Multiple crack initiations associated with recrystallized grain boundaries and carbides were observed. By analysis of the slip traces and crack propagation planes, the operated slip systems were identified to be octahedral for both single crystal substrate and recrystallized grains. Distinct crystallographic fractures dominated, accompanied by recrystallized grain boundary associated crack initiations. This is different from the widely reported solely intergranular cracking at high temperature. Fatigue crack growth rate curves showed evident fluctuation, due to the interaction of fatigue cracks with local microstructures and the crack coalescence mechanism. Both the recrystallized grains and the competition between different slip systems were responsible for the deceleration and acceleration of fatigue microstructurally small crack behavior.

Keywords: single crystal superalloy; recrystallization; fatigue small crack; slip; in situ SEM

---

1. Introduction

Nickel-based single crystal (NBSC) superalloys have been widely used as high temperature materials in aircraft turbines as well as in land-based gas turbines [1], due to its excellent mechanical properties under high temperature. For NBSC superalloys, since the grain boundaries have been removed, grain boundary strengthening elements, such as Hafnium, and Boron, etc. have been removed or greatly reduced to increase the melting temperature of the superalloy [2,3]. Hence, it makes single crystal superalloy susceptible to transverse grain boundaries, for instance recrystallization (RX) [2,3]. Generally, recrystallization is formed in nickel-based single crystal superalloy due to plastic deformation and subsequent exposure to high temperature [2,3]. In the turbine industry, there have been some reports on recrystallization-induced turbine blade accidents in China and other countries [4–7]. It is generally accepted that recrystallized grains may act as potential crack initiation sites and hence would be harmful to the fatigue properties of single crystal superalloys. However, little research has been found in open literature regarding the effect of RX on the fatigue behavior of NBSC superalloy [8] or directionally solidified (DS) alloys [9–11].

Among the limited studies on the fatigue issues of recrystallization, some essential questions have not yet been addressed. For instance, the primary influence factors for the fatigue fracture mechanism of recrystallized NBSC superalloy are still unclear. The conditions which determine if the partial recrystallization grains are fatal or insignificant to fatigue performance remain unknown, to the author's knowledge. Bürgel et al. [8] performed fatigue tests on recrystallized single crystal superalloy at 950 °C, suggesting the insignificant influence of recrystallization on crack initiation life, although the recrystallized sample showed higher crack density. Previous studies on partially recrystallized NBSC superalloy also suggested that 150 μm recrystallization layers could remarkably reduce the low cycle fatigue life at 550 °C [2]. Meng J et al. [5] and Zhang B et al. [7] reported that the recrystallized grains would lead to a remarkable decrease in creep life, due to the early cracking of recrystallized grain boundaries perpendicular to the loading direction. Xie G et al. [6] studied the characteristics of recrystallized grain boundaries in the creep crack initiation behavior of a directionally solidified Ni-base superalloy. The recent studies on directionally solidified DZ4 superalloy indicated that the influence of recrystallization was closely related to the microstructure of recrystallized grains. It may evidently decrease the fatigue life of DZ4 [10,12] or even increase the fatigue life by a dense recrystallized layer composed of refined grains [13]. Zhao Y et al. [11] reported that the effect of recrystallized grains led to an evident decrease in fatigue life in directionally solidified superalloy DZ40M. They suggested the effects of the cracking of recrystallized grain boundaries and twin boundaries on fatigue life. Jia B et al. [12] studied directionally Solidified DZ4 superalloy with different recrystallized layer thicknesses by shot peening and annealing, suggesting that thicker recrystallized layer leads to a larger drop in fatigue life. There are quite limited studies on the fatigue failure mechanism of partially recrystallized single crystal superalloys. The noteworthy research of Sehitoglu H. on single crystals using an in situ technique [14,15] provides good insights into the deformation and slip activities associated with fatigue crack advance. Besides, recent studies [9] have indicated that temperature played an important role in altering the fatigue fracture mode of recrystallization. Nevertheless, the effects of temperature have not yet brought conclusive results. Besides, the authors are unaware of any in situ studies on fatigue crack growth (FCG) in the RX layer of single crystal superalloy. Sehitoglu H. [16] has studied the difference between single crystal and polycrystalline Haynes 230 alloy in terms of tensile curves, slip activity, and local plastic deformation accumulation occurring in the grains and along grain boundaries, which can shed lights on the recrystallization issue. Understanding of fatigue crack initiation and propagation behavior in RX layer can be crucial for the safety evaluation of turbine blades with recrystallization issues.

In this work, the fatigue deformation and fracture behavior of a nickel-based single crystal superalloy with recrystallized grains were studied at 550 °C by in situ SEM. The real time observations were used to identify the operated slip systems and the shear stress responsible for growing fatigue crack tip. The effects of recrystallized grains and the competition between different slip systems on fatigue small crack growth behavior were analyzed based on in situ studies. The underlying mechanism for the variation of crack path and small crack growth rates was discussed accordingly.

2. Material and Experimental

The material used in this study was a nickel-based single crystal superalloy, which was developed for fabricating high-performance gas-turbine blades [17]. The nominal composition of this NBSC alloy is (in wt.%): 0.067 C, 3.9 W, 12.0 Cr, 9.0 Co, 3.6 Al, 5.0 Ta and the remainder is Ni. The slab specimen had a dog-bone shape with a 2.5 mm by 0.4 mm gauge cross section, as shown in Figure 1. In the present experimental study, the specimen/loading axis is along [011], with the transverse orientation (crack growth direction) along [100], as shown in Figure 2. It is known that multiple octahedral slip systems, i.e., {111} <110>, can be activated in the NBSC alloy. Figure 4 shows the duplex octahedral slip planes and the corresponding slip directions, respectively.

The specimen was subjected to shot peening (0.5 MPa, 20 min) on the later edge to introduce plastic deformation. A solution heat treatment was performed at 1220 °C for 4 h, followed by air cooling. To prevent oxidization, the specimens were enclosed in a quartz cuvette and slight argon gas was maintained. It has also been used in the previous studies of recrystallized DZ4 alloy [13].

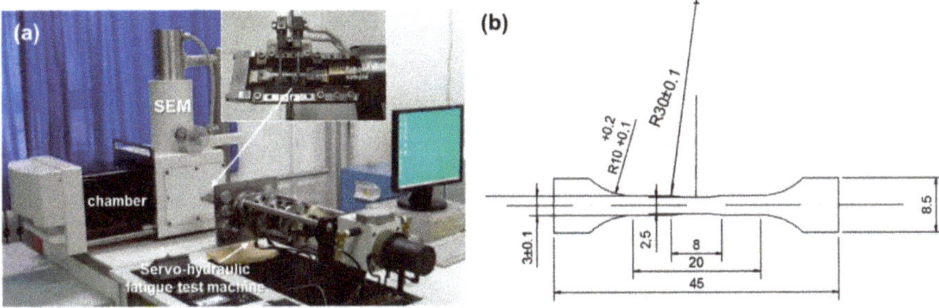

Figure 1. Fatigue test setup and specimen: (**a**) In situ SEM fatigue testing facility; (**b**) Fatigue test specimen geometry (dimensions in mm).

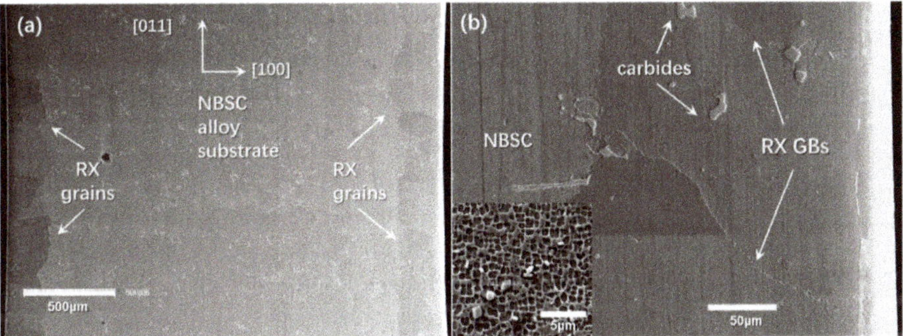

Figure 2. Single crystal superalloy with surface recrystallized grains: (**a**) top view showing recrystallized layer of 200 µm thickness; (**b**) recrystallized grain boundaries and carbides (the insert is high magnification of the γ/γ′ microstructure of single crystal superalloy).

The fatigue specimen was grinded and polished following general procedure until it was a 1 µm diamond paste. To reveal the prevailing microstructure for in situ observations, the specimen was etched in an aqueous solution of 4 g $CuSO_4$ + 20 mL HCl + 20 mL H_2O prior to fatigue tests. No effect of the etching on the crack path was found. The fatigue crack propagation tests were performed in the vacuum chamber of the SEM using a specially designed servo-hydraulic testing system (Shimadzu, Kyoto, Japan). This machine provided pulsating (sinusoidal wave) loads at 10 Hz of ±1 kN maximum capacity and a displacement range of ±25 mm. The signal of the SEM was directly transferred to a computer via a direct memory access type A/D converter, making it possible to sample 960 × 1280 frames of SEM images successively. The SEM was operated at secondary electronic mode with an accelerating voltage of 15 kV.

A constant maximal stress of 750 MPa was adopted throughout the test to study the fatigue crack propagation process at 550 °C. The selected temperature of the local region of the turbine blade where recrystallized grains were found was 550 °C. The waveform utilized was sinusoidal. All fatigue tests were load controlled at a stress ratio of 0.1 with a loading frequency of 2 Hz. Three samples were tested for the present temperature and stress. Images of fatigue crack with different lengths were taken in situ at different cycles of loading and hence the crack growth rate can be calculated from the measured crack lengths. By comparing with the propagation process, the dependence on the local structures and cracking manner can be analyzed and disclosed.

3. In Situ Observation Results

3.1. Fatigue Crack Initiation Mechanism

Figure 3a shows that duplex slip traces were formed in the nickel-based single crystal superalloy substrate, with the loading axis along the vertical direction. The slip traces were measured to be about 35°, inclined with respect to the loading direction. It was due to the fact that two octahedral slip systems have the same value of maximal Schmid factor, as shown in Figure 4. The dominated slips occurred on the two octahedral planes, i.e., ($1\bar{1}\bar{1}$) and (111), by searching all possible slip systems, either octahedral or cube. In Figure 3b, the slip traces interacted with recrystallized grains; however, they failed to enter the recrystallized grains, showing the blocking effect of recrystallized grain boundaries (GBs). After another hundred loading cycles, a crack was noticed at the NBSC/RX interface in Figure 3c, due to the impingement of slip bands against the recrystallized GBs. The intrusion and extrusion formed at the recrystallized GB would finally evolve into a micro-crack along the GB. This phenomenon has been well described in other metals by Zhang Z.F. et al. [18]. Aside from the cracking of the NBSC/RX interface, cracks were also formed at the horizontal recrystallized GB, as shown in Figure 3d, due to the deformation mismatch of adjacent grains with different crystal orientation. Besides, the blocky carbides located at the recrystallized grain boundaries were primary crack initiators due to the stress concentration and the brittleness of carbides. It is favorable for the trend of multiple crack initiation sites [2,3]. Figure 3e,f show two typical examples, which are similar to those reported in the fatigue fracture of the recrystallized directionally solidified DZ4 superalloy. Crystal plasticity finite element simulation studies [19] also indicated that the modulus, strength, and shape of carbides, and grain orientations would affect the maximal cyclic plastic strain occurring in the neighborhood of carbide and thus promote the fracture of carbide, leading to earlier crack initiation.

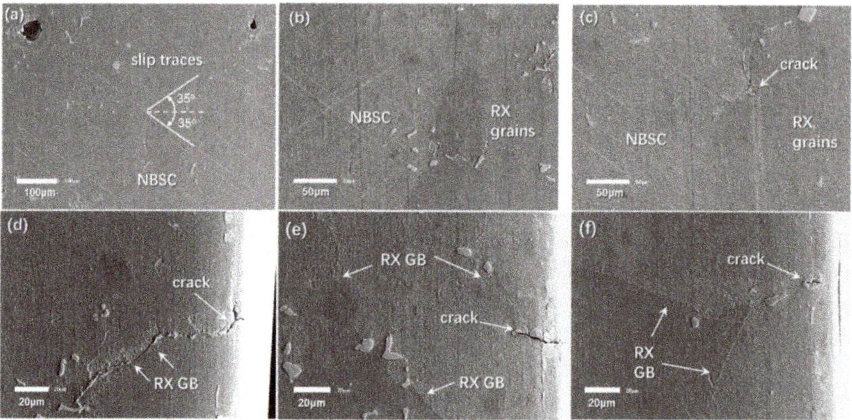

Figure 3. Fatigue deformation and crack initiation features: (**a**) duplex slip traces in single crystal substrate; (**b**) RX grains free of slip traces; (**c**) crack initiation at the interface between RX grains and single crystal substrate; (**d**) crack initiation at recrystallized grain boundary; (**e,f**) crack initiation by fracture of carbides at recrystallized grain boundary. (The loading direction is vertical.)

3.2. Fatigue Crack Propagation Behavior

In spite of multiple crack initiations in the recrystallized layer, most of the cracks appeared to be non-propagating or propagate at extremely slow growth rate in the whole period of fatigue life. Two propagating cracks of importance were identified. The propagation behavior was carefully examined to show the fatigue cracking features and to understand the underlying fracture mechanism.

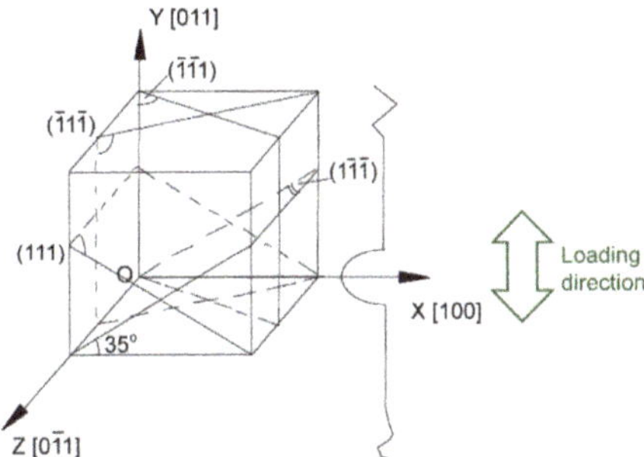

Figure 4. Crystallographic representation and slip systems of nickel-based single crystal superalloy with loading along [011] orientation.

Figure 5 depicts the evolution of the fatigue crack shown in Figure 3e. It is evident that, although the crack nucleation was caused by the carbide at the grain boundary, the subsequent propagation followed an evidently transgranular cracking manner, via the crystallographic slip mechanism, which was different from the widely reported intergranular fracture. Single slip trace was observed in the recrystallized grain, which was related to the grain orientation. The crystallographic crack appeared sharp with small crack opening displacement (COD), due to its shearing dominated nature, as shown in Figure 5d,e. While the crack approached the interface between RX and single crystal substrate, a secondary crack emerged following the same shearing direction, as shown in Figure 5e,f. The secondary crack was also associated with the carbide at the RX/NBSC interface. In the following loading cycles, the crack did not evolve into a primary crack.

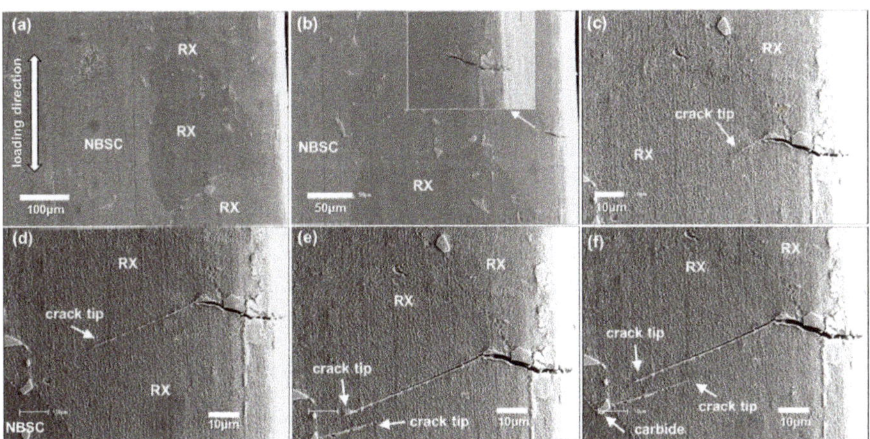

Figure 5. Fatigue crack initiation and propagation at surface recrystallized grains of nickel-based single crystal superalloy: (**a**) 0 cycle; (**b**) 392 cycles; (**c**) 31,036 cycles; (**d**) 62,956 cycles; (**e**) 115,004 cycles; (**f**) 126,276 cycles.

The corresponding fatigue crack growth rate (FCGR) was by the slope (*da/dN*) of crack length increment per cycle, given by:

$$(da/dN)_i = (a_{i+1} - a_i)/(N_{i+1} - N_i) \tag{1}$$

The stress intensity factor (SIF) range for a single edge crack plate of finite width is given by:

$$\Delta K = \Delta\sigma \sqrt{\pi a} F(a/W) \tag{2}$$

where $\Delta\sigma$ is the stress range, a is the crack length defined as the projected length vertical to the loading axis, and W is the width of the specimen. Let $\xi = a/W$, the function of $F(\xi)$ ($\xi = a/W$) is defined as in [20].

$$F(\xi) = 0.265(1 - \xi)^4 + (0.857 + 0.265\xi)(1 - \xi)^{(-3/2)} \tag{3}$$

As adopted in many studies, the crack growth rate (da/dN) was plotted against the stress intensity range (ΔK), as first proposed by Paris and Erdogan [21]. The fatigue crack growth rate curve is shown in Figure 6. The initial propagation of fatigue crack was fast at "b" and showed deceleration afterwards. The fatigue crack propagated at a slow rate, until an evident crystallographic crack tip was formed, shown in Figure 5c. The crack propagated obviously along the crystallographic plane in the next thousand cycles, shown in Figure 5c–e, at a relatively stable propagation rate, as shown in Figure 5d–e. Afterwards, as shown in Figure 5f, a secondary crack nucleating from the carbide at recrystallized GB was formed, following the same direction as the prior crack. In the remaining fatigue life cycles, these two fatigue cracks did not merge with each other and the fatigue cracks were almost non-propagating, or the propagation was negligible, as shown in Figure 5f.

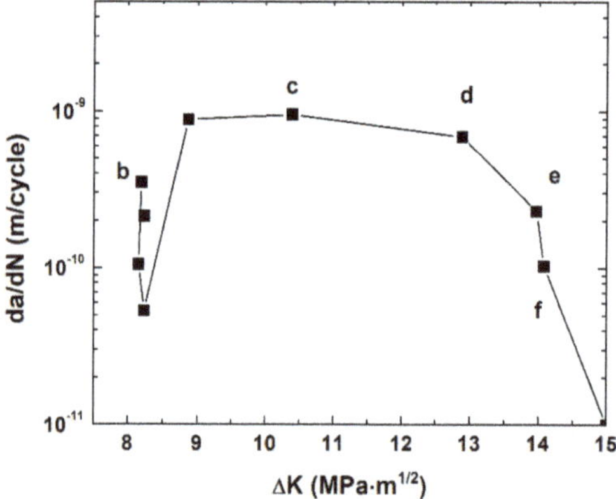

Figure 6. Fatigue crack growth rate of Crack-1 in recrystallized nickel-based single crystal superalloy.

Figure 7 shows the propagation behavior of the primary fatigue crack, which is responsible for the final fracture of the fatigue sample. Figure 7b–j correspond to "b" to "j" in the fatigue crack growth rate curve of primary crack in Figure 8. In Figure 7a, fatigue crack initiated at the RX grain boundary, due to the strain mismatch caused by different orientations of adjacent grains [19]. Of interest is that the following crack propagation exhibited transgranular mode, which was distinct from the previously reported intergranular fracture. Fatigue crack extended into both grains via crystallographic mode, as shown in Figure 7b. In Figure 7c, the primary crack tip "T1" was arrested and a secondary crack "T2" was formed. It is inferred that duplex slip systems can be preferably activated in the RX grain, similar to the [011] oriented single crystal substrate. The fatigue crack "T2" was also arrested, shown in Figure 7d, corresponding to a small FCGR marked as "d" in the FCGR curve in Figure 8. A crack, "T3", was initiated from another RX GB, as shown in the inserted micrograph of Figure 7d. Crack "T4" was formed with a direction parallel to Crack "T1". Crack "T2" become non-propagating for a long time until coalescence with Crack "T4", as shown in Figure 7e, which leads to remarkable increase in the primary crack length. It corresponds to a peak of FCGR, marked as "e" in the FCGR curve in Figure 8. A high magnification examination showed that the primary crack tip penetrated the RX/NBSC interface. In Figure 7f,g, the primary crack showed an evident increase in crack opening displacement and out of plane displacement. It is associated with the fact: (i) the primary crack is long enough to provide a higher ΔK and a larger COD; (ii) the NBSC substrate has different activated slip orientations, causing a deceleration of crack propagation in the substrate, corresponding to "g" in the

FCGR curve in Figure 8. Figure 7h show the high magnification examination of the primary crack tip in the NBSC substrate, revealing that the crack move forward again, depicted by "h" in Figure 8. The fatigue crack was overall along the horizontal, but microscopically following a zig-zag mode via operation of duplex slip systems in Figure 4. In Figure 7i, the fatigue crack was shifted to another octahedral slip plane, which decreased the FCGR, marked as "i" in Figure 8. The slip bands become denser compared with that in Figure 7f, and the COD was remarkably large, with the crack mating surfaces partly contacted with each other, as arrowed in Figure 7f. As shown in Figure 7j, the crack propagation came back to the zig-zag cracking mode via duplex slip systems, as marked by the dashed lines "C3". For the carbides perpendicular to slip bands "C2", the interaction of slip bands against carbide led to the opening of the interface and a crack was formed.

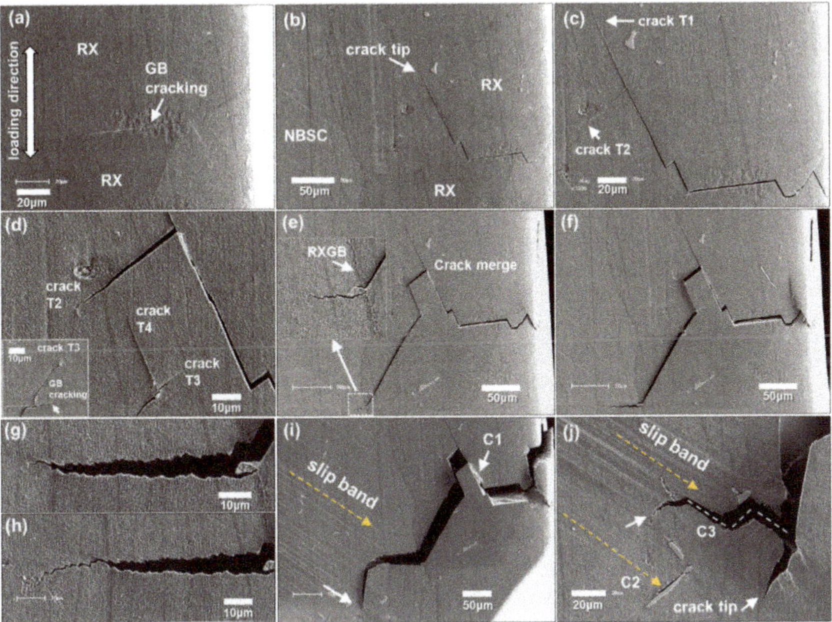

Figure 7. Fatigue propagation behavior of recrystallized single crystal superalloy: (**a**) 115,004 cycles; (**b**) 125,316 cycles; (**c**) 137,266 cycles; (**d**) 149,841 cycles; (**e**) 152,223 cycles; (**f**) 157,161 cycles; (**g**)158,531 cycles; (**h**) 159,858 cycles; (**i**) 162,326 cycles; (**j**) 163,571 cycles.

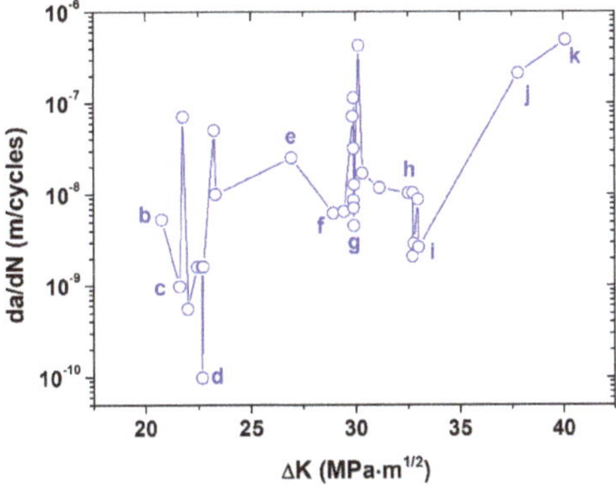

Figure 8. Fatigue crack growth rate curves of the primary crack (da/dN in terms of ΔK).

Figure 9 shows the fatigue crack features before the final rupture of the recrystallized single crystal superalloy in Figure 7. The slip bands in the NBSC substrate became more intense and the cracking of the carbide/NBSC interface ("C4") was more evident than that of Figure 7j. The zig-zag cracking ("C5") was quite pronounced, following $(1\bar{1}1)$ and (111), alternatively. River patterns were clearly observed on the fractured crystallographic plane "C6", indicating cleavage fracture along a slip plane. The octahedral plane was identified to be $(\bar{1}\bar{1}1)$, according to the configuration in Figure 4, by searching all possible octahedral slip systems.

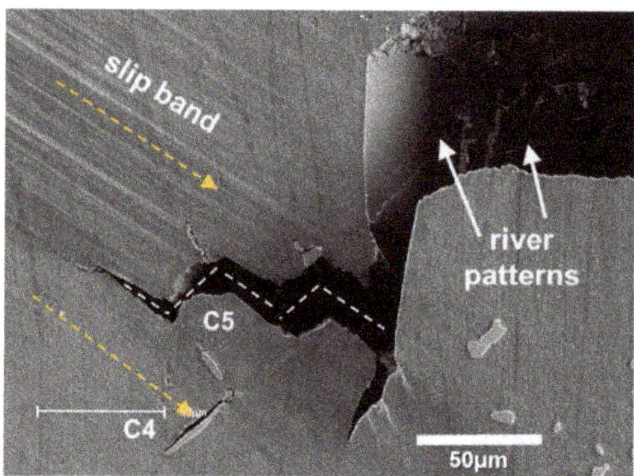

Figure 9. Fatigue crack features before final rupture of recrystallized single crystal superalloy.

4. Discussion

4.1. Effect of Temperature

It is revealed from the above in situ observations that octahedral slip induced cracking is the dominant fatigue fracture mode in the present study. This failure mode was also reported in the PWA1480 superalloy by Telesman et al. [22]. This supports that, at low or intermediate temperatures, the crack growth of NBSC superalloy would occur either on a single octahedral slip plane or on several octahedral slip planes, resulting in crystallographic {111} facets on the fracture surface [17]. Different octahedral slip systems may be operated, controlled by both crystal orientation and test conditions [23,24]. The fatigue failure on specified slip planes is known to be governed by the resolved shear stress, instead of the maximum principle stresses in conventional polycrystalline materials. The dominated deformation mechanism is the shearing of γ/γ' at this temperature, as observed in the studies of low cycle fatigued NBSC specimens at 600 °C [17].

The author's previous studies [2] indicated that intergranular cracks were just occasionally observed, and crystallographic cracking was dominant for fatigue failure of recrystallized NBSC superalloy at 550 °C. At higher temperature, for instance at 850 °C [25], intergranular fracture along recrystallized grain boundaries would be preferred. Figure 10 shows the longitudinal section microstructure of recrystallized NBSC superalloy after conventional strain-controlled fatigue testing at 850 °C. Both oxidation-assisted grain boundary degradation and strength decrease in the grain boundary would lead to intergranular cracking [25]. For this study at 550 °C in vacuum, it is evidently revealed that the recrystallized grain boundary and the associated carbides played an important role in the fatigue crack initiation process. The subsequent crack propagation was overall dominated by crystallographic cracking via one or multiple slip systems. It is expected that, at higher temperature, intergranular cracking will play a more and more important role in the failure analysis of recrystallized single crystal superalloys.

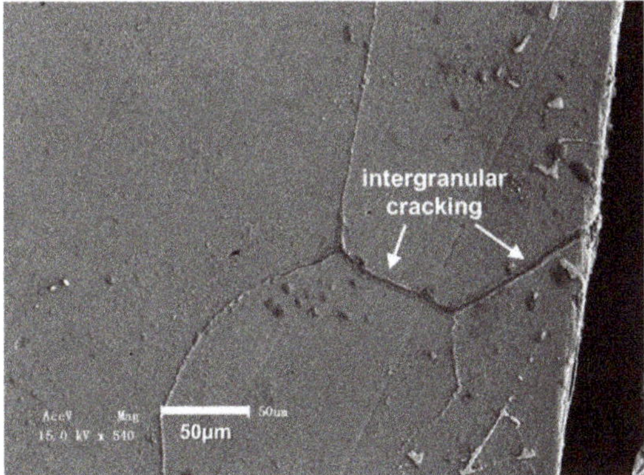

Figure 10. Intergranular cracks observed in the recrystallized single crystal superalloy after fatigue testing at 850 °C [25].

4.2. Microstructurally Small Crack Behavior

In the above in situ observations, the fatigue crack growth behavior in Figures 5 and 7 show evident dependence on the microstructure of recrystallized grains and NBSC substrate. First of all, for the original NBSC superalloy, casting porosity has been reported to be the primary crack initiation site [2,17]. In contrast, for the recrystallized NBSC superalloy, it showed evident multiple crack initiation related to recrystallized GBs and carbides. Hence, the recrystallized grain orientation, misorientation, and carbide property would affect the fatigue crack initiation process.

For the small crack growth behavior within the recrystallized grains, the fatigue crack paths were strongly dependent on the local recrystallized grain orientation, as shown in both Figures 5 and 7. Especially in Figure 7, the fatigue crack path showed an evident bifurcation phenomenon and secondary cracking, which was closely related to the cracking of recrystallized GBs and the blocking effect of the RX/NBSC interface. This process is essentially affected by the local cyclic plastic strain distribution within the recrystallized grains, as indicated by crystal plasticity finite element simulation [19].

Referring to Figures 6 and 8, both FCGR curves showed remarkable fluctuations, which is indicative of microstructural fatigue and small crack behavior. The deceleration of FCGR was usually related to the blocking of the recrystallized GBs or interfaces. While the crack came across the RX/NBSC interface, different slip planes in the NBSC substrate also led to the temporary arrest of small crack growth. Even when the fatigue crack propagated in the [011] substrate, there was competition between different octahedral slip systems, as shown in Figures 7 and 9. The alternate operation of duplex slip planes, i.e., $(1\bar{1}1)$ and (111), were commonly observed, following secondary cracking in Figure 7c, or zig-zag cracking mode in Figure 7g,j. According to Telesman et al. [22], the slip systems with high resolved shear stress intensity are preferential to be operated during cyclic loading. When the operated slip system was not efficient in extending the fatigue crack length, as shown in Figure 7i, the fatigue crack would return to propagate along the duplex slip planes soon, as shown in Figure 7j. For the acceleration of FCGR in this study, it was usually related to the overcoming of microstructural barriers, such as recrystallized GBs or interfaces, as shown in Figure 7b,f. Another important mechanism of accelerated crack growth is via the coalescence of primary cracks with the secondary cracks, as shown in Figure 7e.

5. Conclusions

The fatigue fracture behavior of nickel-based single crystal superalloy with surface recrystallized grains was studied at 550 °C by in situ SEM investigations. The following conclusions were obtained:

(1) Multiple crack initiation was observed at the recrystallized grain boundaries, either by intergranular cracking of grain boundaries, or by the fracture of carbides at the grain boundaries.

(2) The operated slip systems for both the nickel-based single crystal superalloy and the recrystallized grains were identified to be octahedral slips, by analysis of surface slip trace and crystallographic configuration.

(3) The recrystallized nickel-based single crystal superalloy showed distinct fatigue fracture mechanisms at 550 °C. Crystallographic slip deformation is responsible for all fatigue cracks in the present study. The intergranular cracking of the recrystallized grain boundary was due to the impingement of intensive slip bands against the grain boundary. The subsequent fatigue crack propagation was dominated by transgranular cracking manner. At high temperature, commonly reported intergranular cracking would be expected.

(4) Fatigue crack propagation exhibited microstructural small crack behavior. Local microstructural inhomogeneities, such as recrystallized grain boundaries or interfaces, were found to expressly alter the propagation path as well as the crack growth rate. Besides, the competition between different slip systems and the crack coalescence mechanisms were another two important factors responsible for the evident fluctuation in fatigue crack growth rate curve.

Author Contributions: Conceptualization, X.M. and H.-j.S.; Data curation, H.Y., J.T. and H.Z.; Formal analysis, H.Y., J.J., Z.W., X.M., J.T., H.Z. and W.Z.; Investigation, H.Y., X.M., H.-j.S. and W.Z.; Methodology, J.J., Z.W., X.M. and W.Z.; Resources, X.M., H.-j.S.; Supervision, H.-j.S.; Validation, X.M. and J.J.; Writing—original draft, H.Y. and X.M.; Writing—review and editing, X.M. All authors have read and agreed to the published version of the manuscript.

Funding: This research was funded by National Natural Science Foundation of China (No. 11902370), Guangdong Project of Basic and Applied Basic Research (No. 2019B030302011, 2019A050510022, 2019B010943001, 2017B020235001), China Postdoctoral Science Foundation (No. 2019M653173 and 2019TQ0374), Guangdong Education Department Fund (No. 2016KQNCX005), and Fundamental Research Funds for the Central Universities (No. 19lgpy304).

Acknowledgments: This project is supported by the National Science Foundation of China (No. 11902370), Guangdong Major Project of Basic and Applied Basic Research (2019B030302011), International Sci and Tech Cooperation Program of GuangDong Province (2019A050510022), Key-Area Research and Development Program of GuangDong Province (2019B010943001), Guangdong Provincial Basic and Applied Basic Research (2017B020235001), China Postdoctoral Science Foundation (2019M653173 and 2019TQ0374), Guangdong Education Department Fund (2016KQNCX005), and Fundamental Research Funds for the Central Universities (19lgpy304).

Conflicts of Interest: The authors declare no conflict of interest.

References

1. Joyce, M.R.; Wu, X.; Reed, P.A.S. The effect of environment and orientation on fatigue crack growth behaviour of CMSX-4 nickel base single crystal at 650 °C. *Mater. Lett.* **2004**, *58*, 99–103. [CrossRef]

2. Ma, X.; Shi, H.J.; Gu, J.L.; Yang, Z.; Chen, G.F.; Luesebrink, O.; Harders, H. Influence of surface recrystallization on the low cycle fatigue behavior of a single crystal superalloy. *Fatigue Fract. Eng. Mater.* **2015**, *38*, 340–351. [CrossRef]

3. He, Y.H.; Hou, X.Q.; Tao, C.H.; Han, F.K. Recrystallization and fatigue fracture of single crystal turbine blades. *Eng. Fail. Anal.* **2011**, *18*, 944–949. [CrossRef]

4. Li, Z.; Xu, Q.; Liu, B. Experimental investigation on recrystallization mechanism of a Ni-base single crystal superalloy. *J. Alloy. Compd.* **2016**, *672*, 457–469. [CrossRef]

5. Meng, J.; Jin, T.; Sun, X.; Hu, Z. Effect of surface recrystallization on the creep rupture properties of a nickel-base single crystal superalloy. *Mater. Sci. Eng. A* **2010**, *527*, 6119–6122. [CrossRef]

6. Xie, G.; Lou, L.H. Influence of the characteristic of recrystallization grain boundary on the formation of creep cracks in a directionally solidified Ni-base superalloy. *Mater. Sci. Eng. A* **2012**, *532*, 579–584. [CrossRef]

7. Zhang, B.; Lu, X.; Liu, D.L.; Tao, C.H. Influence of recrystallization on high-temperature stress rupture property and fracture behavior of single crystal superalloy. *Mater. Sci. Eng. A* **2012**, *551*, 149–153. [CrossRef]

8. Bürgel, R.; Portella, P.D.; Preuhs, J. Recrystallization in Single Crystals of Nickel Base Superalloys. In *Superalloys 2000*; Pollock, T.M., Kissinger, R.D., Bowman, R.R., Eds.; TMS: Warrendale, PA, USA, 2000; pp. 229–238.

9. Ma, X.F.; Shi, H.J.; Gu, J.L. In-situ scanning electron microscopy studies of small fatigue crack growth in recrystallized layer of a directionally solidified superalloy. *Mater. Lett.* **2010**, *64*, 2080–2083. [CrossRef]

10. Shi, H.J.; Zhang, H.F.; Wu, Y.Q. Effect of recrystallization on low-cycle fatigue behavior of DZ4 directionally-solidified superalloy. *Key Eng. Mater.* **2006**, *306–308*, 175–180. [CrossRef]

11. Zhao, Y.; Wang, L.; Li, H.Y.; Yu, T.; Liu, Y. Effects of recrystallization on the low cycle fatigue behavior of directionally solidified superalloy DZ40M. *Rare Met.* **2008**, *27*, 425–428. [CrossRef]

12. Jia, B.; Li, C.G.; Li, H.Y. Influence of Recrystallization Layer at Surface on Fatigue Behaviors of Directionally Solidified DZ4 Superalloy. *Mater. Eng.* **2008**, *6*, 64–71.

13. Ma, X.; Shi, H.J. In situ SEM studies of the low cycle fatigue behavior of DZ4 superalloy at elevated temperature: Effect of partial recrystallization. *Int. J. Fatigue* **2014**, *61*, 255–263. [CrossRef]

14. Rabbolini, S.; Luccarelli, P.G.; Beretta, S.; Foletti, S.; Sehitoglu, H. Near-tip closure and cyclic plasticity in Ni-based single crystals. *Int. J. Fatigue* **2016**, *89*, 53–65. [CrossRef]

15. Rabbolini, S.; Pataky, G.J.; Sehitoglu, H.; Beretta, S. Fatigue crack growth in Haynes 230 single crystals: An analysis with digital image correlation. *Fatigue Fract. Eng. Mater.* **2015**, *38*, 583–596. [CrossRef]

16. Luccarelli, P.G.; Pataky, G.J.; Sehitoglu, H.; Foletti, S. Finite element simulation of single crystal and polycrystalline Haynes 230 specimens. *Int. J. Solids Struct.* **2017**, *115*, 270–278. [CrossRef]

17. Ma, X.F.; Shi, H.J.; Gu, J.L.; Wang, Z.X.; Harders, H.; Malow, T. Temperature effect on low-cycle fatigue behavior of nickel-based single crystalline superalloy. *Acta Mech. Solida Sin.* **2008**, *21*, 289–297. [CrossRef]

18. Zhang, Z.F.; Wang, Z.G. Dependence of intergranular fatigue cracking on the interactions of persistent slip bands with grain boundaries. *Acta Mater.* **2003**, *51*, 347–364. [CrossRef]

19. Ma, X.; Wei, D.; Han, Q.; Rui, S. Parametric study of cyclic plasticity behavior in a directionally solidified superalloy with partial recrystallization by crystal plasticity finite element simulation. *J. Mater. Eng. Perform.* **2019**. [CrossRef]

20. Wang, X.S.; Fan, J.H. An evaluation on the growth rate of small fatigue cracks in cast AM50 magnesium alloy at different temperatures in vacuum conditions. *Int. J. Fatigue* **2006**, *28*, 79–86. [CrossRef]

21. Paris, P.; Erdogan, F. A critical analysis of crack propagation laws. *J. Basic Eng. Dec.* **1963**, *85*, 528–533. [CrossRef]

22. Telesman, J.; Ghosn, L.J. Fatigue crack growth behavior of PWA 1484 single crystal superalloy at elevated temperatures. *J. Eng. Gas Turbines Power* **1996**, *118*, 399–405. [CrossRef]

23. Henderson, M.B.; Martin, J.W. The influence of crystal orientation on the high temperature fatigue crack growth of a Ni-based single crystal superalloy. *Acta Mater.* **1996**, *44*, 111–126. [CrossRef]

24. Suresh, S. *Fatigue of Materials*, 2nd ed.; Cambridge University Press: New York, NY, USA, 1998.

25. Ma, X.; Jiang, J.S.; Zhang, W.J.; Shi, H.J.; Gu, J.L. Effect of Local Recrystallized Grains on the Low Cycle Fatigue Behavior of a Nickel-Based Single Crystal Superalloy. *Crystals* **2019**, *9*, 312. [CrossRef]

© 2020 by the authors. Licensee MDPI, Basel, Switzerland. This article is an open access article distributed under the terms and conditions of the Creative Commons Attribution (CC BY) license (http://creativecommons.org/licenses/by/4.0/).

Effect of In-Situ Synthesized Boride Phases on the Impact Behavior of Iron-Based Composites Reinforced by B_4C Particles

Fehmi Nair [1] and Mustafa Hamamcı [2,*]

[1] Department of Mechanical Engineering, Faculty of Engineering, Erciyes University, Kayseri 38000, Turkey; fnair@erciyes.edu.tr

[2] Department of Mechanical Engineering, Faculty of Engineering, Iğdır University, Iğdır 76000, Turkey

* Correspondence: mustafa.hamamci@igdir.edu.tr; Tel.: +90-476-223-0010

Received: 4 February 2020; Accepted: 22 April 2020; Published: 25 April 2020

Abstract: The objective of this study is to investigate the impact behavior of iron-based composites reinforced with boron carbide (B_4C) particles and in-situ synthesized iron borides (Fe_2B/FeB). The composite specimens (Fe/B_4C) were fabricated by hot-pressing under a pressure of 250 MPa at 500 °C, and sintered at a temperature of 1000 °C. The effects of the reinforcement ratio on the formation of in-situ borides and impact behavior were investigated by means of different volume fractions of B_4C inside the iron matrix: 0% (un-reinforced), 5%, 10%, 20%, and 30%. Drop-weight impact tests were performed by an instrumented Charpy impactor on reinforced and un-reinforced test specimens. The results of the impact tests were supported with microstructural and fractographical analysis. As a result of in-situ reactions between the Fe matrix and B_4C particles, Fe_2B phases were formed in the iron matrix. The iron borides, formed in the iron matrix during sintering, heavily affected the hardness and the morphology of the fractured surface. Due to the high amount of B_4C (over 10%), porosity played a major role in decreasing the contact forces and fracture energy. The results showed that the in-situ synthesized iron boride phases affect the impact properties of the Fe/B_4C composites.

Keywords: metal matrix composites; powder metallurgy; Fe/B_4C composites; iron boride phases (Fe_2B/FeB); Charpy impact test

---

1. Introduction

Metal matrix composites (MMCs) are preferred in machine elements and constructive structures working under many loading types, with their high specific strength, high wear resistance, and advanced thermal properties. Particle-reinforced composites are advanced materials with the potential to provide improved properties, progressively used in the automotive, aircraft, and space industries, as well as in civil and defense applications. Structures made of particle-reinforced MMCs are dependent on their reinforcement properties, such as particle sizes, particle geometries, and particle contents in the composites when considering the composite characteristics and fabrication process [1,2].

Composite materials are usually exposed to impact damages in different environments. For that reason, problems associated with impact damages are limiting the design criteria for composite materials [3]. The response of the composites can be affected by different failure modes, such as microstructure properties, crack initiation and propagation, and damage type of the reinforcement particles [4,5].

The low-velocity impact and Charpy tests have been used for many years as fast and economical comparison tools to determine the energy absorption, notch sensitivity, and impact fracture behavior of materials for qualitatively comparing different composite structures and assessing their fabrication and service conditions [6–8]. Most of the studies were focused on aluminum- and magnesium-based composites, while little attention has been given to the development of iron matrix composites produced by the powder metallurgy method. The use of iron and its alloys as a matrix in composite materials is very important for wear and corrosion resistance. Besides, they have a higher stiffness, strength, and toughness when compared with aluminum- or magnesium-based composites. MMCs, iron-based composites, and alloyed steels were examined in some studies for fracture characteristics and impact behavior [9–13].

The powder metallurgy method can produce unique microstructures and compositions required for hard composite materials. However, it is well known that powder metallurgy composites are influenced by an inherent porosity after the pressing and sintering. Pores act as stress and crack originators under several impact loadings [5,6,14].

Boron carbide (B_4C) is an excellent material with a wide range of application areas, such as the chemical industry, metallurgy, hardening of metal surfaces, cutting tools and abrasives, and space and armor applications. In some studies that are in close scope, boron carbide with added iron matrix materials has been investigated by means of the structure formation and the growth mechanism of iron borides in boriding treatments. Boriding is a well-known surface hardening process based on the formation of intermetallic Fe_2B and FeB phases, which occurs as a result of the reaction between iron and boron atoms. Regarding composite materials, their strength and hardness can be improved by the presence of intermetallic phases and compounds [15].

The basic advantage of iron borides is that in certain amounts they enhance the hardness and strength of iron-based materials with hard Fe_2B or FeB phases. The formation of iron borides in the internal structure can be achieved by in-situ reactions.

Regarding the MMCs, the interface between reinforcement and the metallic matrix plays an important role in the mechanical properties, as the load has to be transferred from matrix to reinforcement through the interface. If the interface is weak, the composite will perform poorly during service conditions. Besides, microcrack initiation and the damage of materials in particle-reinforced composites is often associated with interfacial bonding, which can occur in the chemical reactions of reinforcement and the matrix through the diffusion of atoms. The in-situ reactions at the interface lead to a strong bond between the iron matrix and the B_4C particles eventually turned into iron borides.

The preferred method to add boron to the iron structure is the use of boron carbide powder containing boron elements. B_4C particles are the main source of FeB and Fe_2B phases, distributed homogeneously and throughout the structure. As a result of the reaction between boron and many elements in the periodic table, a wide range of borides are formed [16,17]. Small additions of boron have a positive effect on the mechanical properties of powder metallurgy steels during the densification stage of sintering [18]. The iron-rich liquid phase is formed between the ceramic particles, and the FeB phase is formed after solidification [19]. At sintering temperatures up to 1100 °C, the Fe_2B and FeB phases begin to form around boron carbide due to the reaction between iron and boron, resulting in an almost complete breakdown of the B_4C particles and the formation of graphite inside a porous and borided zone [16,20]. Increasing the sintering temperature (1000–1100 °C) leads to the formation of the liquid phase, and the sintering process is accelerated [20]. Regarding the solid-state sintering, only solid phases are present at the sintering temperature. However, in liquid phase sintering, small amounts of liquid phase are present. The reaction between Fe and B_4C at 1000 °C leads to the release of free carbon and the formation of the FeB phase according to the reaction $B_4C + 4Fe = 4FeB + C$. The iron particles react with the free carbon and, possibly, with carbon that originated from the boron carbide to form the FeC or Fe_2C carbides. When carbon exceeds the limit of solubility, iron carbide is formed as the second phase [21]. Depending on the duration and temperature of the sintering process, the boron content, and the chemical composition, a single phase (Fe_2B) or two intermetallic phases (Fe_2B, FeB) are formed by the diffusion of boron atoms into the metallic materials [22]. In addition, the Fe-B phase diagram has two eutectics—one is present at 1177 °C, 3.8% boron content, and the second is seen at 1497 °C, 18.5% B [23]. Fe_2B grows in low carbon and low alloy steels with a saw-tooth morphology and possesses a high degree of hardness (1600–1800 HV) [24–26]. Firstly, acicular Fe_2B crystals grow on the metal surface, and then the FeB phase is formed, which is harder (1800–2400 HV), comparatively brittle, and contains more boron atoms [27]. These relevant studies have investigated the diffusion mechanisms of boron into the internal structure of composites. However, their mechanical behavior under the different strains they are exposed to under service conditions have not been examined. There are insufficient studies to clarify the influence of intermetallic boride phases on the impact responses of iron-based composites reinforced by B_4C particles. Therefore, the purpose of this study is to determine the relation of the microstructure and impact behavior of Fe/B_4C composites, as well as their micromechanical properties.

In this context, the main contributions of this paper are: (i) fabrication of Fe/B_4C composites by the hot-pressing powder metallurgy method; (ii) investigation of the microstructural properties influenced by iron boride phases formed in situ; and (iii) evaluating the impact behavior and fractographical morphology of the different volume fractions of reinforcement.

2. Materials and Methods

2.1. Preparation of Composite Test Specimens

Commercial Fe and B_4C powders were used as the matrix and reinforcement, respectively, for the fabrication of composites. From the point of view of the reaction between the iron and boron carbide powders, the selected particle sizes of both powders were nominally similar to promote high diffusion rates between iron and boron carbide. The iron particles, manufactured by AEE (Atlantic Equipment Engineers, (Upper Saddle River, NJ, USA), were mixtures of irregular angular shape and flake geometry with a median particle size (d_{50}) of 28.6 µm. The particles of B_4C powder, manufactured by AEE, showed an irregular polygonal and angular morphology with a median particle size (d_{50}) of 30.8 µm. The particle size distributions of the raw powders were determined by a laser diffraction particle-size analysis (Mastersizer 3000, Malvern Panalytical Ltd, Malvern, UK). The results of the particle size distributions are listed in Table 1. The supplied information about the powders was presented in SEM images, seen in Figure 1.

Table 1. Particle size distributions of the raw powders.

Powder	d_{10} (µm)	d_{50} (µm)	d_{90} (µm)
Fe	8.97	28.6	66.4
B_4C	18.7	30.8	56.4

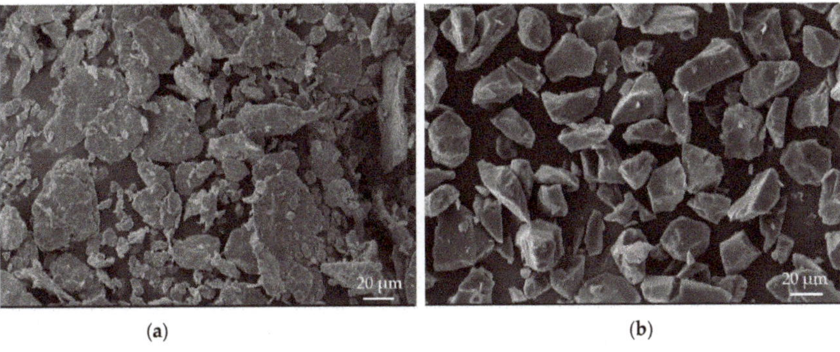

(a)   (b)

Figure 1. Scanning electron microscopy (SEM) images of the raw powders: (a) irregular and flake iron powders; (b) irregular polygonal and angular boron carbide particles.

According to the SEM analysis, no flocculation or aggregates were observed inside the powders. It was found that the Fe powder had a high purity, of 99.8%, and that the B_4C powder had a purity of

99.7%, as measured by a chemical characterization and impurity analysis with Wavelength Dispersive X-ray (WDX) and X-ray Fluorescence (XRF).

Before hot-pressing, mixtures of Fe + B_4C powders were weighed with precision scales to provide four different volume fractions (vol.%) of determined reinforcement percentages (un-reinforced, Fe + 5 vol.%, 10 vol.%, 20 vol.%, and 30 vol.% B_4C) for 55 mm × 55 mm × 10 mm dimensions. The mixing of powders was carried out in different plastic containers for 1 h with a velocity of 100 rpm in the Turbula-T2F mixer (WAB-GROUP, Muttenz, Switzerland). The weights of the matrix and reinforcement powders were determined according to the rule of Mixture Equation (1), which requires that the total volumes of all composite specimens be constant.

$$v_m = \left(\frac{M_m \times 100}{\rho_m \times V_c} \right) v_r = \left(\frac{M_r \times 100}{\rho_r \times V_c} \right) \tag{1}$$

where v is the volume fraction, M is the mass, ρ is the density, V_c is the volume of composite specimen, and subscripts m and r refer to the matrix and the reinforcement, respectively.

2.2. Fabrication of Test Specimens

The AISI H13 hot work tool steel die, which has a square hole with a dimension of 55 mm × 55 mm, was used for the fabrication of composites. The powder mixture was manually laid in the square die after the mixing process was completed. The compacting pressure was applied at room temperature using a uniaxial hydraulic press (max 250 tons). When a pressure of 250 MPa was applied, the heating of the die commenced simultaneously from the outer surface of the die, using a spiral heater. The compacting process shown in Figure 2 was carried out in a protective argon environment. Compacted composites were taken out of the die at room temperature after heating at 500 °C for 30 min under a constant pressure of 250 MPa. Before the sintering process, the compacted composite specimens were cut into pieces of 55 mm × 10 mm × 10 mm by wire erosion to obtain standard impact test specimens. The sintering treatments of the composite specimens were carried out in a high-temperature tubular furnace (Protherm PTF 18/75/300, Protherm, Ankara, Turkey) at 1000 °C for 30 min and 60 min under an argon environment. The heating rate was adjusted to 8 °C/min and then cooled to room temperature under the native conditions in the furnace. The composite specimens were encoded as follows: R0T30—un-reinforced composite (100 vol.% Fe) sintered for 30 min; R5T60—reinforced with a volume fraction of 5 vol.% B_4C and sintered for 60 min, etc.

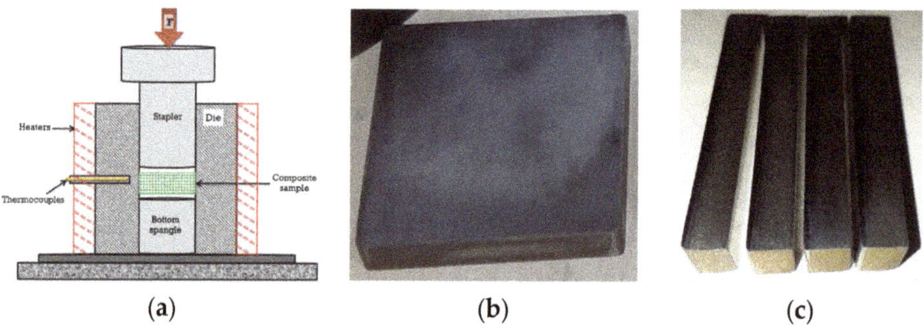

(a)   (b)   (c)

Figure 2. (**a**) Schematic fabrication method of composite specimens under hot-pressing in a square die; fabricated composite specimen: (**b**) compacted Fe + 10% B_4C after hot-pressing (55 mm × 55 mm × 10 mm); (**c**) sliced and sintered (60 min, 1000 °C) un-notched test specimens (55 mm × 10 mm × 10 mm).

2.3. Microstructural and Micromechanical Investigation

Microstructural evaluations were carried out using an optical microscope (Jenavert SL100, Carl Zeiss AG, Oberkochen, Germany) with 25–50× magnification, following the etching of polished surfaces—etched with a 5% nital solution. The formation and distribution of related phases were examined by X-ray diffraction (XRD) analysis, using the Bruker AXS D8 advance diffractometer (Bruker, Billerica, MA, USA) with Cu-Kα radiation (λ = 1.5417 Å), with an applied operating voltage of 25 kV and a current of 35 mA, and by SEM-EDX analysis, using the Carl Zeiss EVO LS10 scanning electron microscope (Carl Zeiss AG, Oberkochen, Germany) with SE1 detectors (secondary electron detectors) under high vacuum conditions, with an accelerating voltage of 25 kV.

The porosity of the sintered composites was determined by the Archimedes method using the Precisa XB 220A precision scale (Precisa Gravimetrics AG, Dietikon, Switzerland). For the rule of mixture, Equation (1) was used to calculate the theoretical density, and porosity was calculated using theoretical and actual density values.

After a metallographic analysis, macro- and microhardness were determined from cross-section surfaces with a Vickers indenter, using the Struers Duramin-5 hardness tester (Struers, Ballerup, Denmark). Five measurements were performed for each specimen to obtain the average hardness by applying a load of 2 kg (HV_2) for 5 s. A 25 g ($HV_{0.025}$) load was also applied to determine the hardness of the iron boride phases at the microscale.

2.4. Charpy Impact Test Setup

Powder metallurgy test specimens were prepared as un-notched, in accordance with the ASTM E23 and ASTM B925 standards. Charpy impact tests were carried out on the low-velocity drop weight test device (CEAST Fractovis Plus, Instron, Norwood, MA, USA) at room temperature. Blocks of composites were cut into slices by wire erosion and freely supported on anvils of the Charpy impact fixture, as shown schematically in Figure 3. The Charpy fixture and impactor are fully suited to the ASTM-E23 standards. The impactor head hit the specimen surface perpendicular to the powder-compacting direction.

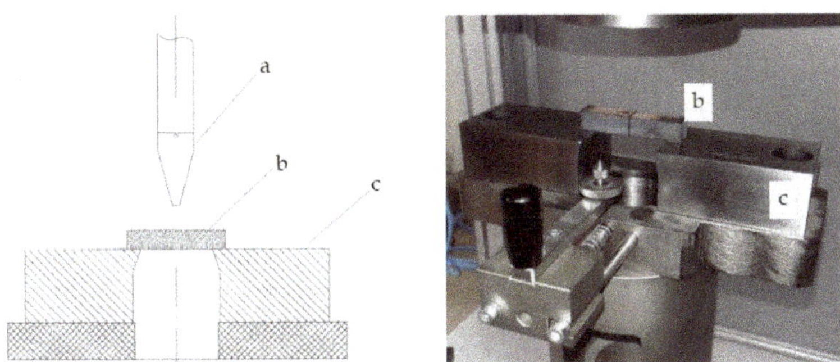

Figure 3. Impact test setup: (**a**) impactor; (**b**) un-notched test specimen; (**c**) Charpy fixture.

Impact parameters were held constant for all impact tests, and the average room temperature was recorded as 21 °C. The weight of the impactor was 5.284 kg, and the height of the specimen surface was 579 mm. The impactor velocity was set at 3.37 m/s, with a resultant energy of 30 J. Energy losses due to air resistance and bearing friction were disregarded due to their small contribution. The anti-rebound system was activated, and repetitive impact tests were implemented on three different specimens with the same compositions under the same test conditions. An acquisition system (DAS16000) was used for monitoring and recording the contact force (N)−time (ms) data coming from the impactor, which was instrumented with a force transducer until the fracture occurred. The contact force values were reported by taking the average of three impact test data.

The kinetic energy change of the impactor head was determined in the time interval at which the contact force started and ended. The difference in the energy value just before the first contact of the impactor head with the test specimen (initial energy of impactor) and after breakage occurred indicates the fracture energy of the specimen.

3. Results and Discussion

3.1. Microstructure Observations

Figure 4 shows the optical micrographs of hot press-compacted Fe/B$_4$C composites (0 vol.%, 5 vol.%, 10 vol.%, 20 vol.%, and 30 vol.% B$_4$C) before the sintering process. The breakage of B$_4$C particles was observed over the 5 vol.% B$_4$C reinforced composites seen in Figure 4c–e, which might have occurred during the mixing or compacting stages.

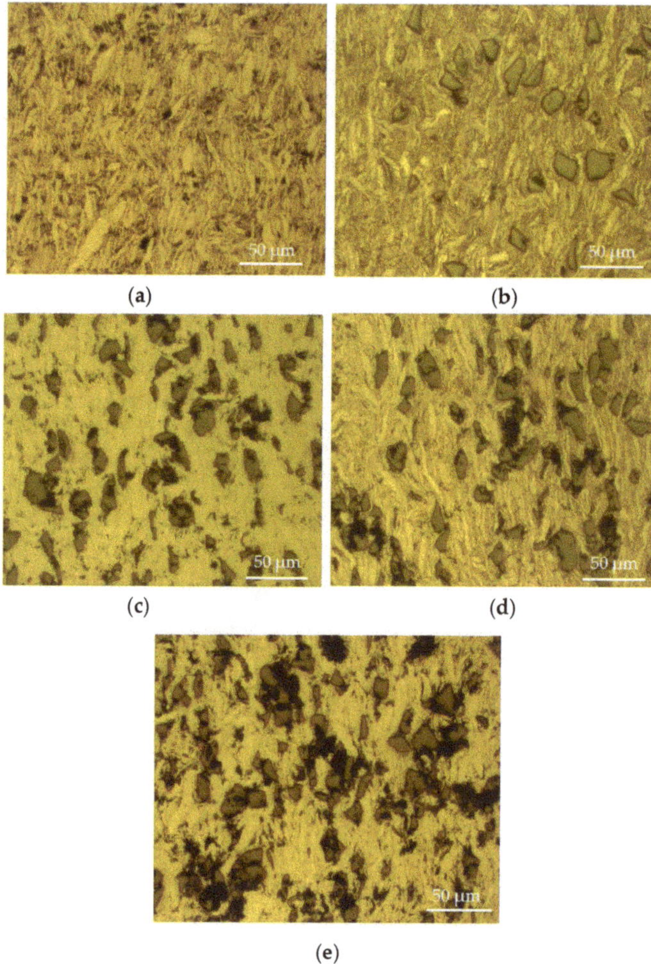

Figure 4. Optical micrographs of unsintered Fe/B$_4$C composites: (**a**) 100 vol.% Fe; (**b**) Fe + 5 vol.% B$_4$C; (**c**) Fe + 10 vol.% B$_4$C; (**d**) Fe + 20 vol.% B$_4$C; (**e**) Fe + 30 vol.% B$_4$C.

Figure 5 shows the optical microstructure images of the composites after the sintering process. The boundaries of the iron particles were more apparent after 60 min of sintering in the un-reinforced specimens (Figure 5b). Porosity formations are seen as dark black areas which increased with the reinforcement.

One of the remarkable microstructure observations was the formation of irregular and acicular (needle-shaped) diffusion zones, as seen in the composite specimens of Figures 5c–f and 6, formed as a result of the diffusion of boron atoms into the iron matrix during sintering. In particular, distinctive formations were observed in 5 vol.% and 10 vol.% B_4C reinforced composites. These in-situ reinforcement phases originated from the B_4C particles and randomly scattered close to the boundaries of the B_4C particles in the matrix. A similar situation was observed by Turov et al. [16].

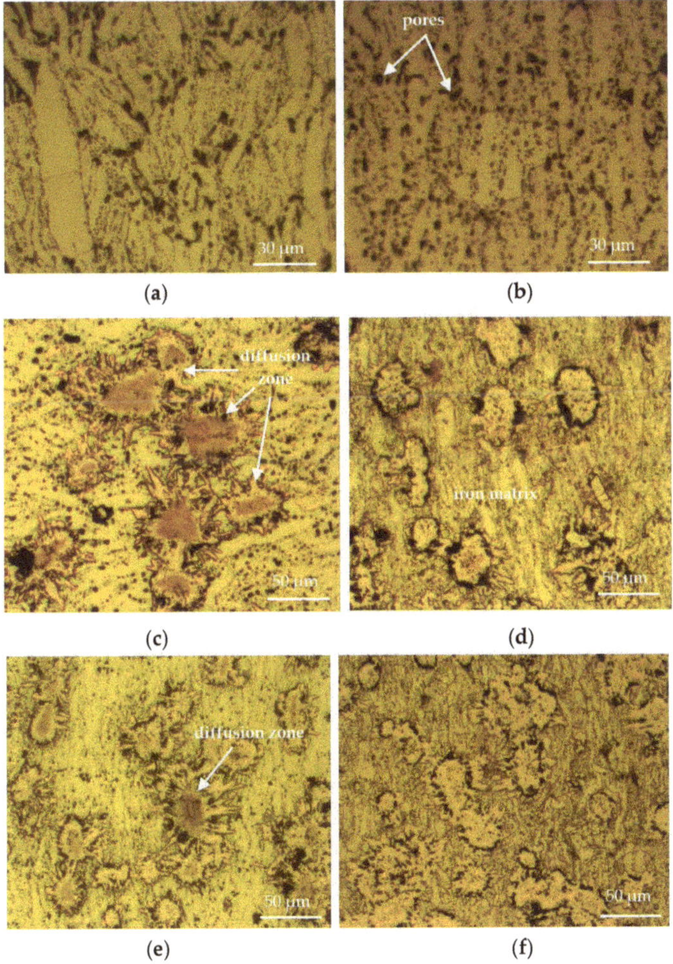

Figure 5. *Cont.*

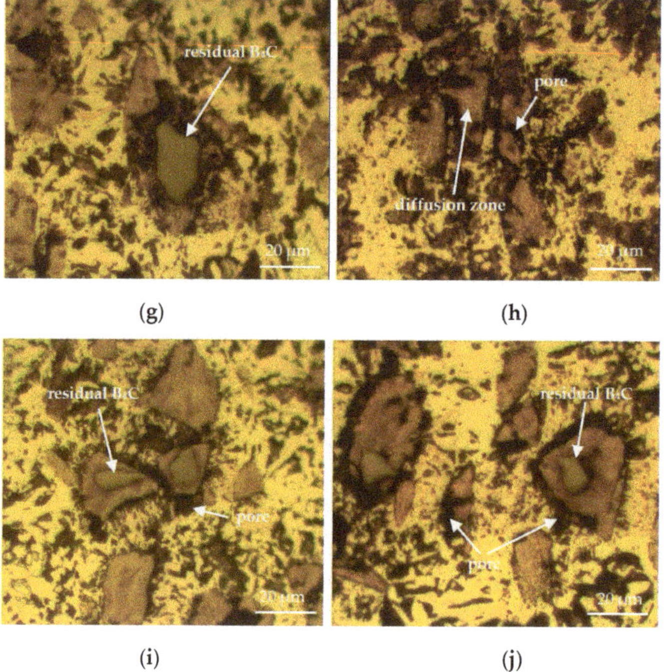

Figure 5. Optical micrographs of sintered Fe/B$_4$C specimens: un-reinforced iron composites sintered for (**a**) 30 min and (**b**) 60 min (R0T30, R0T60); Fe + 5 vol.% B$_4$C sintered for (**c**) 30 min and (**d**) 60 min (R5T30, R5T60); Fe + 10 vol.% B$_4$C sintered for (**e**) 30 min and (**f**) 60 min (R10T30, R10T60); Fe + 20 vol.% B$_4$C sintered for (**g**) 30 min and (**h**) 60 min (R20T30, R20T60); Fe + 30 vol.% B$_4$C sintered for (**i**) 30 min and (**j**) 60 min (R30T30, R30T60).

Figure 5g–j shows the partial decomposition of B$_4$C particles, which remained in residual form at the center of the diffusion zone, above the 5% volume fractions of reinforcement. Larger B$_4$C particles remained in residual form in the structure.

In powder metallurgy, solid state diffusion plays a major role in the formation and growth of interfacial bonding, formed by the dissolution or reaction of the B$_4$C particles and the iron matrix.

B$_4$C particles break down below 1100 °C, and the iron boride phases form at the interface between B$_4$C and iron. Depending on the duration and temperature of the sintering process, the boron content, and the particle size, a single phase (Fe$_2$B) or two intermetallic phases (Fe$_2$B, FeB) are formed by the diffusion of boron atoms into the iron based materials [20–22]. The solubility of boron in steel is higher than 0.004 percent at 1000 °C, and the rate of diffusion is about the same as that of carbon [28]. When the maximum solubility of boron in iron is reached, a solid state solution will be formed, because the solubility of boron in iron is limited. If the amount of boron in iron exceeds the solubility limit, boron cannot enter the iron lattice. The over-reinforcement or diffusion of boron will result in precipitation in the structure, generating an insoluble residue [29].

Residual or unreacted B$_4$C particles act as a barrier to the diffusion of boron, resulting in residual particles. These particles are discontinuous and cause a preferential crack initiation between particle-matrix interfaces under tension or bending forces.

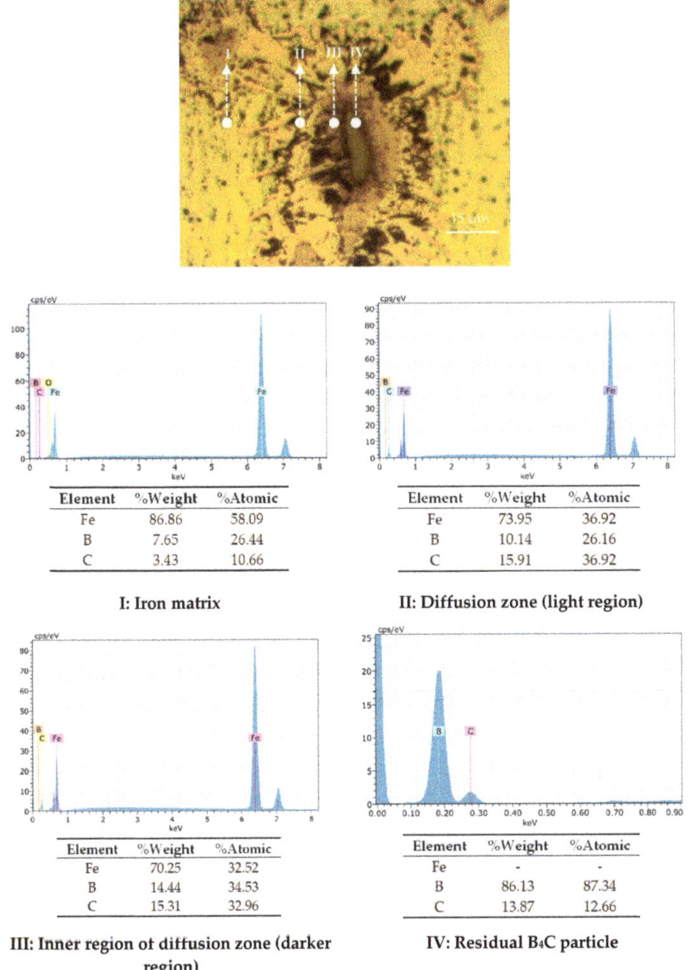

Element	%Weight	%Atomic
Fe	86.86	58.09
B	7.65	26.44
C	3.43	10.66

I: Iron matrix

Element	%Weight	%Atomic
Fe	73.95	36.92
B	10.14	26.16
C	15.91	36.92

II: Diffusion zone (light region)

Element	%Weight	%Atomic
Fe	70.25	32.52
B	14.44	34.53
C	15.31	32.96

III: Inner region of diffusion zone (darker region)

Element	%Weight	%Atomic
Fe	-	-
B	86.13	87.34
C	13.87	12.66

IV: Residual B₄C particle

Figure 6. Energy-dispersive X-ray spectroscopy (EDS) analyses for the Fe + 10 vol.% B_4C (60 min sintered) composite throughout the iron matrix (I) and different points of the diffusion zone (II–IV).

Energy-dispersive X-ray Spectroscopy (EDS) analysis, performed at different points of the diffusion zone (I–IV) (Figure 6), confirmed the presence of FeB and Fe_2B phases, which formed during the reaction between Fe and B_4C. Boron contents decreased from the origin of diffusion (or residual B_4C particles) to the iron matrix. The EDS analysis revealed that the darker regions of the diffusion zone contained more boron (B) atoms, probably due to the existence of an FeB-intensive phase (Figure 6III), whereas the Fe_2B phase is more dominant in the relatively light-colored region (Figure 6II). The XRD analysis, shown in Figure 7, revealed that the composites comprise the Fe_2C and iron boride phases (Fe_2B and FeB). In particular, the intensive Fe_2B phase was observed in the Fe + 30 vol.% B_4C composite.

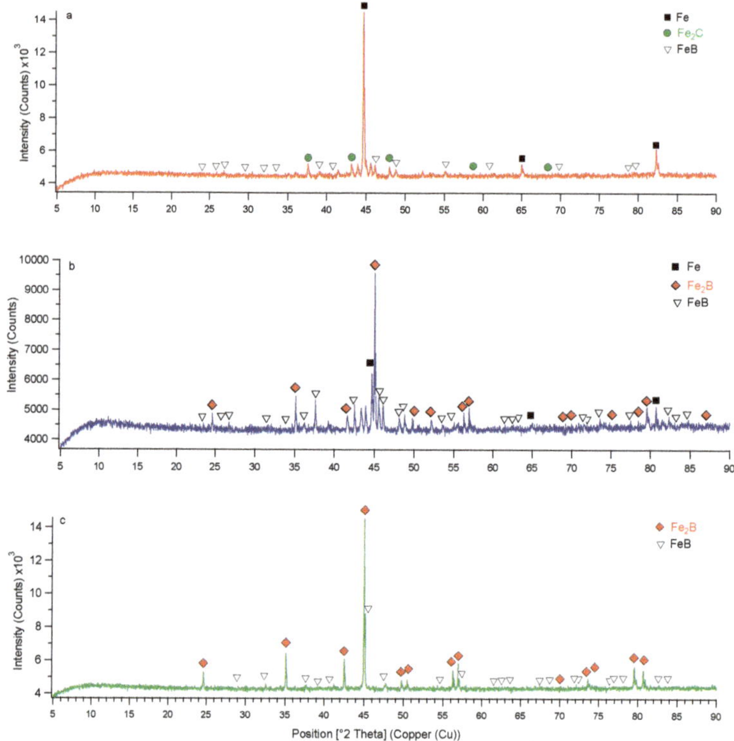

Figure 7. X-ray Diffraction (XRD) spectrum of the 60 min sintered (**a**) Fe + 10 vol.% B$_4$C; (**b**) Fe + 20 vol.% B$_4$C; (**c**) Fe + 30 vol.% B$_4$C composites.

As seen in Figure 8, enhanced porosity was observed as the reinforcement increased, and sintering duration did not have a noticeable effect on porosity. It was concluded that the reinforcement ratio was the effective factor for porosity. The differences between the iron and B$_4$C particle geometries increased the voids. This difference may occur during the compaction stage; in particular, it can be more effective at high reinforcement ratios. Harder B$_4$C powder has a greater effect on the compression behavior of the mixture, as a higher ratio of B$_4$C is more resistant to compressibility than the iron powder. Another reason for the porosity was the dissociation of boron carbide particles from the contacted iron matrix after sintering the Fe-B$_4$C powder mixtures, defined as diffusion porosity [29]. The increase of porosity also reduces the fracture energy, as it reduces the cross-sectional area exposed to fracture.

A gradual increase was seen in the macrohardness of composites with increasing reinforcement ratio (Figure 9). The maximum hardness was measured as 447 HV for the R20T60 specimen. In particular, the increment of hardness became more prominent after the 10 vol.% B$_4$C reinforcement ratio. A similar tendency was observed for the 30 min sintered composites, while the hardness of the 60 min sintered specimens was slightly higher than that of the 30 min sintered composites. The rising tendency of hardness values slowed down after the 20 vol.% reinforcement ratio for the 60 min-sintering composite. In addition to the porosity in the microstructure, possible changes in the grain sizes had an effect on the macrohardness values. Increased sintering duration may also have led to increased boride phases in the internal structure and, consequently, increased hardness.

Hardness differences of the matrix and diffusion zones were demonstrated by the Vickers microhardness measurements (HV$_{0.025}$). The variations of hardness and the traces of the Vickers indenter can be seen in Figure 10. The hardness varies throughout the diffusion zone, with the lowest value in the matrix and the highest one on the the B_4C particles. The average hardness of the iron matrix was measured in the range of 84–102 $HV_{0.025}$, while the diffusion zone was 5–7 times higher than the matrix as 540–640 $HV_{0.025}$. Ozdemir et al. [22] reported that the hardness of boride on the pure iron was over 1700 $HV_{0.01}$, and the hardness of pure iron was about 130 $HV_{0.01}$. Nowacki et al. [25] found that the hardness of Fe-Fe_2B phases changed across a wide range (150–1500 HV_5), in which the proportion of the Fe_2B phase increased (1800 $HV_{0.1}$) as the total hardness of the specimen increased [26].

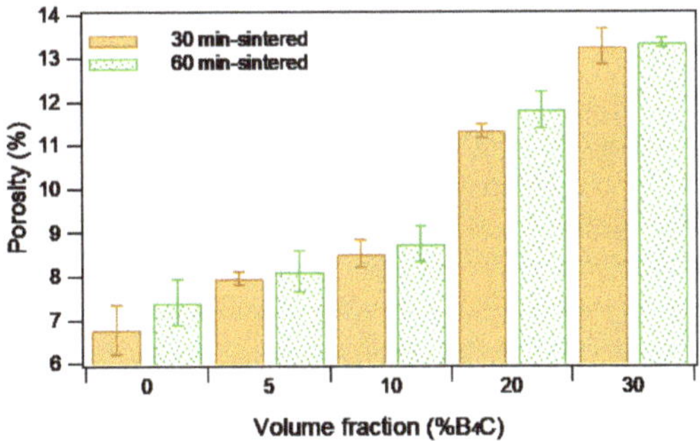

Figure 8. Change in the porosity (%P) after the sintering process as a function of the reinforcement ratio.

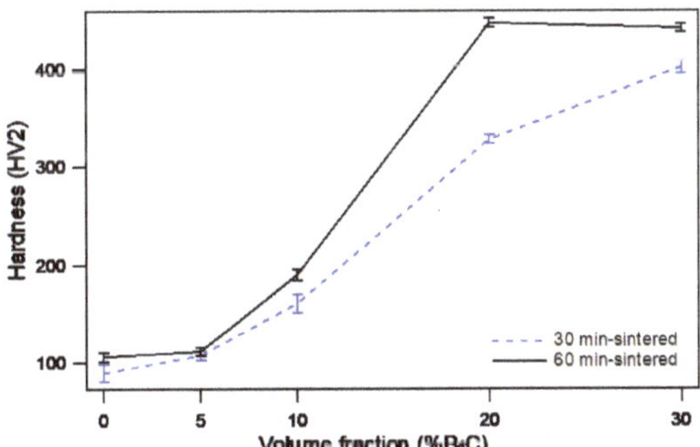

Figure 9. Comparison of the macrohardness of un-reinforced and Fe/B_4C composites with varied durations of sintering.

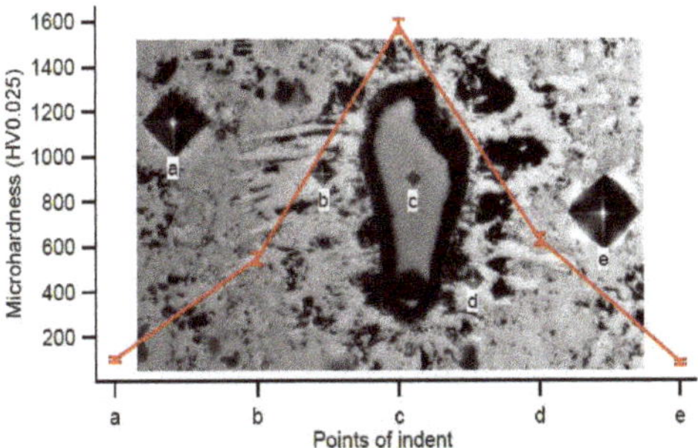

Figure 10. Microhardness distribution of the Fe + 10 vol.% B_4C (30 min sintered) composite throughout the matrix and diffusion zone ($HV_{0.025}$).

3.2. Impact Behavior

The force–time alteration of repeated impact tests followed almost the same trend as the one seen in Figure 11 for 60 min sintered composites. The time axis in the contact force graph shows the time interval between the first contact of the impactor with the specimen surface and the moment when the specimen is broken. This characteristic period, which is defined as the "contact time" or "time to fracture", was considered as the time until the specimen fractured. When the force–time graph was examined, it was clearly seen that the contact time values are different from each other. The contact time is mostly dependent on the composition of the specimens. It is noteworthy that the contact time is significantly reduced in highly reinforced specimens. The fracturing of the specimens at different contact times under the same impact loading is related to the different velocities of crack propagation along cross-sections with the same thickness. When considering the brittle composite, a higher velocity of the crack results in a lower contact time.

In the contact-force time histories it can be seen that the peak levels differ for each experiment. The multiaxial stresses and different crack propagation mechanisms constituted the fundamental differences for the variation of the force–time curves with a different number of peaks. The maximum peak point of the force–time curve was appointed as the maximum force (fracture force) of the tested specimen. At the maximum contact force, a crack occurred and progressed throughout the surface. With some exceptions, the contact force decreased after the maximum peak value in the decreasing cross-sectional area that resisted to the fracture of the specimen.

Lower peak values, occurred before and after the maximum contact forces are noteworthy in Figure 12a–e, respectively. The main reason was the change of direction of the crack's propagation along the cross-section plane. The macrophoto analyses of fractured surfaces demonstrate this non-planar wave-shaped surface. Contactless friction and gaps between the tested specimens, the impactor and the fixture also affected the trend of the curve.

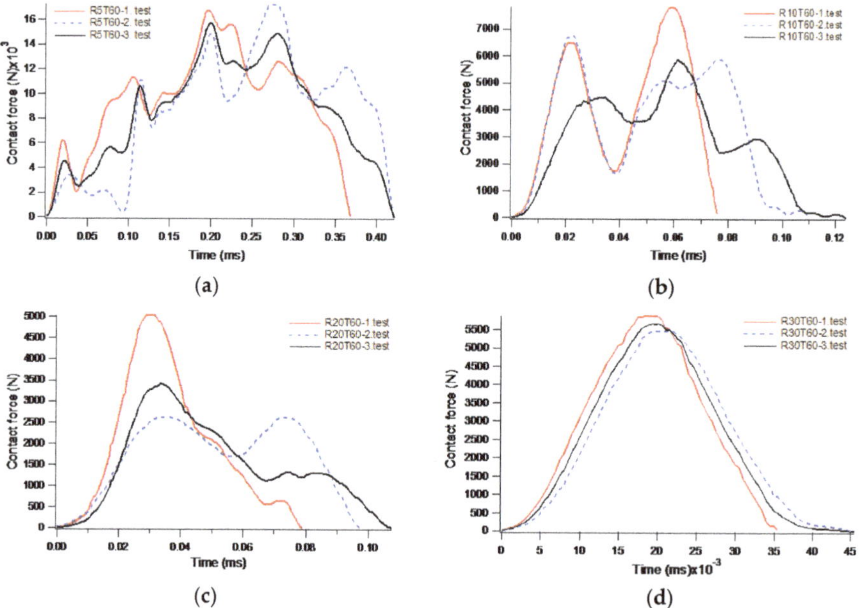

Figure 11. Contact force versus time (F–t) curves after repeating impact tests of composites sintered for 60 minutes: (**a**) Fe + 5 vol.% B$_4$C (R5T60); (**b**) Fe + 10 vol.% B$_4$C (R10T60); (**c**) Fe + 20 vol.% B$_4$C (R20T60); (**d**) Fe + 30 vol.% B$_4$C (R30T60).

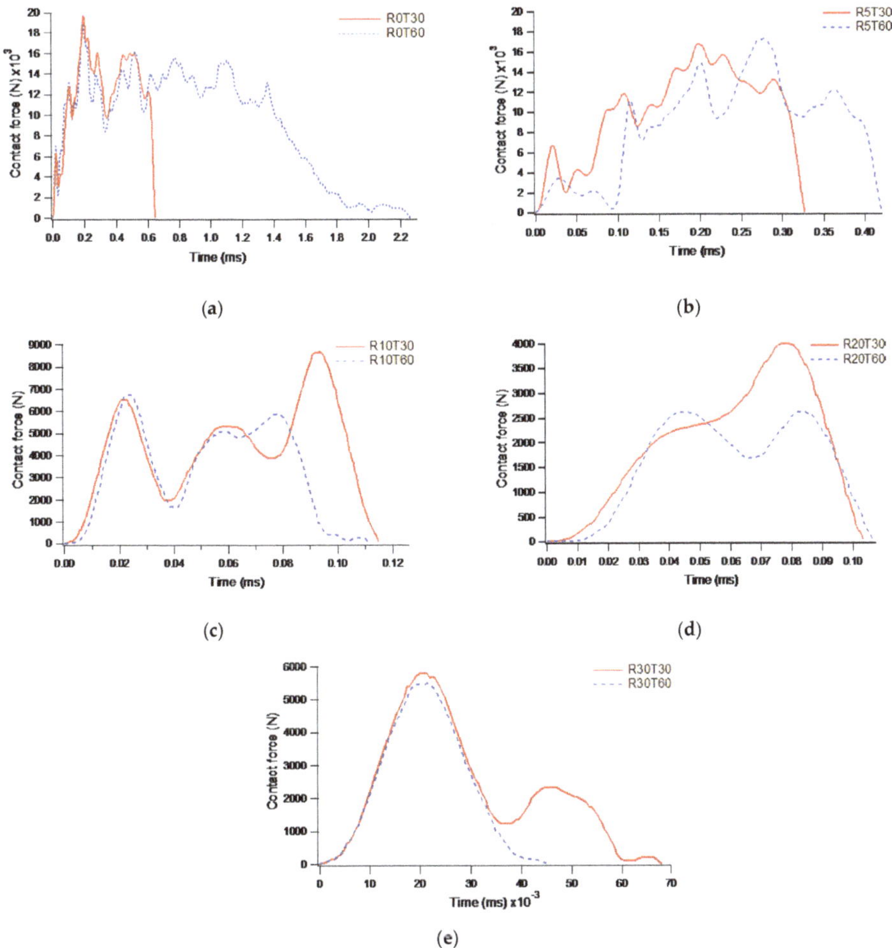

Figure 12. Variation of contact force versus time (F–t) curves for: (**a**) un-reinforced iron specimen sintered for 30 min and 60 min (R0T30, R0T60); (**b**) Fe + 5 vol.% B$_4$C sintered for 30 min and 60 min (R5T30, R5T60); (**c**) Fe + 10 vol.% B$_4$C sintered for 30 min and 60 min (R10T30, R10T60); (**d**) Fe + 20 vol.% B$_4$C sintered for 30 min and 60 min (R20T30, R20T60); (**e**) Fe + 30 vol.% B$_4$C sintered for 30 min and 60 min (R30T30, R30T60).

The effect of sintering duration on the contact force–time behavior (F-t) is presented in Figure 12. A longer duration of sintering caused a negative effect on the fracture force values. While the relatively ductile iron matrix separated as a result of intensive plastic deformation, the breaking mechanism of the harder iron matrix with borided zones was fractured in brittle type. The number of peaks before fracture was reduced when the reinforcement content was increased. Hardness is commonly used for characterizing the brittleness of materials. Composite specimens gain the brittle character with the increase of hardness. The fracturing of brittle materials under dynamic loading results in a higher crack velocity, and fractured surfaces usually start out as relatively smooth surfaces. A brittle fracture can normally be identified by the smoothness of the fractured surface. When the fractured surface is smooth and perpendicular to the applied load, this is strong evidence that the fracture has a brittle component or phase that is also reflected in the force–time curve.

The profile of the contact force curve changed according to the different cases of crack propagation, depending on whether it passed over the borided zone, pores, B_4C particles or the iron matrix. The behavior of the crack was affected by the internal structure of the composites. Since the brittle and ductile regions have different separation mechanisms under loading, the progress of the crack in these regions was also variable, resulting in a fluctuation of the force–time curve. The brittle fracture has been identified by a higher crack propagation velocity when compared with ductile fractures, where the contact times decreased considerably with increasing reinforcement.

A schematical representation of the crack propagation throughout the cross-section of the tested specimens and the effects of the microstructure on the force–time curve is seen in Figure 13.

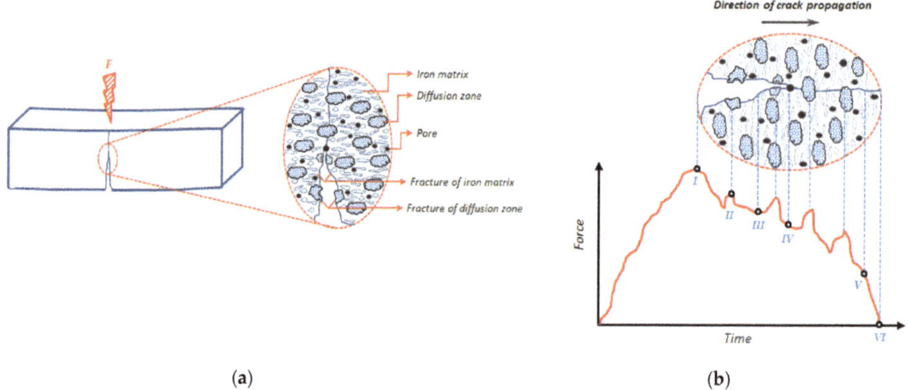

(**a**)   (**b**)

Figure 13. Representation of: (**a**) the propagation of cracks throughout the cross-section of the composite specimen; (**b**) the effect of the microstructure on the contact force–time curve.

The crack, formed at point (I), continued to progress in the structure. When the hard borided zone at point (II) was reached, the resistance to fracture increased momentarily. At point (III), the crack propagation was smoother when it passed the matrix, corresponding to low force values, and these became even lower when it passed through a pore at point (IV). The reduction of the cross-sectional area, as the crack continued to propagate, caused a rapid decrease in the contact force (V) and the test specimen to break completely (VI).

The variation of the impactor energy is presented in Figure 14. According to the graphs, it was seen that the increase of sintering duration had a negative effect on the fracture energy values, particularly at higher volume fractions.

The comparison of the contact force and the impactor energy variations for all the tested specimens sintered under the same conditions is presented in Figures 15 and 16. The maximum force values and impactor energies exhibited different behaviors for different volume fractions. Energy differences were defined by the symbols E1, E2, E3, E4, and E5 for volume fractions of 0%, 5%, 10%, 20%, and 30% B_4C, respectively, for 30 minute sintered specimens, as shown in Figure 15b.

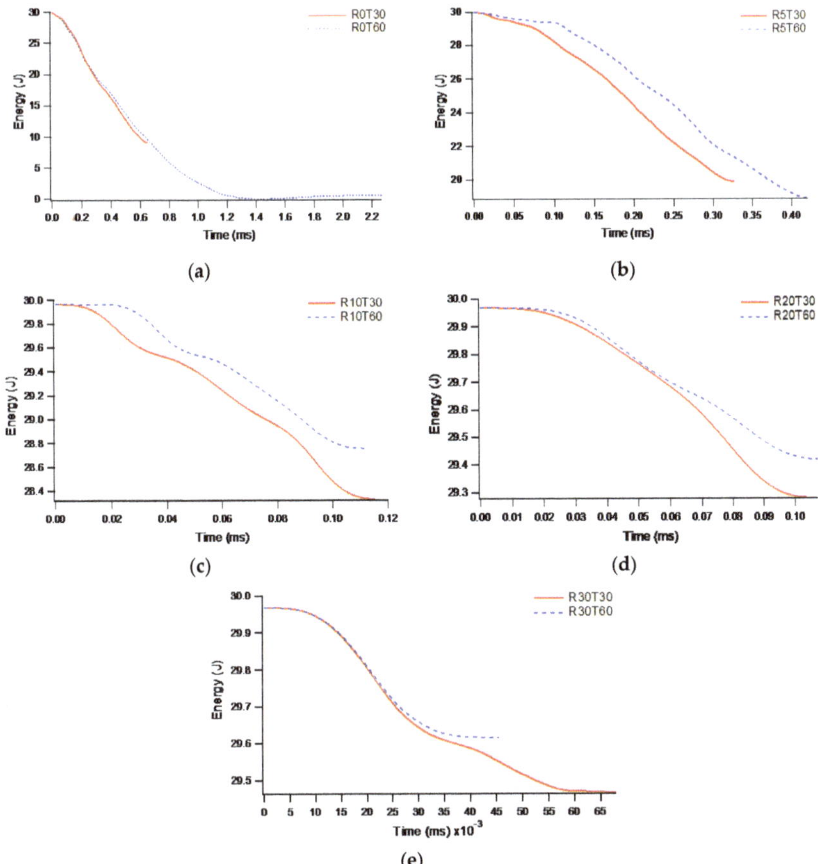

Figure 14. Variation of energy versus time (E–t) curves of impactor for: (**a**) un-reinforced iron specimen sintered for 30 min and 60 min (R0T30, R0T60); (**b**) Fe + 5 vol.% B$_4$C sintered for 30 min and 60 min (R5T30, R5T60); (**c**) Fe + 10 vol.% B$_4$C sintered for 30 min and 60 min (R10T30, R10T60); (**d**) Fe + 20 vol.% B$_4$C sintered for 30 min and 60 min (R20T30, R20T60); (**e**) Fe + 30 vol.% B$_4$C sintered for 30 min and 60 min (R30T30, R30T60).

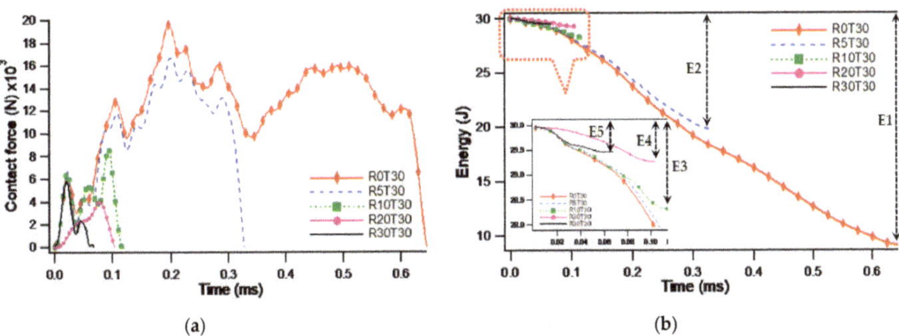

Figure 15. Comparison of 30 min sintered test specimens for: (**a**) contact force–time (F–t); (**b**) energy-time (E–t) curves of impactor; the inset is a magnification of the region indicated by the red dashed line.

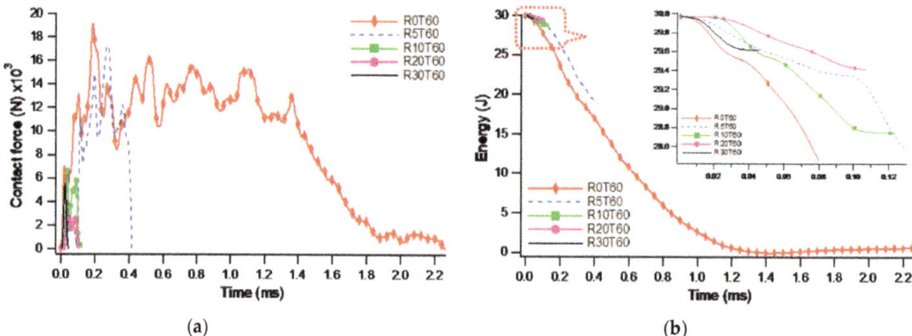

Figure 16. Comparison of 60 min sintered test specimens for: (**a**) contact force–time (F–t); (**b**) energy–time (E–t) curves of impactor; the inset is a magnification of the region indicated by the red dashed line.

Figure 17 shows the category plots of maximum contact forces and fracture energy with respect to the reinforcement ratios and sintering duration. Both the maximum force and the absorbed energy values decreased with increasing volume fractions for each sintering duration.

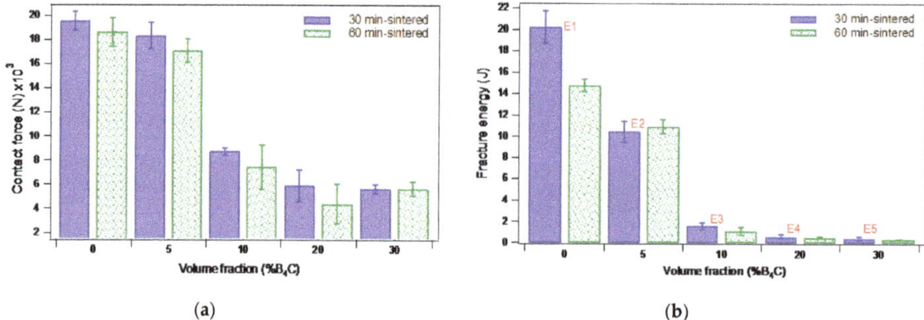

Figure 17. Comparison of un-reinforced Fe and Fe/B$_4$C composites with varied durations of sintering: (**a**) maximum contact forces; (**b**) fracture energy.

Un-reinforced specimens (100 vol.% Fe) had the highest fracture energy for both sintering durations. In particular, the reduction in fracture energy was greater for composites with a volume fraction above 5% B$_4$C.

3.3. Fractographical Analysis

According to the macrophotos of flank and upper surfaces in Figure 18, the fractured surfaces referred to a brittle-type fracture which appeared to be granular and shiny. The propagation of the crack appeared to be of irregular morphology, increased with the reinforcement ratios. The irregular crack propagation was induced by the presence of various obstacles in the structure, such as different hardness zones and the inhomogeneous distribution of boride phases and residual particles. During the propagation, a crack changes direction when faced with these different zones. Regarding the fractured surfaces, the un-reinforced specimen had a planar surface, while the composites showed a wavy profile, as seen in Figure 18's upper surfaces.

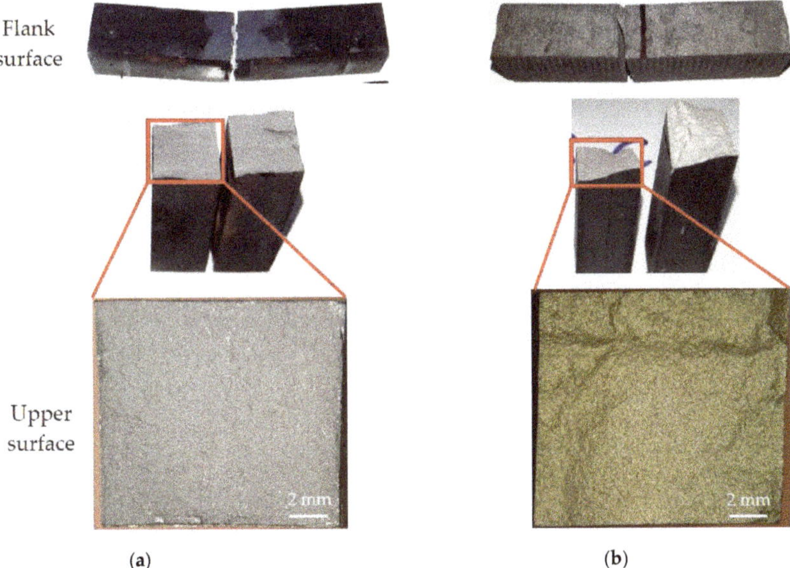

Figure 18. Images of the fractured test specimens from the flank and upper surfaces: (**a**) un-reinforced (R0T60); (**b**) Fe + 20 vol.% B_4C (R20T60) composite.

The cracks have propagated neither in the centreline of the specimens nor throughout the flank surface, as seen in Figure 18. Microcrack initiation and the propagation of cracks were very sensitive to the location of the notch in the composite specimens. The crack first started at the high stress concentration notch, exposed to tensile loading, and then extended to the spreading zone. Non-planar crack propagation and the growth path of the crack also influenced the impact behavior of the composites. The difference between the start point of the crack and the centreline of the specimen, where the impactor contacted the composite on the same axis, caused specimens to sufer damages at different energy levels and contact forces, which was the cause of the standard deviation.

The difference in the fracture morphologies of the iron matrix and the diffusion zone are shown in SEM images (Figures 19–22). The nature of the fracture mode changed from ductile to brittle as the reinforcement ratio increased, even at low volume fractions (5% and 10% B_4C), as shown in Figures 20 and 21a.

The fracture surface of the un-reinforced specimen is characterized by the presence of intensive dimples, as shown in Figure 19. Dimples were observed randomly on the fracture surface, and fracture damage occurred as a result of the combining of these dimples during crack propagation. In addition, oxide formations with a spherical geometry were also observed in the SEM images and proved with EDX analysis, as seen in Figure 19. These oxides, which may be formed during surface polishing processes, are difficult to distinguish from oxides formed during the fabrication process of specimens and come from raw powder.

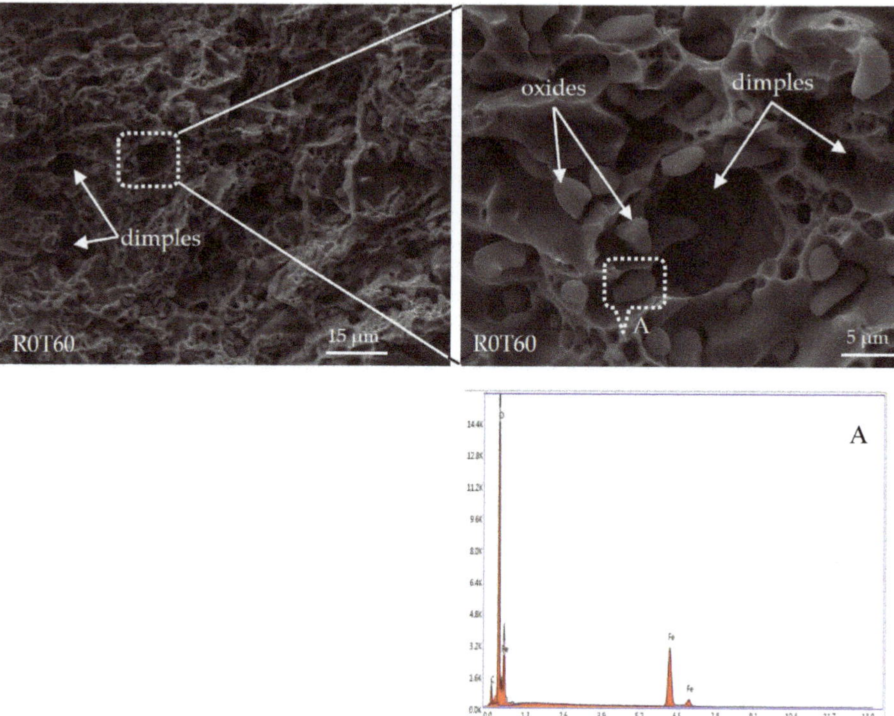

Figure 19. SEM images of the fractured surface from the 60 min sintered un-reinforced test specimen (100 vol.% Fe), and EDX spectrum of the region of A indicated by the white dashed line.

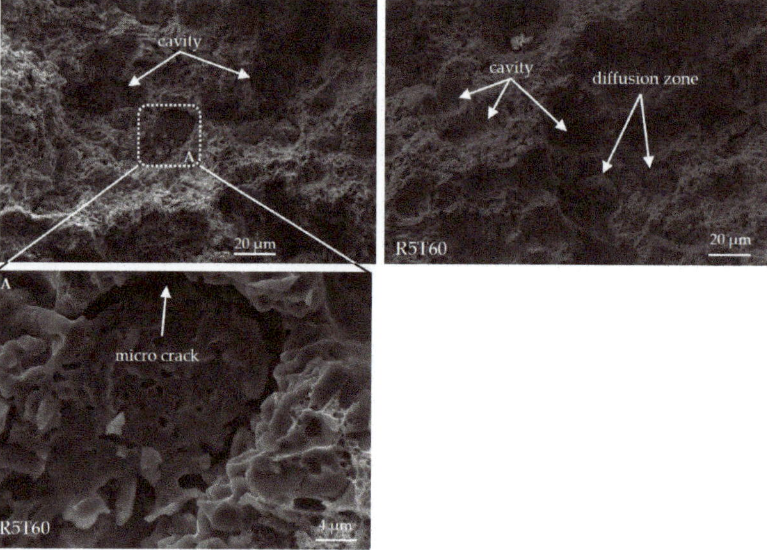

Figure 20. SEM images of fractured surfaces from the 60 min sintered Fe + 5 vol.% B_4C test specimen, A: the magnified region of the white dashed line.

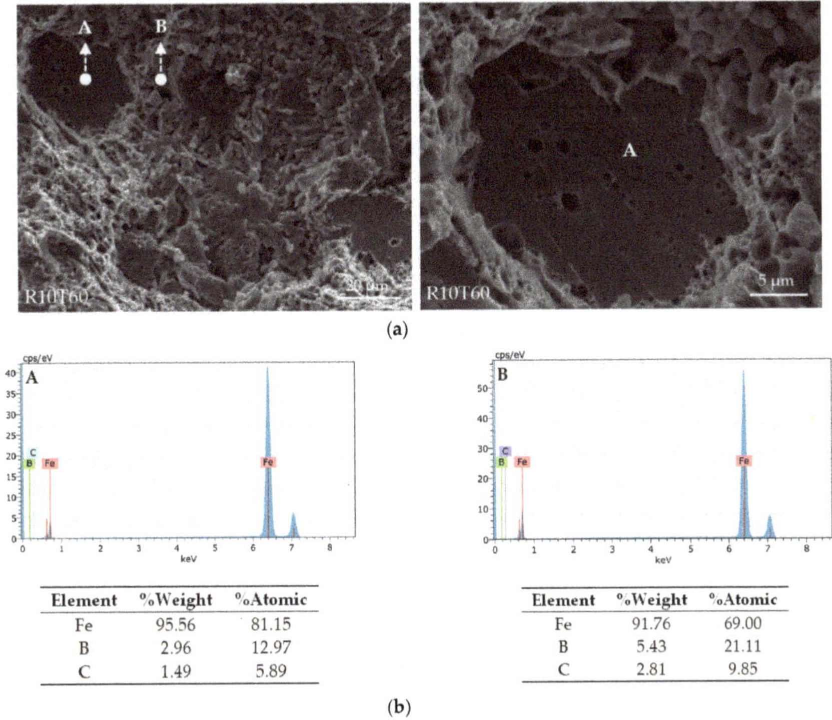

(a)

(b)

Figure 21. (**a**) SEM images of fractured surfaces from the 60 min sintered Fe + 10 vol.% B$_4$C test specimen; (**b**) EDS analysis of the points indicated by the dashed arrow line in (**a**): A iron matrix and B borided zone.

Element	%Weight	%Atomic
Fe	95.56	81.15
B	2.96	12.97
C	1.49	5.89

Element	%Weight	%Atomic
Fe	91.76	69.00
B	5.43	21.11
C	2.81	9.85

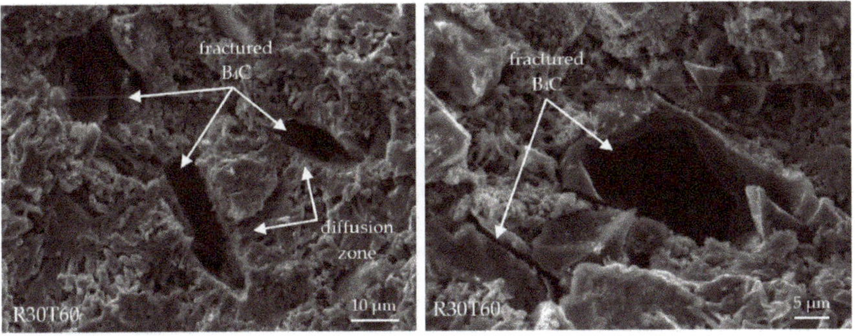

Figure 22. SEM images of fractured surfaces from the 60 min sintered Fe + 30 vol.% B$_4$C test specimen.

It was observed that distinctive cavities occurred when the diffusion zones separated from the matrix surface (Figure 20). If a crack is obstructed by a well-bonded matrix, diffusion zones, or residual particles, it tends to remain constant or to spread around the particle or diffusion zone. If the interfacial adhesion is strong between the matrix and the reinforcement particle interface, damage will occur by breaking the harder regions in the composites. Microcracks were also observed around hard diffusion zones, as confirmed by the SEM images of the R5T60 specimen in Figure 20 (Area A). The magnified region of A in Figure 20 refers to the transgranular fracture of the hard diffusion zone, which encircled the microcrack. In Figure 21, point A refers to a brittle-type fracture with a sharp crack in the iron matrix. The EDS analysis also revealed the difference between the borided zone and the iron matrix on the fractured Fe + 10% B_4C test specimen seen in Figure 21b.

Figure 22 shows the fracture surfaces of the R30T60 specimen. The diffusion zones can be easily distinguished in the fractured surface around residual B_4C particles with their acicular- and columnar-shaped morphology. Residual B_4C particles fractured and broke into pieces under impact loading, as seen in Figure 22. Residual particles and such fracture types were not observed in composites reinforced below 20 vol.% B_4C.

4. Conclusions

In this study, the effects of in-situ synthesized boride phases on the impact behavior of iron-based B_4C particle-reinforced composites were investigated. Fabrication and comprehensive metallographic characterizations were performed. Low-velocity impact tests were conducted and their mechanical behavior was examined in detail.

By analyzing the obtained results, the important findings of this study can be summarized as follows:

(1) Boride phases formed in the internal structure significantly increased the hardness of the composites; on the other hand, the fracture force and impact energy decreased. An increased sintering duration may also lead to increased boride phases and consequently increases hardness, whereas it caused a negative effect on the fracture force values.

(2) B_4C composites reinforced over 20 vol.% had residual B_4C particles, surrounded by the diffusion zones. These B_4C particles fractured and broke into pieces after the impact tests.

(3) The crack propagation and fractured surface morphologies were affected by the distribution and the presence of diffusion zones, porosity, and the residual B_4C particles. When the bonding between the newly formed phases and the matrix was stronger, fracture occurred by the breaking of residual particles.

(4) Residual B_4C particles acted as an impurity factor and caused stress accumulation in the higher volume fractions, while the borided zones led to the strengthening of the interfacial bonding.

(5) The determination of the upper limit of the reinforcement ratio is crucial to achieve the best impact properties, i.e., those that do not allow the formation of residual particles due to an impairment of the mechanical properties.

Author Contributions: The authors collaborated with each other to obtain high-quality research work. F.N. performed the material and procedure selection, analyzed the data, and designed the structure of this paper. M.H. was responsible for the fabrication of test specimens, metallographic preparation, and microstructure characterization of test specimens. He also performed the analysis of the data and designed the structure of this paper. All authors have read and agreed to the published version of the manuscript.

Funding: This research was funded by Erciyes University Scientific Research Projects Coordination Unit, grant number FOA-2014-4997, and the Scientific and Technological Research Council of Turkey (TUBITAK), grant number 117M542.

Acknowledgments: The authors are grateful for the financial support of the Erciyes University Scientific Research Projects Coordination Unit and Scientific and Technological Research Council of Turkey (TUBITAK). The authors are also thankful to Afşin Alper Cerit for help in the development of the experimental setup, and for useful discussions throughout the progress of this work.

Conflicts of Interest: The authors declare no conflict of interest.

References

1. Su, Y.; Ouyang, Q.; Zhang, W.; Li, Z.; Guo, Q.; Fan, G.; Zhang, D. Composite structure modeling and mechanical behavior of particle reinforced metal matrix composites. *Mater. Sci. Eng. A* **2014**, *597*, 359–369. [CrossRef]

2. Safri, S.N.A.; Sultan, M.T.H.; Yidris, N.; Mustapha, F. Low Velocity and High Velocity Impact Test on Composite Materials-A Review. *Int. J. Eng. Sci.* **2014**, *3*, 50–60.

3. Sjoblom, P.O.; Hartness, J.T.; Cordell, T.M. On Low-Velocity Impact Testing of Composite Materials. *J. Compos. Mater.* **1998**, *22*, 30–52. [CrossRef]

4. Straffelini, G.; Molinari, A.; Danninger, H. Impact notch toughness of high-strength porous steels. *Mater. Sci. Eng. A* **1999**, *272*, 300–309. [CrossRef]

5. Straffelini, G. Impact Fracture Toughness of Porous Iron and High-Strength Steels. *Metall. Mater. Trans. A* **2000**, *31*, 1443–1451. [CrossRef]

6. Şahin, Ö.S.; Güneş, A.; Aslan, A.; Salur, E.; Karadağ, H.B.; Akdemir, A. Low-Velocity Impact Behavior of Porous Metal Matrix Composites Produced by Recycling of Bronze and Iron Chips. *Iran. J. Sci. Technol. Trans. Mech. Eng.* **2017**, *43*, 53–60. [CrossRef]

7. Ratto, P.J.J.; Ansaldi, A.F.; Fierro, V.E.; Agüera, F.R.; Villar, H.N.A.; Sikora, J.A. Low Temperature Impact Tests in Austempered Ductile Iron and Other Spheroidal Graphite Cast Iron Structures. *ISIJ Int.* **2001**, *41*, 372–380. [CrossRef]

8. Hufenbach, W.; Ibraim, F.M.; Langkamp, A.; Böhm, R.; Hornig, A. Charpy impact tests on composite structures–An experimental and numerical investigation. *Compos. Sci. Technol.* **2008**, *68*, 2391–2400. [CrossRef]

9. Toktaş, G.; Tayanç, M.; Toktaş, A. Effect of matrix structure on the impact properties of an alloyed ductile iron. *Mater. Charact.* **2006**, *57*, 290–299. [CrossRef]

10. Shahabodin, E.; Saeedi, H.M.; Reza, B.H.; Naser, E. Effect of Iron on the wetting, sintering ability, and the physical and mechanical properties of Boron Carbide composites: A review. *Int. J. Refract. Met. Hard Mater.* **2016**, *57*, 78–92. [CrossRef]

11. Dudrová, E.; Kabátová, M. Fractography of Sintered Iron and Steels. *Powder Metall. Prog.* **2008**, *8*, 59–75.

12. Shanmugasundaram, D.; Chandramouli, R. Tensile and impact behaviour of sinter-forged Cr, Ni and Mo alloyed powder metallurgy steels. *Mater. Des.* **2009**, *30*, 3444–3449. [CrossRef]

13. Ozden, S.; Ekici, R.; Nair, F. Investigation of impact behaviour of aluminium based SiC particle reinforced metal–matrix composites. *Compos. Part A Appl. Sci. Manuf.* **2007**, *38*, 484–494. [CrossRef]

14. Gustafsson, G.; Nishida, M.; Häggblad, H.Å.; Kato, H.; Jonsén, P.; Ogura, T. Experimental studies and modelling of high-velocity loaded iron-powder compacts. *Powder Technol.* **2014**, *268*, 293–305. [CrossRef]

15. Montealegre-Melendez, I.; Arevalo, C.; Ariza, E.; Perez-Soriano, E.M.; Rubio-Escudero, C.; Kitzmantel, M.; Neubauer, E. Analysis of the Microstructure and Mechanical Properties of Titanium-Based Composites Reinforced by Secondary Phases and B_4C Particles Produced via Direct Hot Pressing. *Materials* **2017**, *10*, 1240. [CrossRef] [PubMed]

16. Turov, Y.V.; Khusid, B.M.; Voroshnin, L.G.; Khina, B.B.; Kozlovskii, I.L. Gas transport processes in sintering of an iron-boron carbide powder composite. *Sov. Powder Metall. Ceram.* **1989**, *28*, 618–622. [CrossRef]

17. Üçisik, A.H.; Bindal, C. Fracture toughness of boride formed on low-alloy steels. *Surf. Coat. Technol.* **1997**, *94–95*, 561–565. [CrossRef]

18. Korol'kov, V.V.; Kibak, B. Sintering mechanism of iron powder with microadditions of boron. *Powder Metall. Met. Ceram.* **1997**, *36*, 470–473. [CrossRef]

19. Aizenshtein, M.; Mizrahi, I.; Froumin, N.; Hayun, S.; Dariel, M.P.; Frage, N. Interface interaction in the B_4C/(Fe–B–C) system. *Mater. Sci. Eng. A* **2008**, *495*, 70–74. [CrossRef]

20. Turov, Y.V.; Khusid, B.M.; Voroshnin, L.G.; Khina, B.B.; Kozlovskii, I.L. Structure formation in sintering iron-boron carbide powder composite. *Powder Metall. Ceram.* **1993**, *32*, 465–470.

21. Mizrahi, I.; Raviv, A.; Dilman, H.; Aizenshtein, M.; Dariel, M.P.; Frage, N. The effect of Fe addition on processing and mechanical properties of reaction infiltrated boron carbide-based composites. *J. Mater. Sci.* **2007**, *42*, 6923–6928. [CrossRef]

22. Ozdemir, O.; Usta, M.; Bindal, C.; Ucisik, A.H. Hard iron boride (Fe_2B) on 99.97wt% pure iron. *Vacuum* **2006**, *80*, 1391–1395. [CrossRef]

23. Abenojar, J.; Velasco, F.; Martinez, M.A. Influence of carbon and aluminium additions on the Fe–10% B(wt.) system. *J. Mater. Process. Technol.* **2003**, *143–144*, 28–33. [CrossRef]

24. Campos, I.; Torres, R.; Ramírez, G.; Ganem, R.; Martínez, J. Growth kinetics of iron boride layers: Dimensional analysis. *Appl. Surf. Sci.* **2006**, *252*, 8662–8667. [CrossRef]

25. Nowacki, J.; Klimek, L. Structure and properties of Fe-Fe_2B cermets. *J. Mater. Sci.* **1992**, *27*, 3651–3656. [CrossRef]

26. Nowacki, J.; Klimek, L. The mechanism of reaction sintering of iron-iron boride cermets. *J. Mater. Sci.* **1993**, *28*, 3939–3944. [CrossRef]
27. Martini, C.; Palombarini, G.; Carbucicchio, M. Mechanism of thermochemical growth of iron borides on iron. *J. Mater. Sci.* **2004**, *39*, 933–937. [CrossRef]
28. Busby, P.E.; Warga, M.E.; Wells, C. Diffusion and Solubility of Boron in Iron and Steel. *J. Met.* **1953**, *5*, 1463–1468. [CrossRef]
29. Bagliuk, G. Properties and Structure of Sintered Boron Containing Carbon Steels. In *Sintering—Methods and Products*; Shatokha, V., Ed.; InTech: London, UK, 2012; pp. 249–266.

© 2020 by the authors. Licensee MDPI, Basel, Switzerland. This article is an open access article distributed under the terms and conditions of the Creative Commons Attribution (CC BY) license (http://creativecommons.org/licenses/by/4.0/).

Article

A Numerical Study of the Effect of Isotropic Hardening Parameters on Mode I Fatigue Crack Growth

M. F. Borges, F. V. Antunes *, P. A. Prates and R. Branco

CEEMPRE, Department of Mechanical Engineering, University of Coimbra, 3030-788 Coimbra, Portugal; micaelfriasborges@outlook.pt (M.F.B.); pedro.prates@dem.uc.pt (P.A.P.); ricardo.branco@dem.uc.pt (R.B.)
* Correspondence: fernando.ventura@dem.uc.pt; Tel.: +351-239-790701

Received: 18 December 2019; Accepted: 18 January 2020; Published: 25 January 2020

Abstract: The consideration of plastic crack tip opening displacement (CTOD, δ_p), as a crack driving force has given us the opportunity to predict fatigue crack growth (FCG) rate numerically, and, therefore, to develop parametric studies focused on the effect of loading, geometrical, and material parameters. The objective here is to study the effect of the isotropic hardening parameters of the Voce law on FCG, which are the isotropic saturation stress, Y_{Sat}, and the isotropic saturation rate, C_Y. The increase of these hardening parameters causes δ_p to decrease. However, this effect is much more pronounced for Y_{Sat} than C_Y. The variation is non-linear, and the rate of variation decreases with the increase of isotropic parameters. The increase of Y_{Sat} increases the crack closure phenomenon. Finally, the influence of the isotropic parameters is more relevant for pure isotropic hardening than for mixcd hardening.

Keywords: fatigue crack growth; isotropic hardening; crack tip opening displacement; CTOD; crack closure

---

1. Introduction

Most fatigue crack growth (FCG) studies deal with the effect of load parameters—namely ΔK, stress ratio, K_{max}, and variable amplitude loading. However, the material parameters are also expected to have a major effect on FCG. In real materials, it is not possible to change material properties one-by-one; therefore, analytical or numerical approaches are the solution to study the effect of properties on FCG rate. Table 1 presents some analytical models proposed in the literature [1–10] involving material parameters—namely the elastic parameters (Young's modulus, E, and Poisson's ratio, v), yield stress, Y_0, and isotropic hardening (hardening exponent, n, cyclic strain hardening coefficient, K', and cyclic strain hardening exponent, n'). Young's modulus and Y_0 are included in all of the models. The fracture toughness is included in the models of Nicholls [2] and Chand and Garg [5], while the failure strain is included in other models [3,4]. The isotropic hardening parameters are included in the models of Schwalbe [3] and Chand and Garg [5]. The effect of kinematic hardening parameters is not found in literature models.

Table 1. Fatigue crack growth models that include material properties.

Reference	Model	Comments
Pelloux [1]	$\frac{da}{dN} = \frac{2\Delta K^2}{\pi E Y_0}$	E—Young's modulus Y_0—Yield stress
Nicholls [2]	$\frac{da}{dN} = \frac{\Delta K^4}{4 E Y_0 \leq_c^2}$	-
Schwalbe [3]	$\frac{da}{dN} = \beta \frac{\Delta K^2}{4\pi(1+n)Y_0^2}\left(\frac{2Y_0}{\varepsilon_f E}\right)^{1+n}$	ε_f—failure strain n—hardening exponent
Jablonski [4]	$\frac{da}{dN} = \frac{0.0338(1-v^2)}{\varepsilon_f E Y_0}\Delta K^2$	$n = 10$ v—Poisson's ratio
Chand and Garg [5]	$\frac{da}{dN} = \frac{0.15\Delta K_{eff}^2 Y_0}{nE K_{Ic}^2 (1+R)^{3.8}}$	K_{Ic}—fracture toughness R—stress ratio
Skelton et al. [6]	$\frac{da}{dN} = \frac{\Delta K^2(1-v)}{2\pi E W_c}$	W_c—critical value of density of cumulative energy
Clavel and Pineau [7]	$\frac{da}{dN} = \beta\frac{\Delta K^2}{E Y_0}$	$\beta = 0.25$ Clavel and Pineau $\beta = 0.02\ (1-v^2)$ Liu [8]
Liu and Liu [9]	$\frac{da}{dN} = \frac{0.036}{E Y_0}(\Delta K - \Delta K_{th})^2$	ΔK_{th}—fatigue threshold

Carpinteri [10] performed finite element analyses of crack propagation by means of a strain-based criterion and observed a greater amount of crack extension in the kinematic hardening case for a given remote load, which means that isotropic hardening increases crack growth resistance. Martínez-Pañeda and Fleck [11] modeled advances in a mode I crack under small scale yielding conditions using a cohesive zone formulation. Kinematic hardening significantly raised the level of plastic dissipation. Pommier and Bompard [12] investigated the relative importance of kinematic versus isotropic hardening for modeling crack closure. The material with prevailing kinematic hardening displayed cyclic plasticity at the tip and less crack tip closure than the material with isotropic hardening. The isotropic hardening is often neglected if a material does not exhibit significant cyclic hardening and only kinematic hardening is considered [13–15]. In fact, studies that assume purely isotropic behavior are relatively rare [16–18].

The main objective here is to study the effect of isotropic hardening parameters on FCG rate. The parameters studied were isotropic saturation rate, C_Y, and isotropic saturation stress, Y_{sat}. A numerical analysis based on the finite element method was followed to calculate the plastic crack tip opening displacement (CTOD), δ_P, which was found to be linearly related to FCG rate in numerical [19,20] and experimental [21] studies. This approach has already been used in previous works to predict the effect of a material's yield stress, Y_0 [22], and Young's modulus [23]. All other physical and numerical parameters were kept constant in order to isolate the effect of isotropic hardening parameters.

2. Elastic–Plastic Material Model

The material studied (304L stainless steel) was assumed to have an elastic–plastic behavior. The elastic behavior was isotropic, with $E = 210$ GPa and $v = 0.30$ as parameters of the generalized Hooke's law. The plastic behaviour was described by the von Mises yield criterion coupled with a mixed isotropic–kinematic hardening under an associated flow rule:

$$\sqrt{\frac{3}{2}(\sigma' - X') : (\sigma' - X')} - Y = 0 \tag{1}$$

where σ' is the deviatoric Cauchy stress tensor, X' is the deviatoric backstress tensor described by the Armstrong–Frederick kinematic hardening law [24], and Y is the flow stress described by the Voce isotropic hardening law [25]. The non-linear (exponential) kinematic hardening model proposed by Armstrong and Frederick can be written as:

$$\dot{X'} = C_X \left[X_{Sat} \frac{\Sigma}{\bar{\sigma}} - X' \right] \dot{\bar{\varepsilon}}^p \tag{2}$$

where $\dot{X'}$ is the backstress rate, which represents the translational velocity of the yield surface centre during plastic deformation, $\bar{\sigma}$ is the equivalent stress, and $\dot{\bar{\varepsilon}}^p$ is the equivalent plastic strain rate; C_X and X_{Sat} are the kinematic hardening parameters, respectively representing the saturation rate and the saturation value of the exponential kinematic hardening X, which can be written as [26]:

$$X = X_{Sat}[1 - exp(-C_X \bar{\varepsilon}^p)]. \tag{3}$$

The Voce law is an isotropic hardening model that is often used to describe the behavior of materials that exhibit stress saturation at large strains, and can be written as follows:

$$Y = Y_0 + (Y_{Sat} - Y_0)[1 - exp(-C_Y \bar{\varepsilon}^p)] \tag{4}$$

where Y_{Sat} is the saturation stress, C_Y is the stress saturation rate, and $\bar{\varepsilon}^p$ is the equivalent plastic strain.

The set of material properties that best describe the cyclic behaviour of the SS304L is presented in Table 2, labeled as "Reference". The optimization procedure that was performed to identify these material properties can be found in a previous work [27,28]. Table 2 also presents the numerical changes (−25%, +25%, and +50% of the "Reference" case) in the isotropic parameters Y_{Sat} and C_Y. The case "0.50Y_{Sat}" was not considered because it leads to softening (i.e., $Y_{Sat} < Y_0$). Figure 1 presents the isotropic and mixed hardening stress–strain curves for the materials presented in Table 2, for monotonic uniaxial tension; accordingly, the isotropic hardening curves (see Figure 1a) were obtained from Equation (4) and the mixed hardening curves (see Figure 1b) were obtained from Equation (1). As expected, the level of the mixed-hardening stress–strain curves (see Figure 1b) was higher than that of the isotropic hardening curves (see Figure 1a). Accordingly, the mixed-hardening cases from Table 1 offer greater resistance to plastic deformation than the isotropic hardening cases. Moreover, the sensitivity to numerical changes in Y_{Sat} is higher for isotropic hardening than for mixed-hardening (see Figure 1). Similar trends were obtained for C_Y.

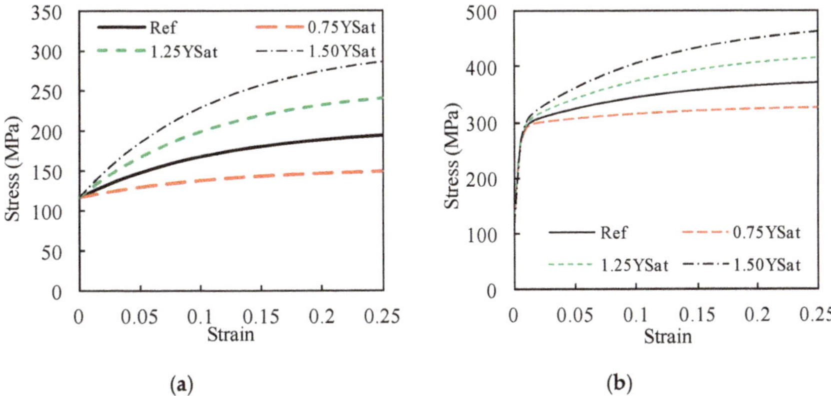

Figure 1. Stress–strain curves for monotonic uniaxial tension. (a) Isotropic hardening. (b) Mixed hardening.

Table 2. The set of material properties.

Material	Hooke's Law		Voce Law			Armstrong–Frederick Law	
SS304L Mixed Hardening [28]	E (GPa)	ν (–)	Y_0 (MPa)	Y_{Sat} (MPa)	C_Y (–)	C_X (–)	X_{Sat} (MPa)
Reference	196	0.3	117	204	9.00	300	176
$0.50Y_{Sat}$	196	0.3	117	204	9.00	300	176
$0.75Y_{Sat}$	196	0.3	117	102	9.00	300	176
$1.25Y_{Sat}$	196	0.3	117	153	9.00	300	176
$1.50Y_{Sat}$	196	0.3	117	255	9.00	300	176
$0.50C_Y$	196	0.3	117	204	4.50	300	176
$0.75C_Y$	196	0.3	117	204	6.75	300	176
$1.25C_Y$	196	0.3	117	204	11.25	300	176
$1.50C_Y$	196	0.3	117	204	13.50	300	176
SS304L Purely Isotropic	E (GPa)	ν (–)	Y_0 (MPa)	Y_{Sat} (MPa)	C_Y (–)	C_X (–)	X_{Sat} (MPa)
Reference	196	0.3	117	204	9.00	0	0
$0.75Y_{Sat}$	196	0.3	117	153	9.00	0	0
$1.25Y_{Sat}$	196	0.3	117	255	9.00	0	0
$1.50Y_{Sat}$	196	0.3	117	306	9.00	0	0
$0.50C_Y$	196	0.3	117	204	4.50	0	0
$0.75C_Y$	196	0.3	117	204	6.75	0	0
$1.25C_Y$	196	0.3	117	204	11.25	0	0
$1.50C_Y$	196	0.3	117	204	13.50	0	0

3. Numerical Model

A compact tension specimen, C(T), was numerically modeled in DD3IMP using in-house code [29,30], with a width, W, equal to 50 mm, and an initial crack length, a_0, of 24 mm. The symmetry conditions of the specimen's geometry allowed for the modeling of only one-quarter of the specimen, reducing the numerical overhead. Frictionless contact was modeled over a symmetry plane placed behind the growing crack front in order to simulate plasticity-induced crack closure. Relative to the specimen's thickness, t, only 0.1 mm were simulated to reduce numerical effort and to simulate a plane stress state with the proper boundary conditions. A remote cyclic load was applied at the hole of the specimen with a magnitude varying between 41.67 N and 4.167 N. The stress ratio, R, was therefore equal to 0.1 and K_{max}, K_{min}, and ΔK were 18.3, 1.83, and 16.5 MPa.m$^{0.5}$, respectively. All simulations were performed with two load cycles between crack increments that occurred at minimum load. As previously mentioned, the numerical parameters were kept constant in order to isolate the effect of isotropic hardening.

An ultra-refined mesh was employed at the crack tip region with the element's dimensions being 8×8 μm^2. The goal of employing this mesh was to accurately quantify strain gradients and local stress. The remaining volume of the specimen's geometry was discretized by a coarser mesh. The finite element mesh was constituted by a total of 7287 three-dimensional (3D) linear isoparametric elements and 14,918 nodes. Each crack increment—i.e., each node release—corresponded to the dimension of the elements in the ultra-refined region (8 μm). The simulations stopped when the total crack propagation, Δa, reached the value of 1272 μm, making 159 crack propagations. This allowed for the stabilization of crack tip fields and the crack closure level. Therefore, stable values of CTOD were able to be obtained. In some numerical simulations, the crack closure effect was removed by eliminating the contact of the crack flanks, which meant that the crack flanks may have overlapped. This is physically impossible; however, it can be numerically made, allowing for the study of crack propagation in a simpler state. All simulations were evaluated at the first node behind the crack tip (8 μm behind the tip).

4. Results

4.1. Typical Curves

Figure 2a shows a common variation of crack tip opening displacement, CTOD, with the applied load, F, in simulations with the crack closure phenomenon. These curves are constituted by several regimes. In regime 1,2, the crack is closed despite the increase of F. At point 2, the opening of the crack occurs. When the crack first opens, it experiences purely elastic deformation, as shown in regime 2,3. Point 3 corresponds to the transition between purely elastic and elastic–plastic behaviour at the crack tip, and is characterized by a plastic crack tip opening displacement, $CTOD_p$, equal to 0.001 µm. This value is purely empirical. At this stage, the plastic deformation increases rapidly until it reaches the maximum value of F at point 4. The unloading of the specimens is similar to the loading process: first, the crack experiences purely elastic deformation in regime 4,5, followed by the elastic-plastic in regime 5,6. At point 6, the crack closes.

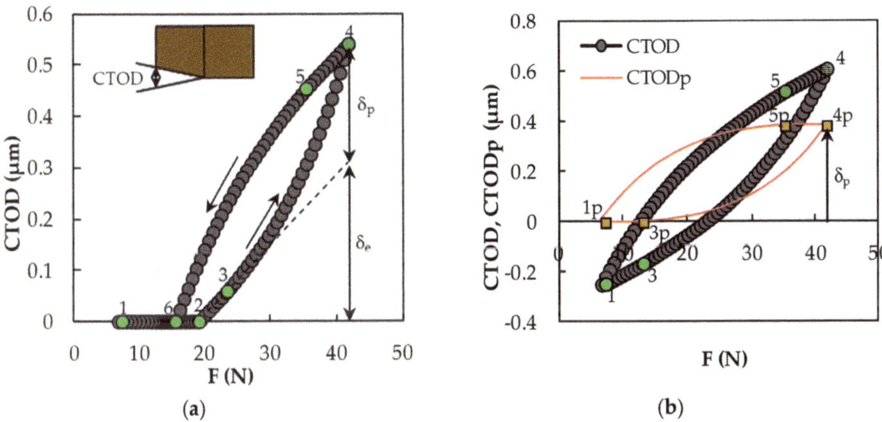

Figure 2. (**a**) Crack tip opening displacement (CTOD) vs. load (with crack closure); (**b**) CTOD and plastic CTOD vs. load (without crack closure); (SS304L mixed hardening).

Figure 2b plots a typical CTOD against the F curve in the simulations without the crack closure phenomenon. In this kind of curve, the crack is already open at the minimum value of F, making points 1 and 2 coincident. Regimes 1–3 and 3–4 are the purely elastic and elastic–plastic regimes during the specimen's loading, respectively. Regimes 4,5 and 5,6 are the same regimes, but during the specimen's unloading. The higher values of CTOD achieved in these simulations are due to the higher effective loads contributing to deformation in the absence of crack closure. This figure also shows the variation of $CTOD_p$ with F. In the purely elastic regimes, $CTOD_p$ remains constant and increases/decreases non-linearly in the elastic–plastic ones.

4.2. Effect of Y_{Sat} and C_Y without Crack Closure

Figure 3a shows three curves of CTOD versus load for different values of Y_{Sat} (cases "1.25Y_{Sat}", "Ref", and "0.75Y_{Sat}"—see Table 2). As expected, reducing the isotropic saturation stress has, as a consequence, higher values of CTOD, and, therefore, higher values of deformation at the crack tip. The increase of deformation is due to the plastic part, as shown in Figure 3b, which was built by extracting the elastic CTOD from the total CTOD presented in Figure 3a. In a uniaxial test, when Y_{Sat} is reached, the increment in plastic deformation is performed at a constant value of stress (equal to Y_{Sat}). Decreasing the value of Y_{Sat} leads to a decrease in the amount of stress required for plastic strain (see Figure 1a); in other words, for a given stress value, the plastic strain will be higher for decreasing values of Y_{Sat}. This can be seen by the ranges of deformations present in both Figure 3a,b, and also by the separation of the curves at lower values of the load. Similarly, for a given load value, the CTOD will be higher for decreasing values of C_Y.

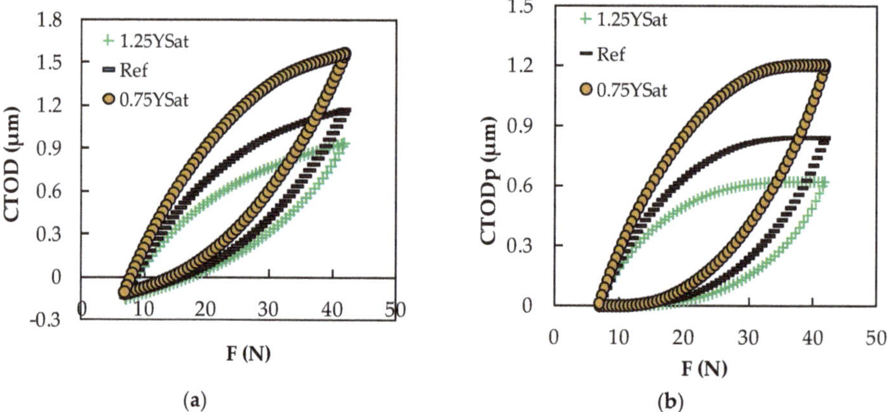

Figure 3. (**a**) Total CTOD vs. load; (**b**) Plastic CTOD vs. load. (SS304L isotropic; without crack closure).

Figure 4a,b show the variation in the plastic CTOD range, δ_p, with Y_{Sat} and C_Y, respectively, for the purely isotropic and mixed hardening models of the SS304L. Both figures show a non-linear variation of δ_p with the isotropic hardening parameters, making it difficult to establish analytical models containing these hardening parameters. The increase of the hardening parameters causes δ_p to decrease with a rate that also decreases with the hardening parameters. The reduction of δ_p is much more pronounced for Y_{Sat} than for C_Y, which can be expected when considering the physical meaning of both parameters. Also, for both figures, the mixed hardening model achieves less δ_p than the isotropic hardening model. This supports the results shown in Figure 1, where mixed-hardening cases (see Figure 1b) offer greater resistance to plastic deformation than isotropic hardening cases (see Figure 1a).

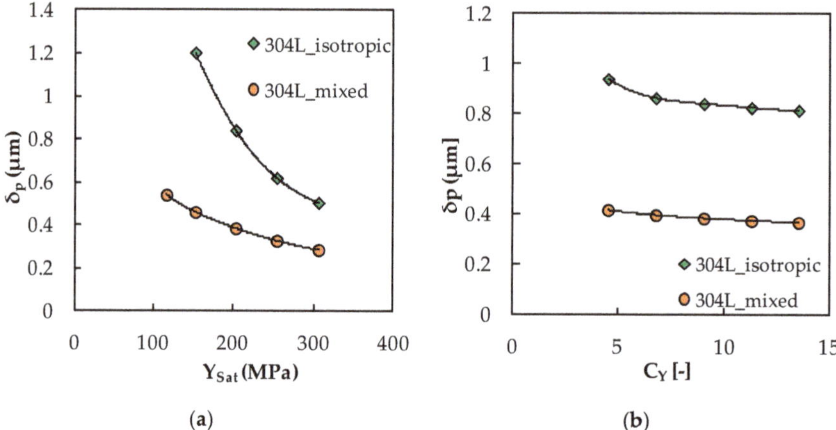

Figure 4. Plastic CTOD range, δ_p, vs. (**a**) saturation stress, Y_{Sat}; (**b**) stress saturation rate, C_Y, (without crack closure).

4.3. Effect of Y_{Sat} and C_Y with Crack Closure

Figure 5a shows three CTOD versus F curves for the purely isotropic SS304L, in simulations with crack closure, in which the values of Y_{Sat} were changed, corresponding to the cases "$0.75Y_{Sat}$", "Ref", and "$1.25Y_{Sat}$" in Table 2. The numerical changes made to Y_{Sat} had a great influence on the range of CTOD values; in other words, the deformation obtained at the crack tip was deeply affected. As was expected, increasing the value of Y_{Sat} reduced the CTOD range and the total deformation due to the reduction of the plastic deformation, as shown in Figure 5b. Also, it can be seen here that the crack opening load, F_{open}, and the crack closure load, F_{close}, were affected by the numerical changes. F_{open} tended to increase and F_{close} to decrease when the value of Y_{Sat} was higher.

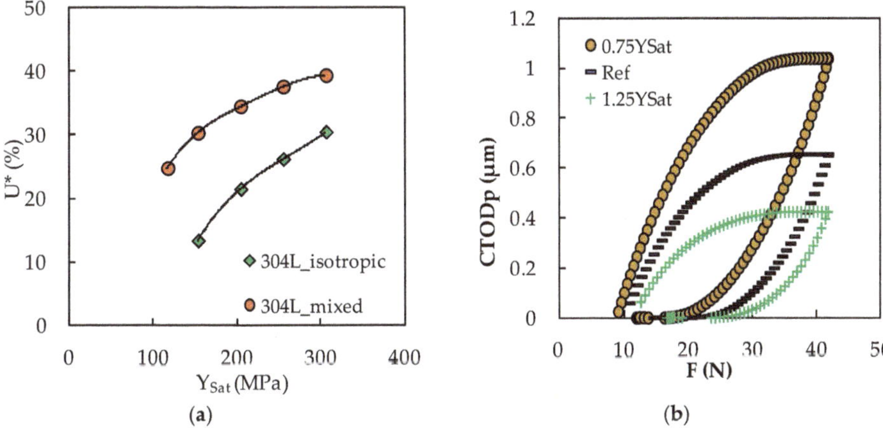

Figure 5. (a) CTOD vs. load; (b) plastic CTOD vs. load (SS304L isotropic; with crack closure).

Figure 6 presents the evolution of U^* with crack growth, Δa. U^* is the percentage of the load cycle during which the crack is closed:

$$U^* = \frac{F_{open} - F_{min}}{F_{max} - F_{min}} \times 100 \tag{5}$$

where F_{max} and F_{min} are, respectively, the maximum and minimum loads. At the beginning of the numerical simulation, i.e., where $\Delta a = 0$, U^* is small and increasing progressively towards a stable value, the residual plastic wake is formed. The stable values are reached at about $\Delta a = 500$ µm, which corresponds to 60 crack increments. Therefore, the 160 crack increments considered in the present study are more than enough to obtain stable values of the crack opening level. Additionally, there is more crack closure for mixed hardening than for isotropic hardening.

The crack closure levels are represented in Figure 7a,b: U^* is a function of Y_{Sat} and C_Y, respectively, for both the purely isotropic hardening and mixed hardening models. As previously stated, the increase of Y_{Sat} caused F_{open} to increase, and, therefore, it was expected that U^* would increase, as shown in Figure 7a. Also, the purely isotropic hardening reduced U^* in comparison with the mixed hardening. Despite the high numerical range of variation of the isotropic parameters, the variation of U^* was relatively small for both hardening conditions (approximately 17% in Figure 7a and 4% in Figure 7b).

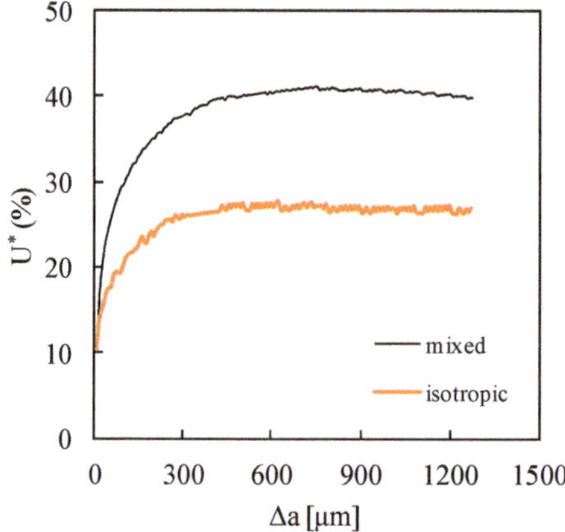

Figure 6. Crack closure level, U^*, vs. crack increment (reference cases).

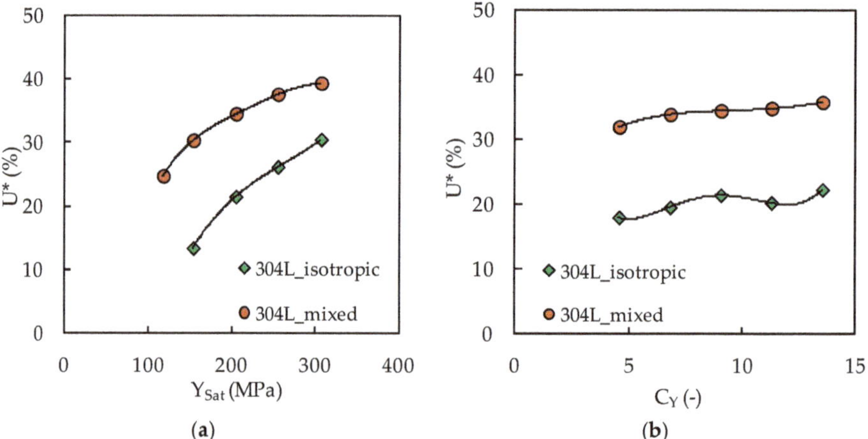

Figure 7. Crack closure level, U^*, vs. (**a**) isotropic saturation stress, Y_{Sat}; (**b**) isotropic saturation rate, C_Y (with crack closure).

The plot of δ_p shown in Figure 8a,b serves as a function of Y_{Sat} and C_Y, respectively. As noted in Figure 7a,b, the purely isotropic hardening model provided less U^* than the mixed hardening model. Therefore, a more effective load range was available in the purely isotropic model, contributing to more plastic deformation, as shown in the results in Figure 8a,b. As in Figure 4a,b, the mixed hardening model achieved less δ_p than the pure isotropic model. The dashed lines indicate the values obtained without contact of crack flanks. The difference in the solid lines is the effect of crack closure, which is relevant in all cases.

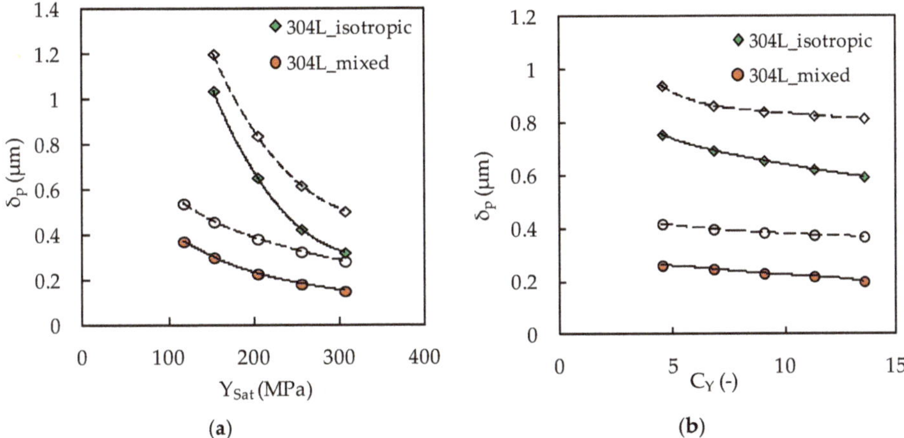

Figure 8. Plastic CTOD range, δ_p, vs. (**a**) saturation stress, Y_{Sat}; (**b**) stress saturation rate, C_Y (with crack closure, solid lines; without crack closure, dashed lines).

5. Conclusions

A numerical approach, based on plastic CTOD, was used here to study the effect of isotropic parameters, which are the isotropic saturation stress, Y_{Sat}, and the isotropic saturation rate, C_Y. The main conclusions are:

The increase of the hardening parameters causes δ_p to decrease. However, this effect is much more pronounced for Y_{Sat} than for C_Y. The variation is non-linear, and the rate of variation decreases with the increase of the isotropic parameters.

The increase of Y_{Sat} causes the increase of the crack closure phenomenon.

The influence of the hardening parameters is more relevant for pure isotropic hardening than for mixed hardening.

Author Contributions: M.F.B.: Treatment of numerical results and writing; F.V.A.: Literature Review; P.A.P.: Modelling of material Behaviour; R.B.: Final revision. All authors have read and agreed to the published version of the manuscript.

Funding: This research was funded by the project no.028789, financed by the European Regional Development Fund (FEDER), through the Portugal-2020 program (PT2020), under the Regional Operational Program of the Center (CENTRO-01-0145-FEDER-028789) and by the Foundation for Science and Technology IP/MCTES through national funds (PIDDAC). The authors also acknowledge the Research Center for Mechanical Engineering, Materials, and Processes (CEMMPRE).

Conflicts of Interest: The authors declare no conflict of interest.

References

1. Pelloux, R.M. Crack extension by alternating shear. *Eng. Fract. Mech.* **1970**, *1*, 697–704. [CrossRef]
2. Nicholls, D.J. The relation between crack blunting and fatigue crack growth rates. *Fatigue Fract. Eng. Mater. Struct.* **1994**, *17*, 459–467. [CrossRef]
3. Schwalbe, K.H. Comparison of several fatigue crack propagation laws with experimental results. *Eng. Fract. Mech.* **1974**, *6*, 325–341. [CrossRef]
4. Jablonski, D.A.; Carisella, J.V.; Pelloux, R.M. Fatigue crack propagation at elevated temperature in solid solution strengthened superalloys. *Metall. Trans. A* **1977**, *8*, 1893–1900. [CrossRef]
5. Chand, S.; Garg, S.B.L. Crack propagation under constant amplitude loading. *Eng. Fract. Mech.* **1985**, *21*, 1–30. [CrossRef]
6. Skelton, R.P.; Vilhelmsen, T.; Webster, G.A. Energia Criteria and cumulative damage during fatigue crack growth. *Int. J. Fatigue* **1998**, *20*, 641–649. [CrossRef]

7.	Clavel, M.; Pineau, A. Fatigue behaviour of two Nickel-base alloys I: Experimental results on low cycle fatigue, fatigue crack propagation and substructures. *Mater. Sci. Eng.* **1982**, *55*, 157–171. [CrossRef]
8.	Liu, H.W. Fatigue crack growth by crack tip cyclic plastic deformation: The unzipping model. *Int. J. Fract.* **1989**, *39*, 63–77. [CrossRef]
9.	Liu, H.W.; Liu, D. A quantitative analysis of structure sensitive fatigue crack growth in steels. *Scr. Met.* **1984**, *18*, 7–12. [CrossRef]
10.	Carpinteri, A. Crack Growth Resistance in Non-Perfect Plasticity: Isotropic Versus Kinematic Hardening. *Appl. Fract. Mech.* **1985**, *4*, 117–122. [CrossRef]
11.	Martínez-Pañeda, E.; Fleck, N.A. Crack Growth Resistance in Metallic Alloys: The Role of Isotropic Versus Kinematic Hardening. *J. Appl. Mech.* **2018**, *85*, 1–6. [CrossRef]
12.	Pommier, S.; Bompard, P. Bauschinger effect of alloys and plasticity-induced crack closure: A finite element analysis. *Fatigue Fract. Eng. Mater. Struct.* **2000**, *23*, 129–139. [CrossRef]
13.	Tong, J.; Lin, B.; Lu, Y.-W.; Madi, K.; Tai, Y.H.; Yates, J.R.; Doquet, V. Near-tip strain evolution under cyclic loading: In situ experimental observation and numerical modelling. *Int. J. Fatigue* **2015**, *71*, 45–52. [CrossRef]
14.	Paul, S.K.; Tarafder, S. Cyclic plastic deformation response at fatigue crack tips. *Int. J. Press. Ves. Pip.* **2013**, *101*, 81–90. [CrossRef]
15.	Bahloul, A.; Bouraoui, C. The overload effect on the crack-tip cyclic plastic deformation response in SA333 Gr 6 C-Mn steel. *Appl. Fract. Mech.* **2019**, *99*, 27–35. [CrossRef]
16.	Camas, D. Numerical Study of The Three-Dimensional Behaviour of Plasticity Induced Crack Closure Phenomenon in Bi-Dimensional Specimens. Ph.D. Thesis, University of Málaga, Málaga, Spain, 2013.
17.	González-Herrera, A.; Zapatero, J. Influence of minimum element size to determine crack closure stress by the finite element method. *Eng. Fract. Mech.* **2005**, *72*, 337–355. [CrossRef]
18.	González-Herrera, A.; Zapatero, J. Numerical study of the effect of plastic wake on plasticity-induced fatigue crack closure. *Fatigue Fract. Eng. Mater. Struct.* **2009**, *3218*, 249–260. [CrossRef]
19.	Antunes, F.V.; Rodrigues, S.M.; Branco, R.; Camas, D. A numerical analysis of CTOD in constant amplitude fatigue crack growth. *Theor. Appl. Fract. Mech.* **2016**, *85*, 45–55. [CrossRef]
20.	Antunes, F.V.; Branco, R.; Prates, P.A.; Borrego, L. Fatigue crack growth modelling based on CTOD for the 7050-T6 alloy. *Fatigue Fract. Eng. Mater. Struct.* **2017**, *40*, 1309–1320. [CrossRef]
21.	Vasco-Olmo, J.M.; Díaz, F.A.; Antunes, F.V.; James, M.N. Characterisation of fatigue crack growth using digital image correlation measurements of plastic CTOD. *Appl. Fract. Mech.* **2019**, *101*, 332–341. [CrossRef]
22.	Antunes, F.V.; Borges, M.F.; Prates, P.; Branco, R.; Oliveira, M. Effect of yield stress on fatigue crack growth. *Frat. Integrità Strutt.* **2019**, *13*, 9–19. [CrossRef]
23.	Borges, M.F.; Antunes, F.V.; Prates, P.A.; Branco, R.; Vojtek, T. Effect of Young's modulus on Fatigue Crack Growth. *Int. J. Fatigue* **2019**, 105375. [CrossRef]
24.	Frederick, C.O.; Armstrong, P.J. A mathematical representation of the multiaxial Bauschinger effect. *Mater. High Temp.* **2007**, *24*, 1–26. [CrossRef]
25.	Voce, E. The relationship between stress and strain for homogeneous deformation. *J. Inst. Met.* **1948**, *74*, 537–562.
26.	Prates, P.A.; Oliveira, M.C.; Fernandes, J.V. On the equivalence between sets of parameters of the yield criterion and the isotropic and kinematic hardening laws. *Int. J. Mater.* **2015**, *8*, 505–515. [CrossRef]
27.	Prates, P.A.; Pereira, A.F.G.; Sakharova, N.A.; Oliveira, M.C.; Fernandes, J.V. Inverse strategies for identifying the parameters of constitutive laws of metal sheets. *Adv. Mater. Sci. Eng.* **2016**. [CrossRef]
28.	Antunes, F.V.; Ferreira, M.S.C.; Branco, R.; Prates, P.; Gardin, C.; Sarrazin-Baudoux, C. Fatigue crack growth versus plastic CTOD in the 304L stainless steel. *Eng. Fract. Mech.* **2019**, *214*, 487–503. [CrossRef]
29.	Menezes, L.F.; Teodosiu, C. Three-dimensional numerical simulation of the deep drawing process using solid finite elements. *J. Mater. Proc. Technol.* **2000**, *97*, 100–106. [CrossRef]
30.	Oliveira, M.C.; Alves, J.L.; Menezes, L.F. Algorithms and strategies for treatment of large deformation frictional contact in the numerical simulation of deep drawing process. *Arch. Comput. Method Eng.* **2008**, *15*, 113–162. [CrossRef]

 © 2020 by the authors. Licensee MDPI, Basel, Switzerland. This article is an open access article distributed under the terms and conditions of the Creative Commons Attribution (CC BY) license (http://creativecommons.org/licenses/by/4.0/).

Article

Fatigue Property and Improvement of a Rounded Welding Region between the Diaphragm Plate and Closed Rib of an Orthotropic Steel Bridge Deck

Datao Li *, Chunguo Zhang and Pengmin Lu *

Key Laboratory of Road Construction Technology and Equipment, MOE, Chang'an University,
Xi'an 710064, China; zcguo2008@163.com
* Correspondence: ldt1688@chd.edu.cn (D.L.); lpmin@chd.edu.cn (P.L.)

Received: 31 December 2019; Accepted: 20 January 2020; Published: 21 January 2020

Abstract: By means of finite element modeling (FEM) and fatigue experiments, we study the fatigue performance of the rounded welding region between the diaphragm plate and closed rib of orthotropic steel bridge deck in this work. A local sub-model of the rounded welding region from the orthotropic steel bridge deck was developed to analyze the stress distributions. Based on the analysis results we designed the fatigue specimen for the fatigue test of this detailed structure. The fatigue experimental results revealed that the crack initiates from the weld toe of the rounded welding region and the stress concentration at the rounded welding region is the main mechanism of fatigue crack initiation. In addition, we propose three improvements to reduce the stress concentration of the rounded welding region, and the local structure optimization scheme of the diaphragm–rib weld can effectively improve the fatigue resistance of the detailed weld structure.

Keywords: fatigue performance; rounded welding region; finite element modeling (FEM); structure optimization; reinforcing plate

---

1. Introduction

The orthotropic steel bridge decks are used in most of the world's major long span bridges with an important character of low dead weight. The orthotropic steel deck consists of a deck plate supported in mutually perpendicular directions; all these elements are connected by welded connections. Owing to the cyclic load stress caused by a high number of vehicles, fatigue cracks in orthotropic steel bridges significantly occur at partial-penetration fillet-welded connections [1–7]. Usually, enhancing the fatigue strength of welded joints, such as the rib-to-rib welded details, rib-to-crossbeam welded joints, and field splice joints of longitudinal ribs [3,8–10] is an effective method to improve the fatigue resistance of the orthotropic steel decks.

Till now, numerous works have focused on the fatigue behaviors of structure details in orthotropic bridges decks. Taking orthotropic decks of the Williamsburg Bridge for instance, Tsakopoulos and Fisher [11] studied fatigue resistance of local welding structure between rib and diaphragm, and the observations are recommended in the 1994 AASHTO LRFD Bridge Design Specifications. Several modifications are proposed for orthotropic deck design and some modifications are adopted in the 2000 Interim AASHTO LRFD Specifications. In addition, Connor and Fisher [12] explored an identical method to examine how fatigue stress range is defined and determined during the testing, then the fatigue resistance of welded rib-to-web connections in steel orthotropic bridge decks is obtained. Furthermore, Xiao et al. [13] evaluated the stress distributions and fatigue life of the rib-deck welded joints in orthotropic steel decks using finite element modeling (FEM). Ya et al. [14] further measured the fatigue behavior of rib-deck welded joints on the orthotropic steel bridge deck. Moreover, the effects of fabrication procedures on fatigue resistance of welded joints in orthotropic steel decks have also been discussed [15]. Miki [16] used the FE sub-models of rib-to-deck joint derived from the global model of a real bridge to investigate the notch stresses at the weld root, the observations show the increase of the weld penetration results in the higher fatigue resistance. In addition, the effects of concrete cracking on the stress range response of welded joint in rib-to-floor beam can be evaluated through the FE analysis [17].

Although numerous efforts have been invested into the weld fatigue of the orthotropic steel bridge deck, the investigations on fatigue behaviors of the rounded welding region between diaphragm plate and closed rib are still under-researched. Facing to the fatigue behavior of this structure detail, we may think about a question, i.e., how does the fatigue behavior of this detail perform under cyclic loading? In this study, we attempt to answer this question using experimental method and FEM. In addition, this study is organized as follows: Firstly, we design fatigue specimens based on the simulated stress distributions of the detailed structures. Then, we perform the fatigue tests of specimens (nine specimens) and obtain the fatigue strength. Finally, we propose three schemes to improve the fatigue strength of this detail, and find that scheme 2 can effectively reduce the stress concentration of the rounded welding region.

2. Experiment

2.1. Design of Fatigue Specimen

According to the general code for design of highway bridges and culverts [18], we could analyze the stress distribution of the detailed structure in the orthotropic steel bridge deck. Firstly, the standard load of a vehicle for the stress analysis is chosen as 550 kN, the axle-loads for the middle and rear are 2×120 kN and 2×140 kN, respectively. In addition, the wheel contact area of two tires is assumed as a rectangle of 600 mm $\times$ 200 mm. The thickness of the asphalt pavement layer is about 55 mm and the load distribution angle is 45°. Thus, the actual load area on the deck is 710 mm $\times$ 310 mm, as shown in Figure 1. Moreover, we choose a bridge deck (3750 mm $\times$ 4800 mm) as the stress analysis object (Figure 2). Based on the aforementioned standard, an assumed truck wheel load along the transverse direction of the bridge deck is used to determine the maximum stress of our interest point (Figure 2). Furthermore, an FE model consisting of a diaphragm and eight closed ribs is created in ANSYS software. In this model, Shell63 element is applied in the FE model, and it contains about 83800 elements and 83500 nodes. For the FE model, the bottom of the diaphragm is fully fixed. At one side of the closed rib, the freedoms in the Y and Z directions are restrained. For the other end, the freedom in the Y direction is fixed. The location of the rear axle (2×140 kN) is just on the top of the diaphragm (Figure 1). In addition, for the FE model, we can find a similar mesh generation in Ref. [13]. Moreover, in Figure 1, we can see the rounded welding region of the orthotropic steel bridge deck as shown in the red-squared part in the A direction. Usually, the fatigue crack initiates in the weld toe of the rounded welding region.

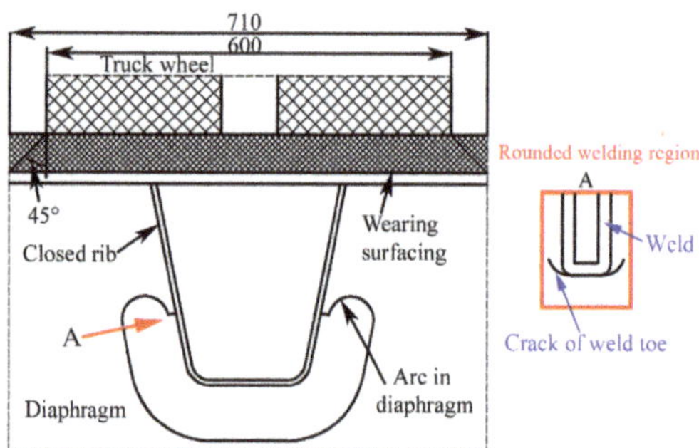

Figure 1. Fatigue crack of the rounded welding region between a closed rib and a diaphragm (in mm).

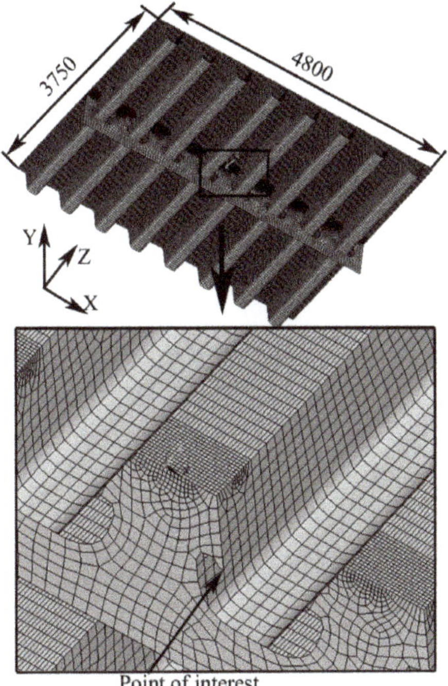

Figure 2. finite element (FE) model and the interest point of the rounded welding region.

Moreover, Figure 3a exhibits a simulated maximum principal stress–time curve of the interest point when the wheel loads are running along the transverse direction (Z direction in Figure 2) of the bridge deck. And the wheels load just locates on the diaphragm and closed rib when the maximum principal stress reaches the maximum value as shown in the curve (Figure 3b). Furthermore, a sub-model is developed to simulate the stress distribution around the interest point. Then, the boundary conditions applied to the sub-model are obtained from the FE model as shown in Figure 3b as the maximum principal stress reaches the maximum value (Figure 3a). In the sub-model, the element (Solid185, it is a 3-dimensional solid element with eight nodes in ANSYS software) sizes of the weld are about 2~3 mm (Figure 4). In addition, a schematic diagram of the diaphragm–rib structure of an orthotropic steel bridge deck and the weld details of interest regions are shown in Figure 5.

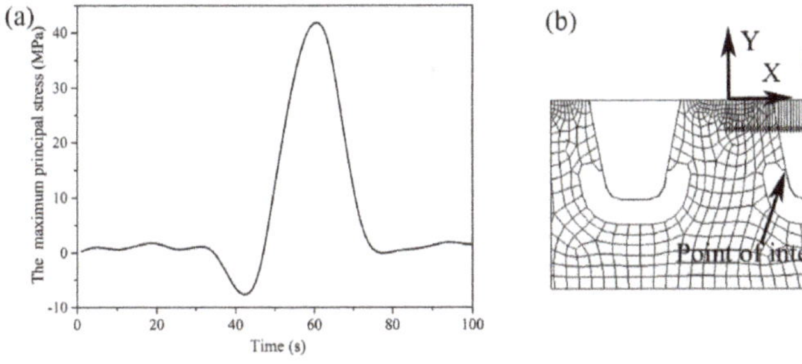

Figure 3. (**a**) Maximum principal stress–time curve of the interest point, and (**b**) the position of wheel load as the maximum principal stress of the interest point reaches its peak.

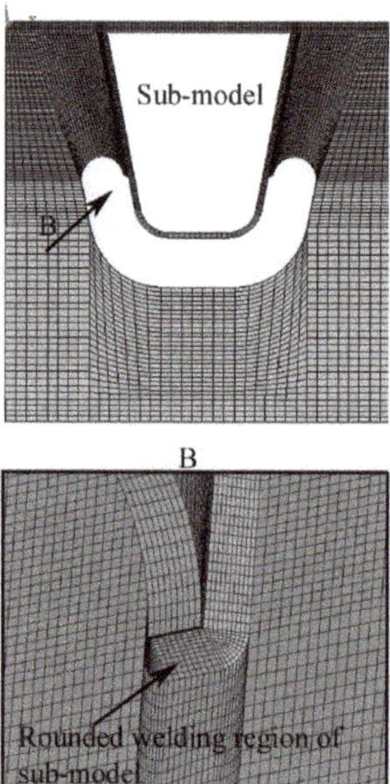

Figure 4. Sub-model of the rounded welding region.

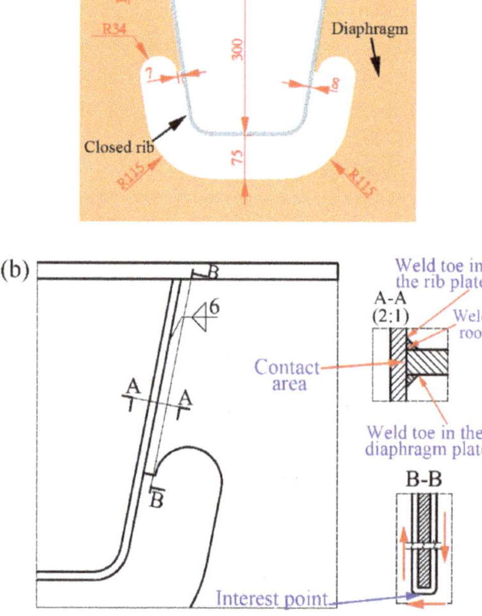

Figure 5. (**a**) Structure schematic diagram (in mm) and (**b**) the fillet weld between closed rib and diaphragm (in mm).

As shown in Figure 5b, double-sided fillet welds are adopted for welding the diaphragm and the closed rib, and the weld width is about 6 mm. The contact elements are created on the contact area between the diaphragm and the closed rib. In addition, the cross section of the fillet weld is a triangular shape (partial view A-A in Figure 5b), the weld toes (in the diaphragm plate and the closed rib plate, respectively) and the weld root are the potential fatigue-crack-initiating points affected by the weld size, plate thickness and weld penetration [19]. Moreover, the maximum principal stress distributions in the welding direction (welding direction is the red arrow route of the local B-B structure in Figure 5b) are shown in Figure 6, and the simulated stresses are from the FE model of the sub-model (Figure 4). The simulated results also indicate that the stress at the weld toe in the closed rib plate is significantly higher than those at weld toe in the diaphragm plate and the weld root. Therefore, the weld toe in the closed rib plate is considered as the interest point, and it is a potential fatigue-crack-initiating point during service condition of the bridge. In addition, the symmetrical principal stress curves in the weld toe in the rib plate may be caused by the boundary conditions applied to the sub-model. Moreover, the node position of weld toe in the rib plate is longer than the weld root and the weld toe in the diaphragm as shown in Figure 5b. Thus, the peak stress region of the weld toe in the rib plate does not line up with the peak stress regions for the weld root and the weld toe in the diaphragm plate.

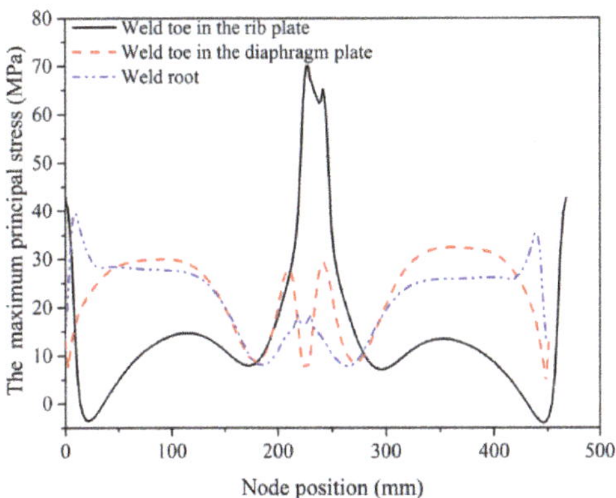

Figure 6. Stress–ode position curves of the rounded welding region in the sub-model (Node position is along the welding direction).

2.2. Fatigue Specimen

Based on the simulated stress distribution around the rounded welding region, we design a fatigue specimen of the structure detail with a rounded welding region (Figure 7). The material of fatigue specimen is Q345qD; it is a common steel material served as the bridge deck in China. In addition, the mechanical properties of the specimen material are shown in Table 1. Manual welding is employed to perform the welding, and the welding process parameters are listed in Table 2. Moreover, weld toes in the rib plate, the diaphragm plate, and the weld root are similar with weld details of the sub-model in Figure 5. In our simulation, the applied load is set as 50 kN. As shown in Figure 8, the maximum stress at the weld toe in the closed rib plate is higher than that at the weld root and the weld toe in the diaphragm plate. Because the weld length of our designed fatigue specimen is about half the length in sub-model (see node positions in Figures 6 and 8), the stress curves of the specimen welds are similar to the stress curves obtained from the sub-model (Figure 6). The similarity of the two stress distribution curves indicates the specimen can be used for the fatigue strength test of this weld detail.

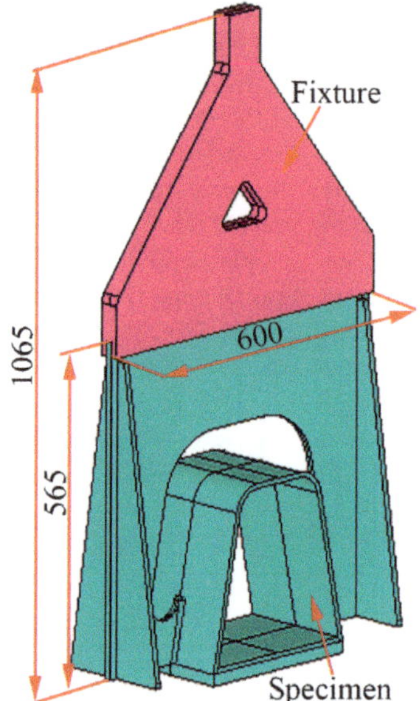

Figure 7. Specimen and fixture (in mm).

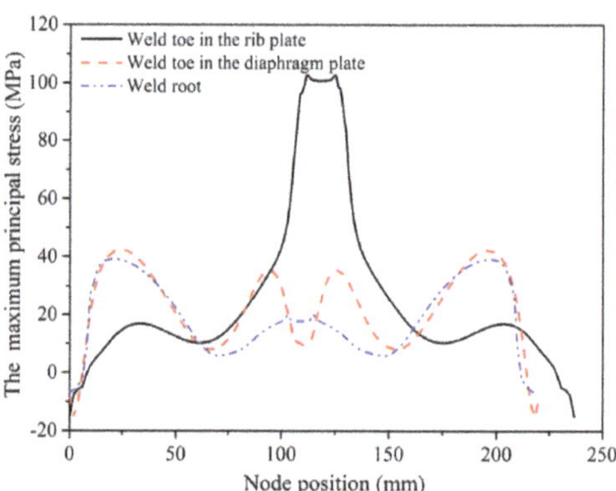

Figure 8. Stress–node position curves of the rounded welding region in the specimen.

Table 1. Mechanical properties of Q345qD.

Q345qD	Yield Strength (MPa)	Tensile Strength (MPa)	Elongation (%)	V notch Impact Energy (−20 °C, J)
	345	510	21	34

Table 2. Welding procedure.

Welding Processes	Manual Welding
Electrode diameter (mm)	Φ 4.0
Welding current (A)	160
Arc voltage (V)	24
Travel speed (cm/min)	30~40

2.3. Fatigue Tests and Results

In the rounded welding region of the fatigue specimen, strain rosettes are placed at a distance of approximately 1.5 t (t is the thickness of the closed rib) far away from the weld toe (Figure 9). In Figure 9a, the strain gauges 7 and 8 are used to determine the offset load caused by fabrication process. In Figure 9b, the strain rosettes 1–3 and 4–6 are distributed symmetrically, they can test the stress distribution of the weld toe in the rib plate. In addition, Figure 9c shows the photo of the fatigue specimen and the testing field. The fatigue tests are conducted by a hydraulic servo fatigue testing machine (SD-500, Changchun Research Institute for Mechanical Science CO. LTD, Changchun, China). During the testing, the load mode is the constant amplitude with a sine waveform at a frequency of 2 Hz, and the averaging stress ratio is about 0.06. Moreover, cameras with 40 magnifications are used to monitor and measure the crack initiation and propagation. In the fatigue test, fatigue cracks of the specimens firstly initiate from the rounded welding region between the diaphragm and closed rib, and the fatigue test results are summarized in Table 3. In addition, the stress range is measured by the strain rosettes 2 and 5 [20].

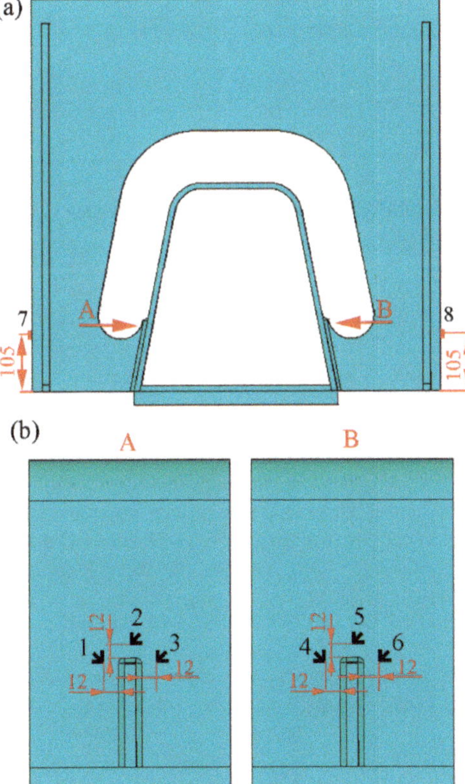

Figure 9. *Cont.*

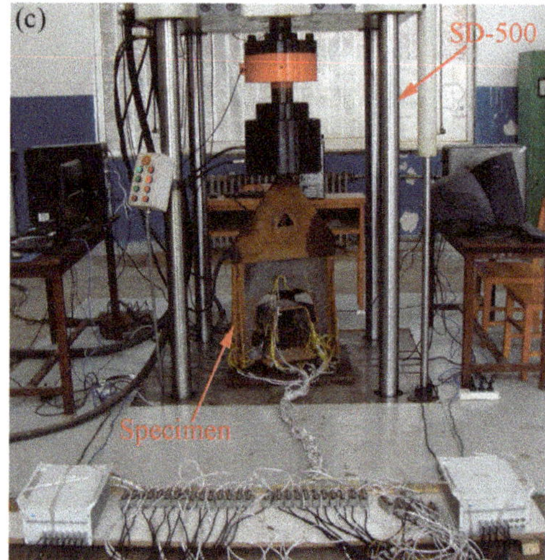

Figure 9. Schematic of the strain rosette location: (**a**) Strain gauges on the diaphragm, (**b**) Strain rosettes on the closed rib in A and B directions, respectively (in mm), (**c**) Photo of and the testing field and fatigue specimen.

Based on the fatigue results in Table 3, Δσ-N equation is fitted by the least square method and the fatigue life curve with 50% confidence bound is given as following:

$$\lg N = 12.33 - 3.4\,(\lg \Delta\sigma) \tag{1}$$

where fatigue cycles N are 2 million, the stress range Δσ is about 59.32 MPa.

The fatigue life equation with 97.7% confidence bound is obtained by subtracting 2*s* (*s* is the standard deviation of lgN, its value is 0.38) from the mean line of Equation (1):

$$\lg N = 11.57 - 3.4\,(\lg \Delta\sigma) \tag{2}$$

where fatigue cycles N are 2 million, the stress range Δσ is about 35.46 MPa. The Δσ-N curves under two confidence bounds are plotted in Figure 10. Moreover, almost all cracks initiate from the rounded welding region between the diaphragm and the closed rib. We do not find a similar structure detail in the design codes for steel structures [21–24]. Therefore, Equation (2) could provide a reference for the fatigue design of this detailed structure.

Table 3. Fatigue test results.

Specimen ID	Minimum Stress (MPa)	Maximum Stress (MPa)	Stress Range Δσ (MPa)	Number of Cycles N at Failure (Crack Reaches 30 mm)
SY-3-1-1	5.83	71.71	65.88	1409107
SY-3-1-2	5.83	104.65	98.82	1059280
SY-3-1-3	5.83	88.18	82.35	736535
SY-3-1-4	5.83	104.65	98.82	435607
SY-3-1-5	5.83	115.63	109.80	409925
SY-3-1-6	5.83	104.65	98.82	331072
SY-3-1-7	5.83	104.65	98.82	205567
SY-3-1-8	5.83	104.65	98.82	241068
SY-3-1-9	5.83	154.06	148.23	82632

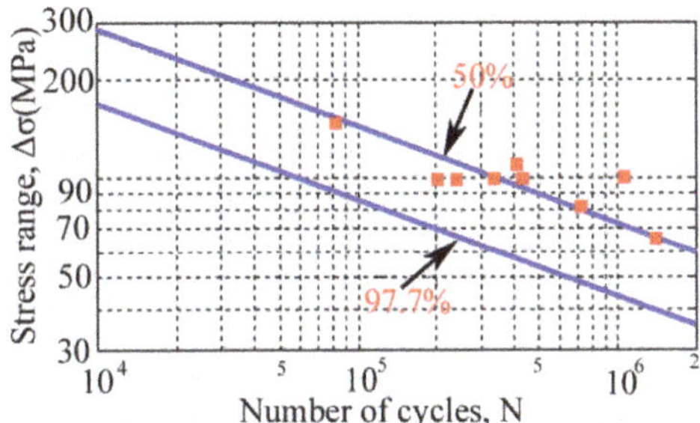

Figure 10. Δσ-N curves of fatigue specimens.

3. Discussion

3.1. Fatigue Crack

The fatigue crack of the tested specimens initiates from the rounded welding region (Figure 11a), then the crack propagates along the weld toe (see red arrows in Figure 11b–d). Based on these fatigue tests observations, the weld toe in the closed rib plate is most susceptible to failure, therefore, the initiated position of the crack and the crack propagation path can indirectly demonstrate the validity of these fatigue specimens.

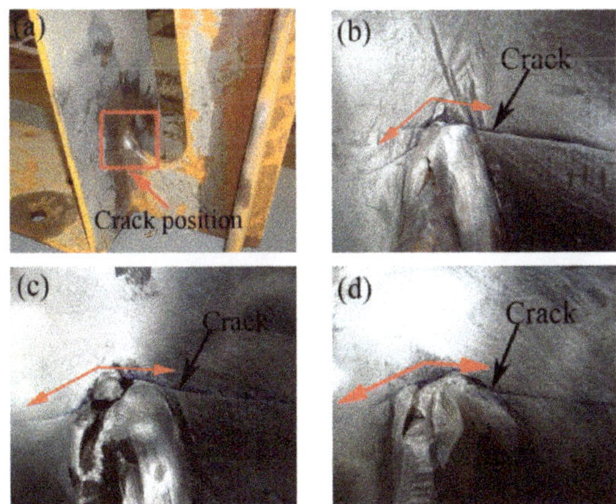

Figure 11. Cracks in the fatigue specimen: (**a**) Crack position in the specimen and Cracks of specimens (**b**) SY-3-1-2, (**c**) SY-3-1-5, (**d**) SY-3-1-6, respectively.

3.2. Hardness Measurements around the Crack

In order to analyze the mechanism of crack initiation, the hardness distribution of the welding region around the fatigue crack is measured by using a Vickers hardness tester [25–27] (Figure 12). The route of indentations successively crosses the parent metal (closed rib plate), heat-affected zone, weld toe (around the fatigue-crack-initiating point) and weld of specimen (from left to right in Figure 12b), and the distance between indentations is about 0.5 mm. Before the test, the surface of the specimen is well polished. In addition, the hardness distribution along the route reveals that the hardness fluctuates at the weld region (Figure 12b). However, the hardness value near the fatigue-crack-initiating point is about 200 HV, the value is slightly higher than the parent material's hardness (~180 HV). This phenomenon indicates the welding process has no significant effects on material hardness at the weld toe in the closed rib plate. But the maximum stress of the rounded welding region still locates at the weld toe, then we get the conclusion that the stress concentration in the rounded welding region is the main mechanism of the fatigue crack initiation.

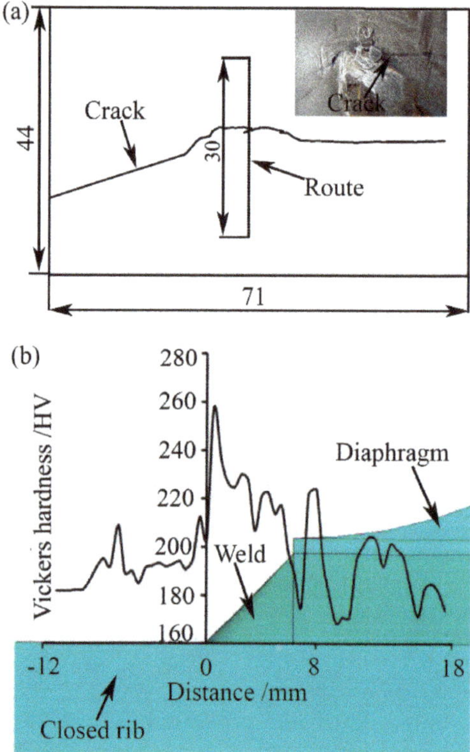

Figure 12. Vickers hardness test: (**a**) Sliced specimen (in mm) and (**b**) hardness distribution along the route.

3.3. Improvement of the Local Structure: Scheme 1

The previous analysis clearly states that the stress concentration is the major mechanism of fatigue crack initiation. Furthermore, we infer the stress concentration results from the size mutations between the diaphragm and the closed rib (see the arc in the diaphragm in Figure 1). Aiming to reduce the stress concentration, a reinforcing plate is incorporated onto the closed rib by welding. The details of the plate and the welding position are marked out in Figure 13a. With the purpose of reducing the peak value of the maximum principal stress in the rounded welding region, a parametric model in ANSYS software is developed to optimize the design parameters ($h1$, $h2$ and t, where t is the thickness of the reinforcing plate). Finally, the optimized results of the $h1$, $h2$ and t are 5 mm, 84 mm and 4 mm, respectively, and the maximum stress at the weld toe in the closed rib plate is 45.135 MPa in scheme 1 (Figure 13b). This value represents a decrease of 35.54% compared with the original peak stress. However, the stress at the weld toe in the diaphragm plate of the optimized reinforcing plate increases from 32.397 to 56.49 MPa (Figure 13b). The peak stress of the weld root is 39.65 MPa, and this value is very close to its original value. In summary, the stress concentration in the rounded welding region is reduced by welding the reinforcing plate. However, the stress at the weld toe in the diaphragm plate becomes larger than the stress at the weld toe in the rib plate. This means the expected failure location will change.

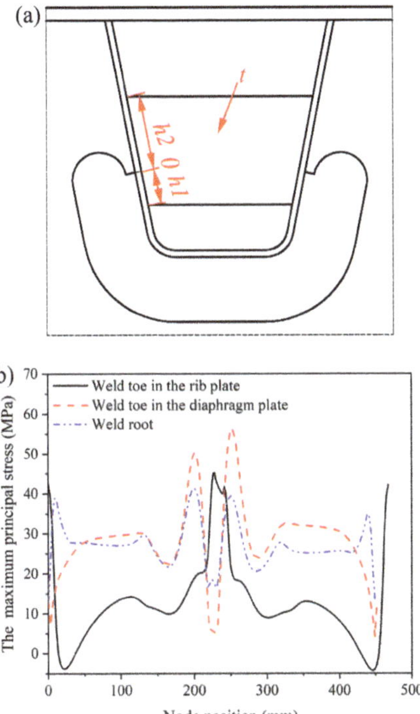

Figure 13. (**a**) Reinforcing plates in improvement scheme 1 and (**b**) stress–node position curves of the rounded welding region.

3.4. Improvement of the Local Structure: Scheme 2

In scheme 2, two reinforcing plates are installed on both sides of the closed rib by welding. The details of the reinforcing plates and the welding positions are identified in Figure 14a. Make sure the complete contact between the reinforcing plate and the closed rib, fusion-through welding is required. In this scheme, the plate thickness $t1$ is larger than $(a1 + a2)/2$. Variables $a1$, $a2$, $t1$, and b are considered as design variables, while the state variable is the peak value of the maximum principal stress in the rounded welding region.

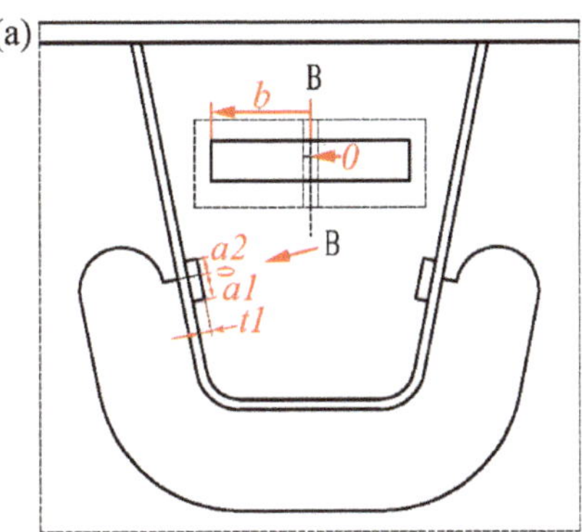

Figure 14. *Cont.*

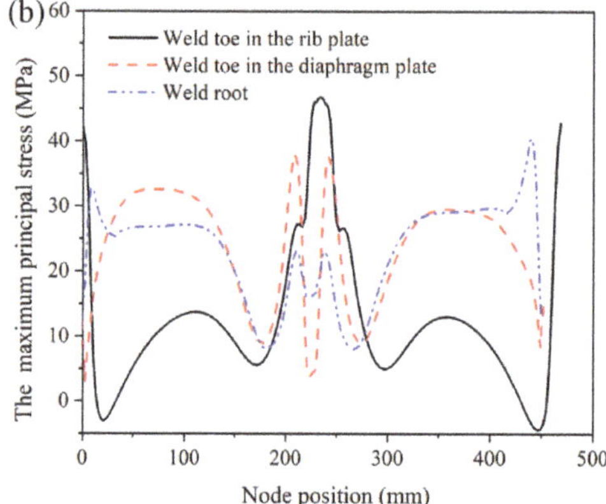

Figure 14. (a) Reinforcing plates in improvement scheme 2 and (b) stress—node position curves of the rounded welding region.

Furthermore, the optimized results of the *a1*, *a2*, *t1*, and *b* are 13 mm, 14 mm, 14 mm, and 135 mm, respectively, and the maximum stress at the weld toe in the closed rib plate is 46.711 MPa in scheme 2 (Figure 14b). This value represents a decrease of 33.29% compared with the original maximum stress. The maximum stress at weld toe in the diaphragm plate is 37.99 MPa and increases by 5.593 MPa (an increase of 17.26%) as compared with the original value. The peak stress at weld root is 40.51 MPa and increases by 1.151 MPa (an increase of 2.9%). In this scheme, the stress concentration in welded region is reduced, while the peak stresses at the weld toe in the diaphragm plate and weld root also increase.

3.5. Improvement of the Local Structure: Scheme 3

In scheme 3, two reinforcing plates are welded on both sides of the closed rib, respectively. We can see the details of reinforcing plates and the welding positions in Figure 15a. An original point is chosen (marked as "0") for optimizing the dimensions of the reinforcing plate. Fusion-through welding is also required, thus the plate thickness *t2* should be larger than *b1*. The variables *a3*, *a4*, *t2*, and *b1* are considered as design variables, and the state variable still is the peak value of the maximum principal stress in the rounded welding region.

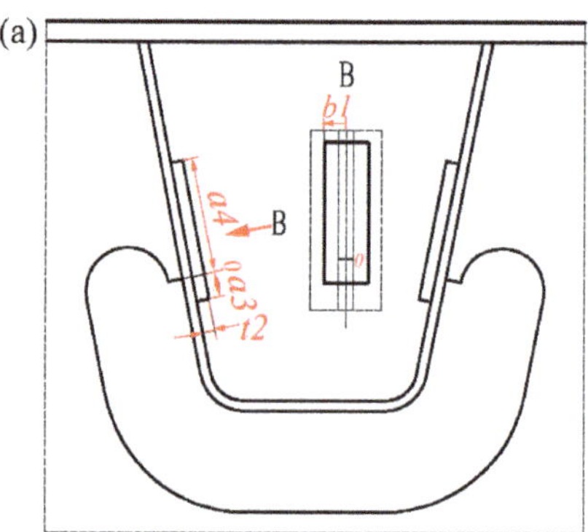

Figure 15. *Cont.*

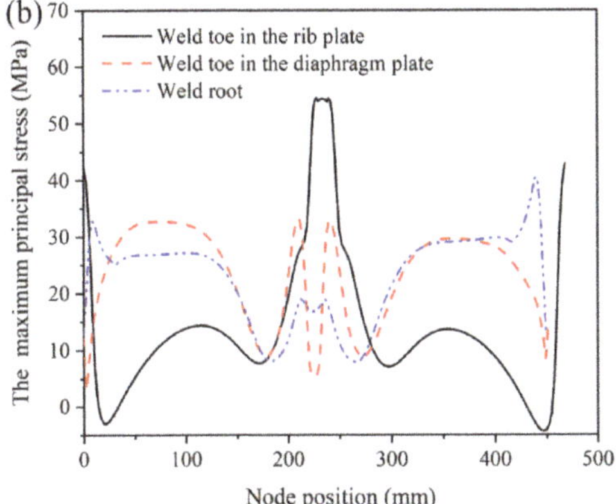

Figure 15. (**a**) Reinforcing plates in improvement scheme 3 and (**b**) stress—node position curves of the rounded welding region.

Finally, the optimized results of the *a3*, *a4*, *t2*, and *b1* are 12 mm, 15 mm, 11.5 mm, and 21 mm, respectively, and the peak stress at the weld toe in the rib plate is 52.8 MPa in scheme 3 (Figure 15b). This value means a decrease of 24.6% compared with the original peak stress. The peak stress at weld toe in the diaphragm plate is 32.673 MPa and increases 0.276 MPa (an increase of 0.85%) as compared with the original value. The peak stress at weld root increases from 39.359 to 40.532 MPa. In this scheme, the reinforcing plate reduces the stress concentration in welded region, but the peak stress at the weld toes in the diaphragm and the weld root does not significantly increase.

3.6. Comparisons of Three Improvement Schemes

Following the above descriptions, distribution curves of the maximum principal stress for the weld toe in the closed rib plate, the weld toe for the diaphragm plate and the weld root of three improvement schemes are plotted in Figure 16a–c, respectively. Moreover, the maximum principal stresses of the rounded welding regions with and without reinforcing plate are listed in Table 4. For the rounded welding region, three improvement schemes can effectively reduce the stress concentration at the weld toe in the closed rib plate. However, the scheme 2 is better than the schemes 1 and 3. Because both schemes 1 and 2 reduce the peak stress more than scheme 3, but scheme 1 increases the stress at the weld toe in the diaphragm plate to be more than the stress at the weld toe in the closed rib plate. Thus, we consider scheme 2 is the best and effectively improve the fatigue resistance of the rounded welding region of the diaphragm–rib structure.

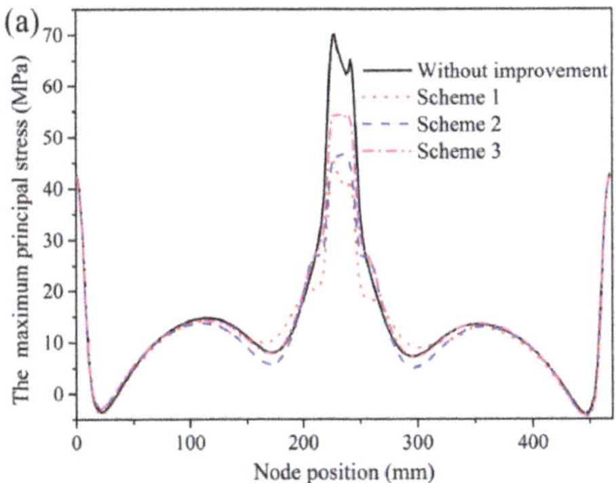

Figure 16. *Cont.*

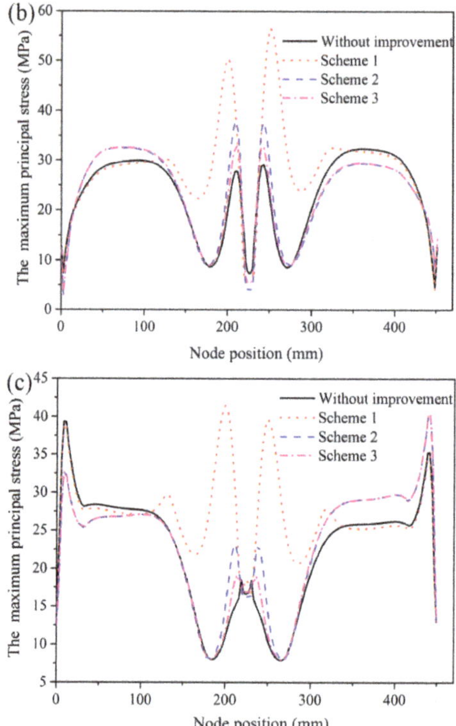

Figure 16. Stress–node position curves in the rounded welding region before and after improvement: Stress curves of (**a**) weld toe in the closed rib plate, (**b**) weld toe in the diaphragm plate and (**c**) weld root.

Table 4. Peak value of the maximum principal stress in the rounded welding region before and after improvement.

Schemes Locations	Without Improvement	Scheme 1	Scheme 2	Scheme 3
Weld toe in the closed rib plate (MPa)	70.023	45.135	46.711	52.8
Weld toe in the diaphragm plate (MPa)	32.397	56.49	37.99	32.673
Weld root (MPa)	39.359	39.65	40.51	40.532

4. Conclusions

In this study, the fatigue performance of the rounded welding region between diaphragm plate and closed rib of an orthotropic steel bridge deck is investigated using experimental methods and FEM. The main conclusions are summarized as follows

(1)   The fatigue crack of the rounded welding region of the orthotropic steel bridge deck initiates from the weld toe in the closed rib plate.

(2)   The fatigue strength equation for the tested specimens is $\lg N = 11.57 - 3.4(\lg \Delta\sigma)$ with 97.7% confidence bound. When the fatigue cycles N is 2 million, the stress range $\Delta\sigma$ is 35.46 MPa. This equation can provide a reference for the fatigue design of this detailed structure.

(3)   The stress concentration of the rounded welding region is the main mechanism of fatigue crack initiation. Two optimized reinforcing plates welded in the closed rib effectively reduce the stress

concentration of the rounded welding region, and the fatigue strength of this welded structure can be enhanced.

Author Contributions: Writing—original draft preparation, D.L.; writing—review and editing, C.Z.; supervision, P.L. All authors have read and agreed to the published version of the manuscript.

Funding: This research was funded by the Project of Jiangxi Provincial Communication Department, grant number 2010C00003. This research was also funded by the National Natural Science Foundation of China, grant number 11902046 and the Fundamental Research Funds for the Central Universities, CHD, grant number 300102259302.

Conflicts of Interest: The authors declare no conflict of interest.

References

1. Freitas, S.T.D.; Kolstein, H.; Bijlaard, F. Composite bonded systems for renovations of orthotropic steel bridge decks. *Compos. Struct.* **2010**, *92*, 853–862. [CrossRef]
2. Kainuma, S.; Yang, M.; Jeong, Y.S.; Inokuchi, S.; Kawabata, A.; Uchida, D. Experiment on fatigue behavior of rib-to-deck weld root in orthotropic steel decks. *J. Constr. Steel Res.* **2016**, *119*, 113–122. [CrossRef]
3. Huang, Y.; Zhang, Q.; Bao, Y.; Bu, Y. Fatigue assessment of longitudinal rib-to-crossbeam welded joints in orthotropic steel bridge decks. *J. Constr. Steel Res.* **2019**, *159*, 53–66. [CrossRef]
4. Choi, J.H.; Kim, D.H. Stress characteristics and fatigue crack behaviour of the longitudinal rib-to-cross beam joints in an orthotropic steel deck. *Adv. Struct. Eng.* **2008**, *11*, 189–198. [CrossRef]
5. Zhang, Q.H.; Cui, C.; Bu, Y.Z.; Liu, Y.M.; Ye, H.W. Fatigue tests and fatigue assessment approaches for rib-to-diaphragm in steel orthotropic decks. *J. Constr. Steel Res.* **2015**, *114*, 110–118. [CrossRef]
6. Cui, C.; Zhang, Q.; Luo, Y.; Hao, H.; Li, J. Fatigue reliability evaluation of deck-to-rib welded joints in OSD considering stochastic traffic load and welding residual stress. *Int. J. Fatigue* **2018**, *111*, 151–160. [CrossRef]
7. Zhang, Q.; Bu, Y.; Li, Q. Review on fatigue problems of orthotropic steel bridge deck. *J. China J. Highw.* **2017**, *30*, 14–30.
8. Deng, Y.; Liu, Y.; Feng, D.; Li, A. Investigation of fatigue performance of welded details in long-span steel bridges using long-term monitoring strain data. *Struct. Control. Health Monit.* **2015**, *22*, 1343–1358. [CrossRef]
9. Alemdar, F. Experimental Study of Fatigue Crack Behavior of Rib-To-Rib Butt Welded Connections in Orthotropic Steel Decks. *Lat. Am. J. Solids Struct.* **2018**, *15*, e128.
10. Zhuang, M.; Miao, C.; Chen, R. Analysis for Stress Characteristics and Structural Parameters Optimization in Orthotropic Steel Box Girders based on Fatigue Performance. *KSCE J. Civ. Eng.* **2019**, *23*, 2598–2607. [CrossRef]
11. Tsakopoulos, P.A.; Fisher, J.W. Full-scale fatigue tests of steel orthotropic decks for the Williamsburg Bridge. *J. Bridge Eng.* **2003**, *8*, 323–333. [CrossRef]
12. Connor, R.J.; Fisher, J.W. Consistent approach to calculating stresses for fatigue design of welded rib-to-web connections in steel orthotropic bridge decks. *J. Bridge Eng.* **2006**, *11*, 517–525. [CrossRef]
13. Xiao, Z.; Yamada, K.; Ya, S.; Zhao, X. Stress analyses and fatigue evaluation of rib-to-deck joints in steel orthotropic decks. *Int. J. Fatigue* **2008**, *30*, 1387–1397. [CrossRef]
14. Ya, S.; Yamada, K.; Ishikawa, T. Fatigue evaluation of rib-to-deck welded joints of orthotropic steel bridge deck. *J. Bridge Eng.* **2010**, *16*, 492–499. [CrossRef]
15. Sim, H.; Uang, C.; Sikorsky, C. Effects of fabrication procedures on fatigue resistance of welded joints in steel orthotropic decks. *J. Bridge Eng.* **2009**, *14*, 366–373. [CrossRef]
16. Miki, C. Fatigue damage in orthotropic steel bridge decks and retrofit works. *Int. J. Steel Struct.* **2006**, *6*, 255–267.
17. Zhang, Q.; Liu, Y.; Bao, Y.; Jia, D.; Bu, Y.; Li, Q. Fatigue performance of orthotropic steel-concrete composite deck with large-size longitudinal U-shaped ribs. *Eng. Struct.* **2017**, *150*, 864–874. [CrossRef]
18. JTGD60-2004. *General Code for Design of Highway Bridges and Culverts*; China Communications Press: Beijing, China, 2004.
19. Zhang, C.; Lu, P.; Hu, X.; Song, X. Effect of buffer layer and notch location on fatigue behavior in welded high-strength low-alloy. *J. Mater. Process. Technol.* **2012**, *212*, 2091–2101. [CrossRef]
20. Hobbacher, A. *Recommendations for Fatigue Design of Welded Joints and Components*; Springer: Berlin/Heidelberg, Germany, 2009.

21. GB50017-2003. *Code for Design of Steel Structures*; China Architecture & Building Press: Beijing, China, 2003.
22. TB10002-2. *China Railway Major Bridge Reconnaissance Design of Railway Bridge*; China Railway Publishing House: Beijing, China, 2005.
23. Nussbaumer, A.; Borges, L.; Davaine, L. *Fatigue Design of Steel and Composite Structures: Eurocode 3: Design of Steel Structures, Part 1–9 Fatigue; Eurocode 4: Design of Composite Steel and Concrete Structures*; John Wiley & Sons: Hoboken, NJ, USA, 2012.
24. *AASHTO LRFD Bridge Design Specifications*, 6th ed.; Parts I and II; ASCE: Atlanta, GA, USA, 2012.
25. Ghorbal, G.B.; Tricoteaux, A.; Thuault, A.; Louis, G.; Chicot, D. Comparison of conventional Knoop and Vickers hardness of ceramic materials. *J. Eur. Ceram. Soc.* **2017**, *37*, 2531–2535. [CrossRef]
26. Moreira, F.D.L.; Kleinberg, M.N.; Arruda, H.F.; Freitas, F.N.C.; Parente, M.M.V.; De Albuquerque, V.H.C.; Rebouças , P.P. A novel Vickers hardness measurement technique based on Adaptive Balloon Active Contour Method. *Expert Syst. Appl.* **2016**, *45*, 294–306. [CrossRef]
27. Guo, B.; Zhang, L.; Cao, L.; Zhang, T.; Jiang, F.; Yan, L. The correction of temperature-dependent Vickers hardness of cemented carbide base on the developed high-temperature hardness tester. *J. Mater. Process. Technol.* **2018**, *255*, 426–433. [CrossRef]

© 2020 by the authors. Licensee MDPI, Basel, Switzerland. This article is an open access article distributed under the terms and conditions of the Creative Commons Attribution (CC BY) license (http://creativecommons.org/licenses/by/4.0/).

Article

Modelling of Fracture Toughness of X80 Pipeline Steels in DTB Transition Region Involving the Effect of Temperature and Crack Growth

Jie Xu [1], Wei Song [2,*], Wenfeng Cheng [3], Lingyu Chu [1], Hanlin Gao [1], Pengpeng Li [1] and Filippo Berto [4]

[1] School of Materials Science and Physics, China University of Mining and Technology (CUMT), Xuzhou 221116, China; j.xu@cumt.edu.cn (J.X.); chulingyu0zi@163.com (L.C.); TS19180050P31@cumt.edu.cn (H.G.); lpp0424@yeah.net (P.L.)
[2] School of Mechanical & Electrical Engineering, Xuzhou University of Technology, Xuzhou 221018, China
[3] SINOPEC Oil & Gas Pipeline Inspection Co., Ltd., Xuzhou 221008, China; chengwf.gdcy@sinopec.com
[4] Department of Mechanical and Industrial Engineering, Norwegian University of Science and Technology (NTNU), 7491 Trondheim, Norway; filippo.berto@ntnu.no
* Correspondence: swingways@hotmail.com; Tel.: +86-152-0450-6099

Received: 21 November 2019; Accepted: 16 December 2019; Published: 23 December 2019

Abstract: This work presents an investigation of the effects of temperature and crack growth on cleavage fracture toughness for weld thermal simulated X80 pipeline steels in the ductile-to-brittle transition (DBT) regime. A great bulk of fracture toughness (crack tip opening displacement—CTOD) tests and numerical simulations are carried out by deep-cracked single-edge-notched bending (SENB) and shallow-cracked single-edge-notched tension (SENT) specimens at various temperatures ($-90\,°C$, $-60\,°C$, $-30\,°C$, and $0\,°C$). Three-dimensional (3D) finite element (FE) models of tested specimens have been employed to obtain computational data. The results show that temperature exerts only a slight effect on the material hardening behavior, which indicates the crack tip constraint (as denoted by Q-parameter) is less dependent on the temperature. The measured CTOD-values give considerable scatter but confirm well-established trends of increasing toughness with increasing temperature and reducing constraint. Crack growth and 3D effect exhibited significant influences on CTOD-CMOD relations at higher temperatures, $-30\,°C$ and $0\,°C$ for the SENT specimen.

Keywords: fracture toughness; coarse grained heat affected zone (CGHAZ); X80 pipeline steels; weld thermal simulation; finite element analysis (FEA)

1. Introduction

Significant effects of crack size and loading mode (bending vs. tension) on fracture toughness values have been revealed from fracture mechanics testing of ferritic structural steels [1]. Previous numerical investigations [2,3] illustrate the strong dependence of crack-tip fields on specimen geometry and remote loading. According to experimental studies by, e.g., Sorem et al., [4] Joyce and Link [5], and others, significant elevations in fracture toughness (characterized as J_C or K_{IC}) for shallow-cracked specimens and/or subjected to tensile loading have been shown. With increasing loads in specific objects, such as a cracked specimen or a structural component, the crack-tip plastic zone is increasingly affected by the nearby traction free boundary according to small-scale yielding (SSY) theory. Due to the crack tip stress relaxation, the constraint level of specimens decreases and further contributes to the apparently increased toughness of shallow-cracked and tension-loaded geometries from fracture mechanics testing [6–8]. The stress field surrounding the crack is influenced by the crack tip constraint, which cannot be characterized by single fracture mechanics indicator. The second parameter, such as

T-stress [9] or *Q*-parameter [10,11], has been proposed and developed to further describe the crack-tip stress fields and quantify constraint levels for various geometries and loading modes.

High-strength low-alloyed (HSLA) steels increasingly used for high-pressure pipeline operation and offshore structural installation. The installation of pipelines used for transporting oil and gas sometimes takes place in severe environments, such as in the low-temperature region, where the pipelines must have low-temperature toughness [12]. Thus, the major motivation for the improvement of HSLA steels has been provided by the demands for higher strength as well as improved toughness, ductility, and weldability at low temperatures [13,14]. Though the HSLA steels own the excellent properties of tensile strength and ductile to brittle transition (DBT), according to the Charpy-impact test investigation, the ductile brittle transition (DBT) on the basis of microscopic mechanism occurs with decreases of temperature. On the other hand, X80 steel pipelines are exposed in extreme low-temperature environments, and it is meaningful to characterize the fracture toughness with temperature variations during the processing of DBT. The balance of high strength and toughness can be deteriorated by welding thermal cycles, producing local poor toughness in the welded joints [15]. The heat-affected zone (HAZ) of a weldment is in many cases considered to be the weakest part and is crucial in the failure of steel structures because of its heterogeneous microstructure produced during the welding process [16,17]. Therefore, treatment of brittle fracture in weldment and HAZ is challenging. In the 1990s, there was a significant focus on characterizing the local stress fields in weldment and HAZ. SINTEF/NTNU developed the so-called *J-Q-M* theory (see, e.g., Zhang et al. [18–20]) where both constraint effects due to geometry and material mismatch were included in characterization of the local stress field ahead of the crack tip. In addition, some researchers have investigated the related HAZ properties of high-strength steels on the basis of physical simulation, welding heat input effect on HAZ in S960QL steel, and the matching effect on fatigue crack growth behavior of high-strength steels GMA welded joints [21–24].

This paper mainly focuses on the single-edge-notched tension (SENT) and single-edge-notched bending (SENB) specimens, which are usually used to characterize pipeline steels with thin-walled thickness, as specified recently by a so-called SENT methodology to identify a SENT specimen(s) to match the crack tip constraint of cracked pipe sections. The effect of constraint on the fracture toughness of weld thermal simulated X80 pipeline steels in the ductile-to-brittle transition relationship is clarified by combination of experimental assessment and numerical simulation. The fracture toughness comparison of X80 steel SENB specimens ($a/W = 0.5$) under different temperature have been presented in [25]. The material is determined by the increasing demand due to a great account of applications for manufacturing high-strength pipes for the oil and gas industry. Fracture toughness (as denoted by crack tip opening displacement (CTOD)) tests are performed at different temperatures, $-90\,°C$, $-60\,°C$, $-30\,°C$, and $0\,°C$. Both the traditionally used deeply cracked SENB specimens with $a/W = 0.5$ and SENT specimens with $a/W = 0.3$ are used to characterize the crack tip constraint effect on the fracture toughness in the ductile-to-brittle transition region. 3D nonlinear finite element models are employed to analyze the crack-tip stress fields of tested specimens by considering the effects of the constraint and without/with crack growth on fracture toughness. The numerical analysis is compared with the experimental results.

2. Experimental Details

2.1. Material Description

The material used in our study is the HSLA X80-grade steel, which has a minimum yield strength of 555 MPa (tensile strength 625 MPa). The nominal outer diameter of the pipe is 510 mm, and the nominal wall thickness is 14.6 mm. The typical chemical composition of this material is listed in Table 1.

Table 1. Chemical composition of the X80 steel (wt. %).

Steel	C	Si	Mn	P	S	Others
X80	0.04~0.07	~0.25	≤1.8	≤0.01	≤0.001	Mo, Ni, Cu, Ti, Nb, V, Al

2.2. Weld Thermal Simulation Technique

The weld thermal simulation technique is used to obtain the tested specimens. The single welding cycle simulation is intended to represent the coarse-grained HAZ (CGHAZ) of the welds [26], which has been used in the preparation of weld thermally simulated specimens. In this study, the specimens were heated to the maximum temperature of 1350 °C by resistance-heating in a computer-controlled Gleeble weld thermal simulator (Figure 1a). The whole temperature vs. time history applied during the weld thermal simulation can be seen in Figure 1b. The thermal history is described as a sequence of heating and cooling intervals. The simulation was performed by heating the sample to 1350 °C for 2 s followed by controlled cooling; the cooling intervals were 2 s in the 1350 °C–1200 °C range, and 15 s between 800 °C–500 °C. Thus, the T8/5 is 15 s. The synthetic CGHAZ microstructure was thereafter produced in a certain region in the specimen where the fatigue pre-crack is introduced after being machined. The prior austenite grain size of CGHAZ was measured to be about 50 μm.

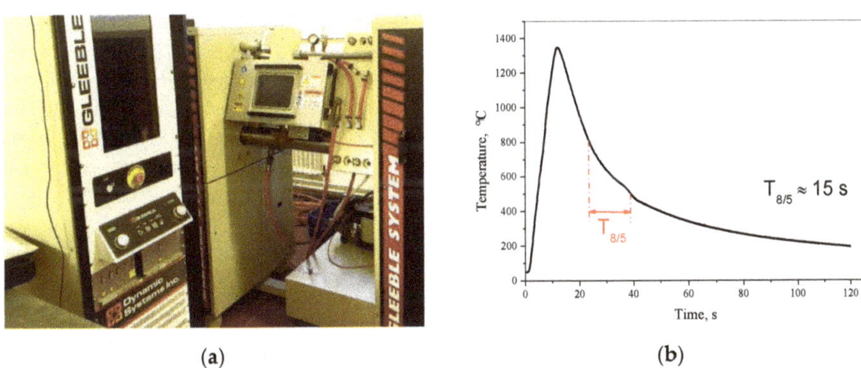

(a)   (b)

Figure 1. (**a**) Gleeble weld thermal simulator; (**b**) temperature vs. time history during the weld thermal simulation.

2.3. The True Stress-Strain Curves

To characterize the material flow properties in 3D finite element model, the true stress-strain curves of material (CGHAZ in X80) were measured by smooth round bar tensile tests at four temperatures as shown in Figure 2. The results show that yield strength here slightly increases with decreasing temperatures. There is also a weak trend of increasing work hardening with decreasing temperatures.

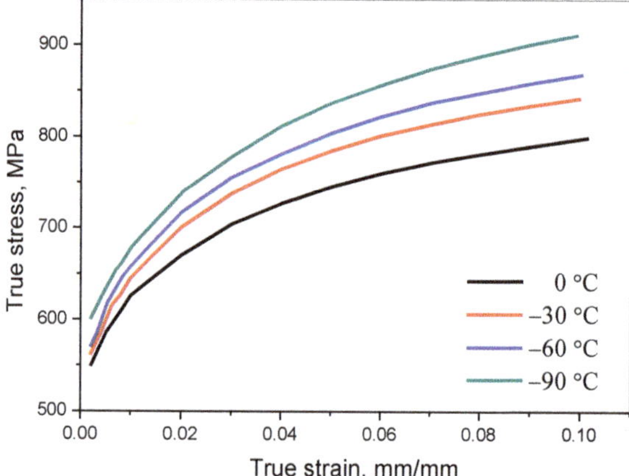

Figure 2. True stress-strain curves in coarse-grained heat affected zone (CGHAZ) of X80 at various temperatures [25].

2.4. Specimen Configurations and Test Program

The geometrical configurations are schematically drawn in Figure 3 for SENB and SENT specimens, which are directly extracted from the X80 pipeline with specimen length along the pipeline longitudinal direction and crack propagation following the pipe thickness orientation, as shown in Figure 3a. For all specimens, a thickness of $B = 10$ mm and width of $W = 10$ mm with crack length (denoted by a), to width ratio of $a/W = 0.5$ for SENB and $a/W = 0.3$ for SENT specimens have been considered. The span of the specimen, S, is chosen to be four times of width, W, for SENB ($S/W = 4$) and $L/W = 3$ for SENT.

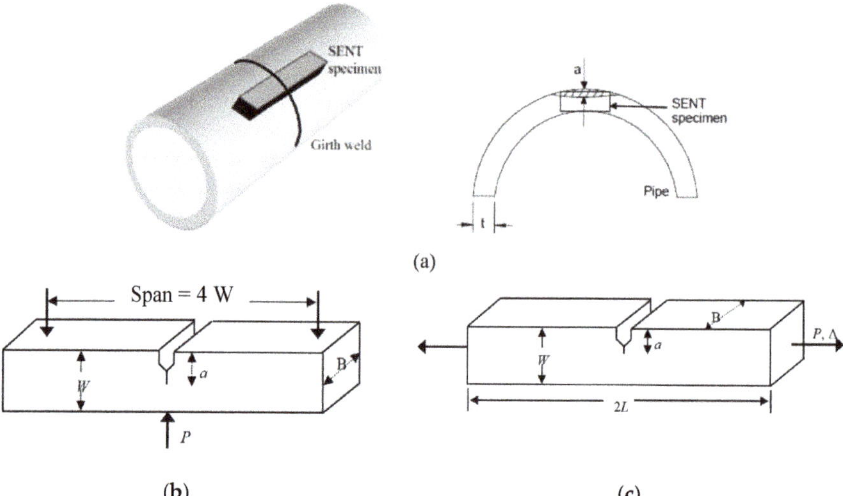

Figure 3. Specimen configurations. (a) Schematic plot of the relationship between single-edge-notched tension (SENT) and pipe; (b) single-edge-notched bending (SENB) with $a/W = 0.5$; (c) SENT with $a/W = 0.3$.

The SENB specimens are prepared and tested according to the standard of BS 7448 [27], while the SENT fracture specimens are machined in accordance with the "Recommended Practice DNV-RP-F108" [28]. Double clip gauge was used to digitally record the load-CMOD (crack mouth opening displacement) curves during the tests. The CTOD values are determined at the maximum load through measured load-CMOD records. The tensile and bending tests were performed at four different temperatures, −90 °C, −60 °C, −30 °C, and 0 °C. The testing rate is 0.5 mm/min of crosshead displacement for each specimen. For each specimen geometry, 10 parallel tests have been carried out at each temperature. After each test, the fatigue pre-cracking length and ductile crack extension that occurred during the test were measured using an optical microscope. The notches were located in the center of the weld, as can be seen in Figure 3.

3. Numerical Procedures

3.1. 3D Finite Element Models

3D finite element models were built using ABAQUS [29] for SENB and SENT specimens as shown in Figure 4. Due to symmetry, one-quarter of the specimen is modeled for finite element analysis considering the geometrical symmetry. A typical mesh configuration of elements surrounding the crack front is used with a small notch (with a notch root radius of $r = 2$ μm) in front of the crack tip. A 3D continuum element with eight-node, full integration (ABAQUS: C3D8), is used for FE calculations. The X80 steel true stress-strain curves obtained from the smooth round bar tensile tests at corresponding temperatures are applied for 3D model calculations. Meanwhile, the nonlinear geometric effect (NLG) is considered in all the finite element analyses. The CTOD-value is extracted from the displacement of a node in front of the initial crack tip [30].

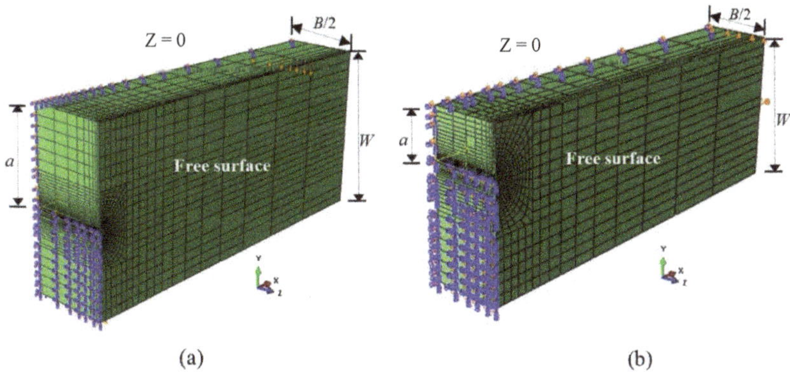

Figure 4. 3D FE models (1/4 model). (**a**) SENB with $a/W = 0.5$; (**b**) SENT with $a/W = 0.3$.

3.2. The MBL (Modified Boundary Layer) Model

In calculating the Q-parameter (quantitative characterization of crack tip constraint), the MBL model solution with $T = 0$ (here, T is the elastic T-stress, which is defined as constant stress acting parallel to the crack and its magnitude is proportional to the nominal stress in the vicinity of the crack) is adopted herein to represent the reference stress field for each case. Due to symmetry, only one-half of the model has been used in the MBL model, as shown in Figure 5. The global finite element mesh for the MBL model is drawn in Figure 5a. Similar models have been used in other studies [31–33]. Details of the mesh in the local region of the crack tip can be seen in Figure 5b. The MBL is a plane strain model with the same mesh arrangement in front of the crack tip (with a notch root radius of $r = 2$ μm) as in 3D models.

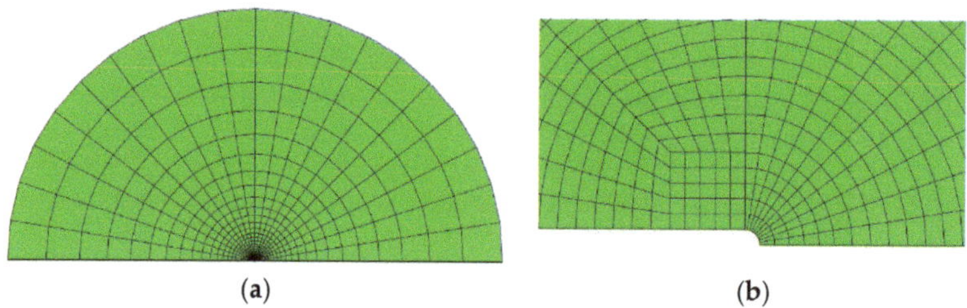

(a) **(b)**

Figure 5. Finite element mesh of the modified boundary layer (MBL) model. (**a**) Global mesh; (**b**) local mesh around crack tip region.

4. Results and Discussion

4.1. Measured and Calculated Load-CMOD Curves

The measured and calculated load-CMOD curves for all specimens at various temperatures are plotted in Figure 6. Only mid-thickness values of CMODs are extracted through the specimen thickness for all cases in this subsection. It can be seen that numerical simulations of load-CMOD curves are in good accord with experimental results for all temperatures. The material becomes quite brittle, which can be clearly seen from the flat fracture surface, and no significant crack growth has been observed from optical microscope observations for both the SENB and SENT specimens at lower temperatures, for example −90 °C and −60 °C. For the SENB specimens at −30 °C and 0 °C, small subcritical (mostly less than 0.2 mm) crack growth has been observed; whereas evident crack growth ($\Delta a > 0.2$ mm) have been observed for the SENT specimens at −30 °C and 0 °C. For the sake of simplicity, no crack growth has been considered in the 3D models in this subsection.

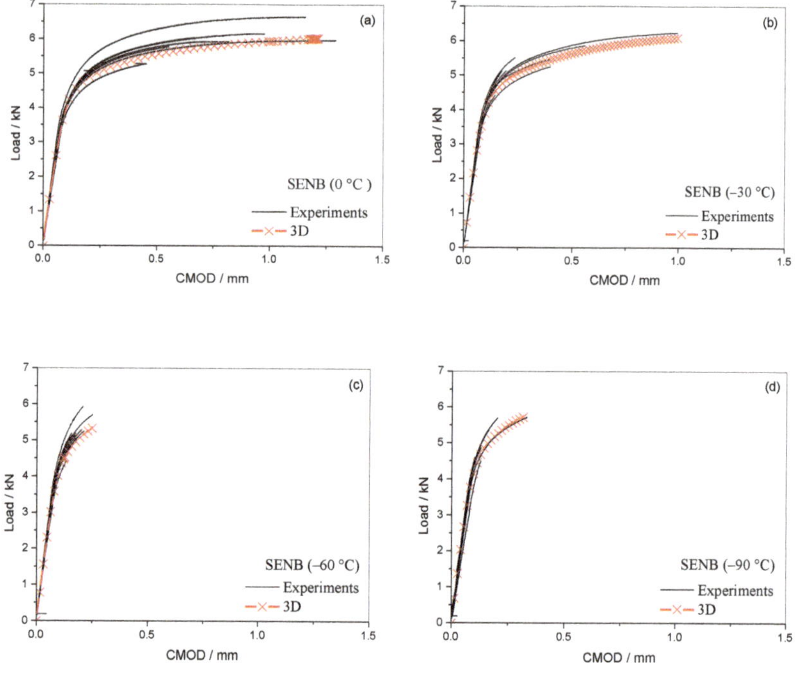

Figure 6. *Cont.*

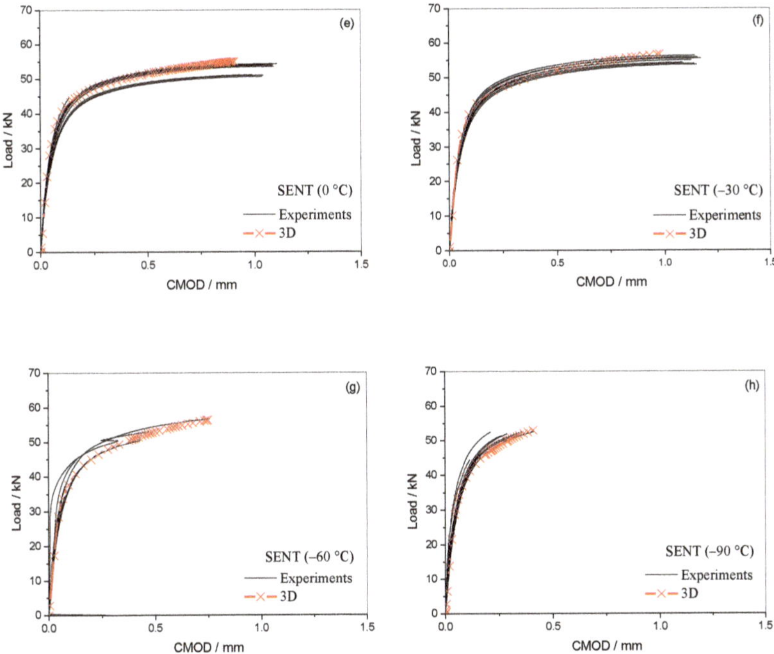

Figure 6. Comparisons of measured and calculated load-CMOD curves for SENB [25] and SENT specimens at various temperatures. (**a**) 0 °C for SENB specimens, (**b**) −30 °C for SENB specimens, (**c**) −60 °C for SENB specimens, and (**d**) −90 °C for SENB specimens, (**e**) 0 °C for SENT specimens, (**f**) −30 °C for SENT specimens, (**g**) −60 °C for SENT specimens, and (**h**) −90 °C for SENT specimens.

According to test demands of SENT and SENB specimen, 10 parallel tests have been carried out at each temperature. In addition, it can be argued that the transferability of the true $\sigma - \varepsilon$ curve from round thermal simulated tensile bar to the fracture mechanics specimens is quiet well. One thing that should be noted is that the average crack depth for both the SENB and SENT specimens at each temperature is used in these 3D models.

4.2. Measured CTOD-Values and Calculated Q-CTOD Relations at Different Temperatures

The results of fracture toughness (CTOD-value) as a function of temperature and their related average curves for SENT and SENB specimens are presented in Figure 7. It shows that the average CTOD values of SENT specimens at each temperature are obviously higher than that of SENB specimens. To illustrate the scatter degree of fracture toughness test data in Figure 7, the statistical characteristics are conducted quantitatively. The related analysis results for SENB and SENT specimens are summarized in Table 2. According to the average CTOD variations in statistical characteristics, these data increase with the temperature elevation from −90 °C to 0 °C for these two specimen types. In addition, from the perspective of standards variation coefficient, it stands for the ratio of the standard deviation to the mean. The higher the coefficient of variation, the greater the level of dispersion around the mean. It can be seen from Table 2 that the dispersion of −30 °C test data is the largest for SENB specimens, and −60 °C test data dispersion is the largest for SENT specimens. Thus, a definite temperature-dependence on dispersion characteristic cannot be drawn from these test data.

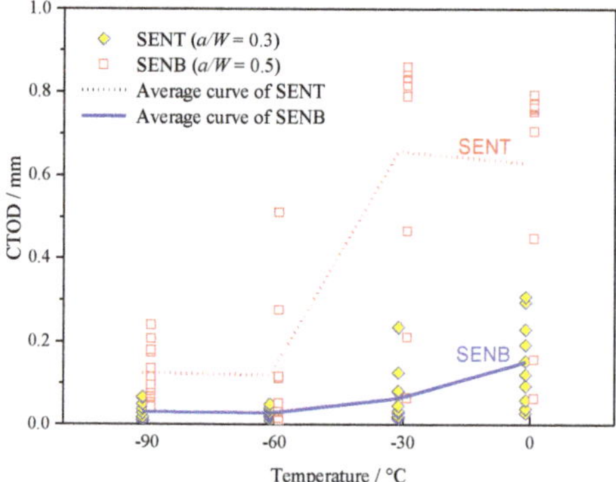

Figure 7. Crack tip opening displacement (CTOD) vs. temperature for HAZ in X80 pipeline steels.

Table 2. Statistical characteristics of test data for SENB and SENT specimens.

Specimens	Statistical Characteristics	−90 °C	−60 °C	−30 °C	0 °C
SENB	Average values (mm)	0.031	0.029	0.065	0.152
	Standard deviation	0.021	0.011	0.064	0.096
	Standard deviation coefficient	0.703	0.392	0.979	0.632
SENT	Average values (mm)	0.125	0.119	0.658	0.63
	Standard deviation	0.055	0.151	0.284	0.098
	Standard deviation coefficient	0.439	1.274	0.432	0.155

It shows that the fracture toughness tends to be scattered at each temperature. Also, the scatter of fracture toughness increases rapidly with increasing temperatures (upper transition region), for instance −30 °C and 0 °C, where a ductile mechanism is involved and cleavage instability may intervene after a certain amount of ductile crack growth as have been observed from the tests, especially for SENT specimens.

Additionally, the average fracture toughness values are higher for SENT specimens with a shorter crack of $a/W = 0.3$ compared to the SENB specimens with $a/W = 0.5$ at each test temperature. Moreover, this difference becomes larger with increasing temperatures.

The detailed effects of temperature and specimen geometry (as quantitatively characterized by crack tip constraint—Q-parameter) on fracture toughness will be studied in the following. As has been known, the J-Q methodology gives a direct measurement of the crack-tip stress field of interest that is related to a reference field [10,11], and can therefore describe the evolution of constraint ahead of the crack tip throughout the loading to large-scale yielding (LSY), where J sets the deformation level and Q is a stress triaxiality parameter. In this paper, the Q-parameter has been used to quantify the crack tip constraint for each specimen at each temperature.

The Q-parameter was originally defined as follows [10],

$$Q = \frac{\sigma_{\theta\theta} - (\sigma_{\theta\theta}{}^{Ref})_{T=0}}{\sigma_0}, \; x/(J/\sigma_0) = 2, \; \theta = 0. \tag{1}$$

where $\sigma_{\theta\theta}$ is the opening stress component of interest, $(\sigma_{\theta\theta}{}^{Ref})_{T=0}$ is the reference stress component characterized by MBL model solution with $T = 0$, σ_0 is the yield stress, and x denotes the distance from the crack tip along the crack plane ($\theta = 0$).

CTOD is selected as the crack driving force in our stud; the following constrain effect definition of Q has been used [27,28]:

$$Q = \frac{\sigma_{\theta\theta}{}^{specimen} - \left(\sigma_{\theta\theta}{}^{Ref}\right)_{T=0}}{\sigma_0}, \text{ at } x/\text{CTOD} = 4, \theta = 0. \tag{2}$$

where $\sigma_{\theta\theta}{}^{specimen}$ is the opening stress component of the specimen at a certain temperature, $\left(\sigma_{\theta\theta}{}^{Ref}\right)_{T=0}$ is the reference stress component at the same temperature, and other parameters are the same as defined in Equation (1). Only the distribution of the crack tip opening stress (σ_{22} at $\theta = 0$) has been studied. In the following, the results of crack tip opening stress distribution at different CTODs are presented in Figure 8 for specimens at 0 °C.

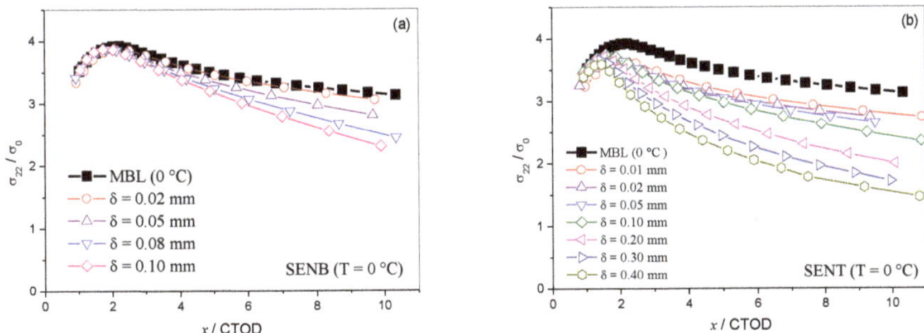

Figure 8. Opening stress distributions ahead of the crack tip, $T = 0$ °C. (**a**) SENB; (**b**) SENT.

For SENT specimen as shown in Figure 8b, it demonstrates that the opening stress distribution in front of the crack tip is nearly parallel to the reference stress field. However, global bending causes the slope of the opening stress distribution to gradually deviate from the reference field for SENB specimen, but still remainsquite similar, as can be seen in Figure 8a. Similar observations have also been found for specimens at other temperatures while the results are not included herein for the sake of simplicity.

The calculated Q-CTODs relations are displayed in Figure 9 for SENB and SENT specimens at various temperatures. Only the mid-thickness layer was used to compute Q-parameter herein. The Q-parameter stands for constraint effect decreases with the increases of CTODs. Meanwhile, the Q-parameter for SENB specimen for each temperature level is considerably larger than that of SENT at same CTODs, which means the crack tip constraint of SENB specimen is higher than that of SENT as has been observed. Nearly constant Q-parameters have been computed for all temperatures considered at the same CTOD values. The results present a weak dependence of temperature on the constraint ahead of the crack tip that can be expected for SENT specimen compared with that of SENB specimen.

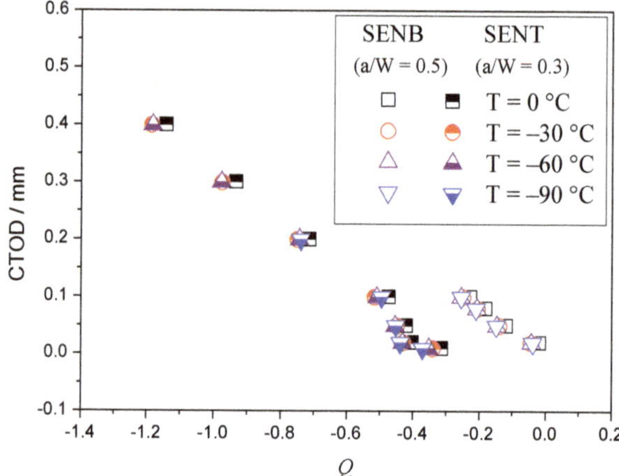

Figure 9. Q vs. CTODs for both SENB and SENT specimens, $T = 0\ °C$, $-30\ °C$, $-60\ °C$, and $-90\ °C$.

4.3. Measured and Calculated CTOD-CMOD Relations

As has been shown in Section 4.1, the experimental results are in good accordance with numerical simulations for the load-CMOD curves for all specimens at each temperature; how the 3D finite element models work for predicting the local fracture parameter as denoted by CTOD will be discussed in this subsection.

Figures 10 and 11 draws the CTODs vs. CMODs relationship obtained from both experiments and numerical calculations for all specimens at various temperatures without considering crack growth. Still, only mid-thickness values of CTODs versus CMODs are extracted herein. It can be seen that the tested results of CTOD-CMOD relations for the SENB specimens (Figure 10) can be quite well predicted by 3D FEA results at all temperatures.

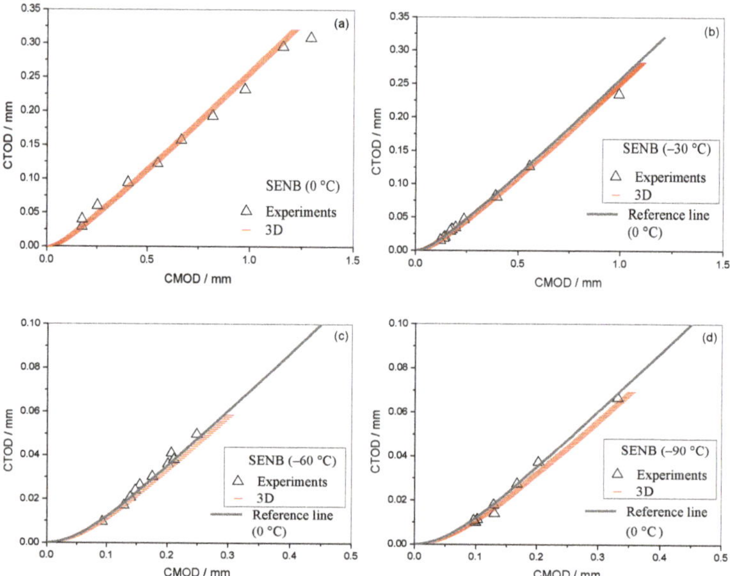

Figure 10. CTOD vs. CMOD relations from experiments and 3D FE analyses for SENB specimens without crack growth at different temperatures, (**a**) 0 °C, (**b**) −30 °C, (**c**) −60 °C, and (**d**) −90 °C [25].

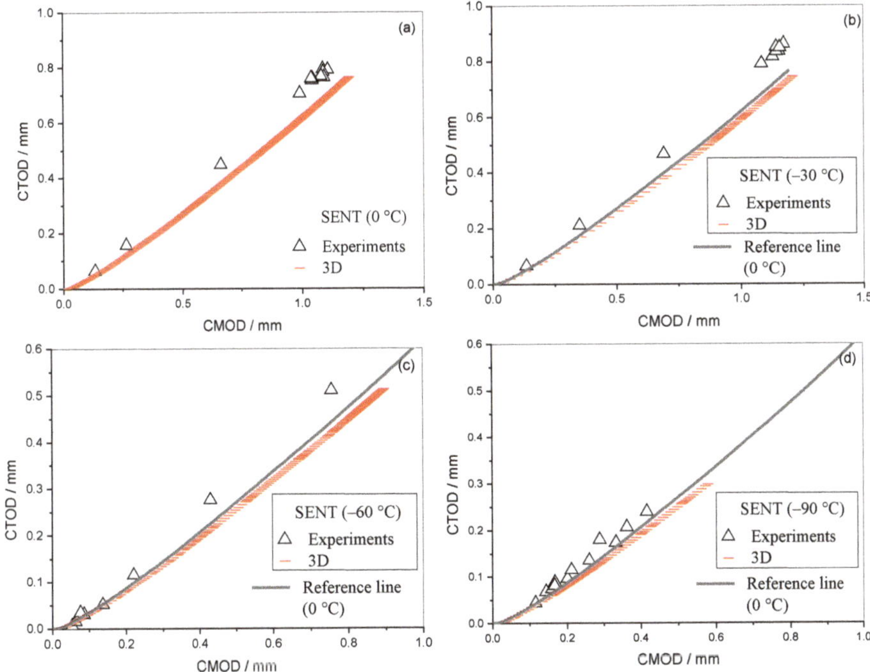

Figure 11. CTOD vs. CMOD relations from experiments and 3D FE analyses for SENT specimens without crack growth at different temperatures, (**a**) 0 °C, (**b**) −30 °C, (**c**) −60 °C, and (**d**) −90 °C.

As for the SENT specimens (Figure 11), a certain difference between experiments and 3D FE simulations can be observed at all temperatures. For small CTODs (less than 0.2 mm), quite good accordance can be obtained between experiments and simulations for all temperatures. However, the predicted CTOD values from the 3D models start to deviate from experimental results with the increases of CTODs, especially for the cases at higher temperatures, 0 °C and −30 °C. This is a remaining issue and more efforts are needed in further work, for example, the influences of local inhomogeneity of microstructure in HAZ and ductile crack propagation can be considered to further modify the models so as to improve the validity and accuracy of simulations with respect to the experiments. In this respect, it is noted that no standard for CTOD measurements in SENT specimens currently exists. The results in this paper warrant the need for further work in order to arrive at a method for experimental measurements of CTOD in SENT specimens, which could eventually form the basis for a standard document.

4.4. The Effect of Crack Growth on the CTOD-CMOD Relations

In this subsection, the effect of crack growth on CTOD-CMOD relations for the SENT specimen at 0 °C and −30 °C are considered in 3D models. Only mid-thickness values for both the CTODs and the CMODs are plotted in Figure 12. The complete Gurson model (see Reference [34]) is involved in calculating crack growth in 3D models. It can be seen that a significant elevation in fracture toughness for SENT specimens by considering crack growth. Also, a wonderful agreement between experiments and 3D FE simulations can be achieved by considering the influence of crack growth on fracture toughness for SENT specimens at both 0 °C and −30 °C. It can be concluded that the ductile crack extension is dependent on the temperature. In the low temperature range, it almost occurs at the cleavage fracture. The ductile crack extension mechanism involved in the upper transition zone of DBT, which cannot be ignored to predict CTOD-value for pipeline steel integrity assessment.

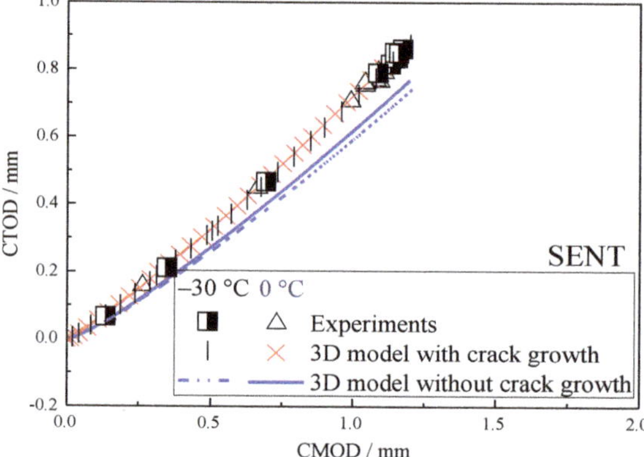

Figure 12. Effect of crack growth on the CTOD vs. CMOD relations for the SENT specimens under 0 °C and 30 °C.

4.5. The Influence of 3D Effect on the Fracture Toughness

In order to find the influence of 3D effect on fracture toughness, the calculated CTOD-CMOD relations at different layers through the specimen thickness compared with experiments are displayed in Figure 13. Two cases of SENT specimens considering crack growth at 0 °C and −30 °C are selected for this question. It can be seen that CTOD-CMOD values change considerably through the specimen thickness. The greater the distance from the specimen mid-thickness, the greater the deviation from experiments. In addition, the predicted CTOD-CMOD relations near the mid-thickness layer are coincident well with the experiments.

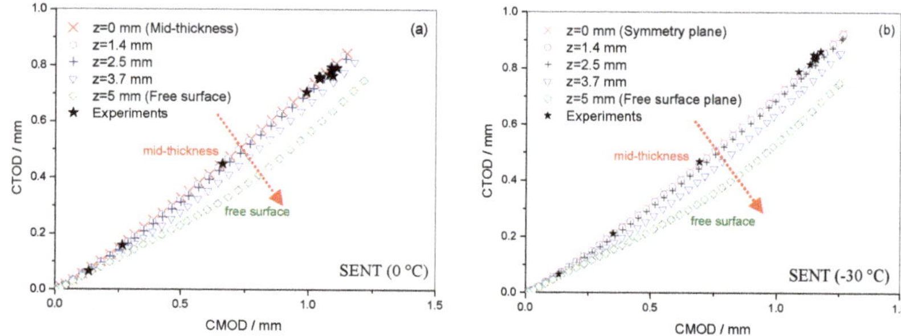

Figure 13. CTOD vs. CMOD relations at different layers through the specimen thickness for the SENT specimens considering crack growth at (**a**) 0 °C; (**b**) −30 °C.

5. Conclusions

In this paper, the HAZ of X80 high-strength steel is studied by experimental and simulation methods. The following conclusion can be drawn as follows:

(1) The HAZ of X80 hardening behavior exhibits a slight effect of temperature variations, which indicates the crack tip constraint is less dependent on the temperature as also observed from 3D FEA results.

(2) The predicted load-CMOD curves from 3D models are in good accordance with experimental results at all temperatures. As for the local fracture parameter, as depicted with the CTOD-CMOD relationship, the experimental data for the SENB specimens can be quite well simulated by 3D simulations without considering crack growth. For the SENT specimens, a good agreement

between experiments and numerical simulations can also be obtained by considering the effect of crack growth in the 3D models.

(3) The tested CTOD-values show considerable scatter but confirm well-established trends of increasing toughness with increasing temperature and reducing constraint.

(4) Cleavage fracture can be clearly observed for SENB specimens at all tested temperatures, while ductile crack growth can be seen for SENT specimens at −30 °C and 0 °C.

(5) From 3D finite element analyses, it has been found out that the CTODs change considerably through the specimen thickness. The predicted CTODs near the mid-thickness layer is coincident well with the experiments for the SENT specimens at 0 °C and −30 °C. The greater the distance from the specimen mid-thickness, the greater the deviation from experiments.

Author Contributions: Conceptualization, J.X.; Methodology, H.G.; Software, L.C.; Validation, W.S.; Investigation, W.C.; Writing-Review & Editing, F.B. All authors have read and agreed to the published version of the manuscript.

Funding: This research was funded by National Natural Science Foundation of China (Project No. 51301197) and the Natural Science Foundation of Jiangsu Province (Project No. BK20130182) and the Fundamental Research Funds for the Central University (Project No. 2011QNA07) as well as the National Key R&D Program of China (2018YFB2001200).

Acknowledgments: The first author sincerely appreciates Zhiliang Zhang from Norwegian University of Science and Technology for his valuable comments. Erling Østby and Bård Nyhus from SINTEF are also gratefully acknowledged for their kind help.

Conflicts of Interest: The authors declare no conflict of interest.

References

1. Mathias, L.L.; Sarzosa, D.F.; Ruggieri, C. Effects of specimen geometry and loading mode on crack growth resistance curves of a high-strength pipeline girth weld. *Int. J. Press. Vessels Pip.* **2013**, *110*, 12–22. [CrossRef]

2. McMeeking, R.M.; Parks, D.M. On Criteria for J-Dominance of Crack-Tip Fields in Large-Scale Yielding. In *ASTM STP 668 Elastic-Plastic Fracture*; Landes, J.D., Ed.; ASTM International: Philadelphia, PA, USA, 1979; pp. 175–194.

3. Shih, C.F.; German, M.D. Requirements for a one parameter characterization of crack-tip fields by the HRR singularity. *Int. J. Fract.* **1981**, *17*, 27–43.

4. Sorem, W.A.; Dodds, R.H.; Rolfe, S.T. Effects of crack depth on elastic-plastic fracture toughness. *Int. J. Fract.* **1991**, *47*, 105–126. [CrossRef]

5. Joyce, J.A.; Link, R.E. Ductile-to-brittle transition characterization using surface crack specimens loaded in combined tension and bending. In *ASTM STP 1321 Fatigue and Fracture Mechanics*; Underwood, J.H., Macdonald, B., Mitchell, M., Eds.; ASTM International: Philadelphia, PA, USA, 1997; Volume 28, pp. 243–262.

6. Dodds, R.H., Jr.; Shih, C.F.; Anderson, T.L. Continuum and micromechanics treatment of constraint in fracture. *Int. J. Fract.* **1993**, *64*, 101–133.

7. Dodds, R.H., Jr.; Nevalainen, M. Numerical investigation of 3-D constraint effects on brittle fracture in SE (B) and C (T) specimens. *Int. J. Fract.* **1996**, *74*, 131–161.

8. Naumenko, V.P.; Limanskii, I.V. Fracture resistance of sheet metals and thin-wall structures. Part 1. critical review. *Strength Mater.* **2014**, *46*, 18–37. [CrossRef]

9. Betegon, C.; Hancock, J.W. Two-parameter characterization of elastic-plastic crack-tip fields. *J. Appl. Mech.* **1991**, *58*, 104–110. [CrossRef]

10. O'Dowd, N.P.; Shih, C.F. Family of crack-tip fields characterized by a triaxiality parameter-I. Structure of fields. *J. Mech. Phys. Solids* **1991**, *39*, 989–1015. [CrossRef]

11. O'Dowd, N.P.; Shih, C.F. Family of crack-tip fields characterized by a triaxiality parameter-II. Fracture application. *J. Mech. Phys. Solids* **1991**, *40*, 939–963. [CrossRef]

12. Bose-Filho, W.W.; Carvalho, A.L.M.; Stragwood, M. Effects of alloying elements on the microstructure and inclusion formation in HSLA multipass welds. *Mater. Charact.* **2007**, *58*, 29–39. [CrossRef]

13. Das, S.K.; Sivaprasad, S.; Das, S.; Chatterjee, S.; Tarafder, S. The effect of variation of microstructure on fracture mechanics parameters of HSLA-100 steel. *Mater. Sci. Eng. A* **2006**, *431*, 68–79. [CrossRef]

14. Bose Filho, W.W.; Carvalho, A.L.M.; Bowen, P. Micromechanisms of cleavage fracture initiation from inclusion in ferritic welds: Part, I. Quantification of local fracture behavior observed in notched test pieces. *Mater. Sci. Eng. A* **2007**, *460–461*, 436–452. [CrossRef]

15. Lambert-Perlade, A.; Gourgues, A.F.; Besson, J.; Sturel, T.; Pineau, A. Mechanisms and modeling of cleavage fracture in simulated heat-affected zone microstructures of a high-strength low alloy steel. *Metall. Mater. Trans. A* **2004**, *35*, 1039–1053. [CrossRef]

16. Mohseni, P.; Solberg, J.K.; Karlsen, M.; Akselsen, O.M.; Østby, E. Investigation of mechanism of cleavage fracture initiation in intercritically coarse grained heat affected zone of HSLA steel. *Mater. Sci. Technol.* **2012**, *28*, 1261–1268. [CrossRef]

17. Moeinifar, S.; Kokabi, A.H.; Hosseini, H.R.M. Influence of peak temperature during simulation and real thermal cycles on microstructure and fracture properties of the reheated zones. *Mater. Des.* **2010**, *31*, 2948–2955. [CrossRef]

18. Zhang, Z.L.; Hauge, M.; Thaulow, C. Two-parameter characterization of near-tip stress fields for a bi-material elastic-plastic crack. *Int. J. Fract.* **1996**, *79*, 65–83. [CrossRef]

19. Thaulow, C.; Hauge, M.; Zhang, Z.L.; Ranestad, O.; Fattorini, F. On the interrelationship between fracture toughness and material mismatch for cracks located at the fusion line of weldments. *Eng. Fract. Mech.* **1999**, *64*, 367–382. [CrossRef]

20. Ren, X.B.; Zhang, Z.L.; Nyhus, B. Effect of residual stresses on the crack-tip constraint in a modified boundary layer model. *Int. J. Solids Struct.* **2009**, *46*, 2629–2641. [CrossRef]

21. Sisodia, R.P.S.; Gáspár, M. Physical simulation-based characterization of HAZ properties in steels. part 1. high-strength steels and their hardness profiling. *Strength Mater.* **2019**, *51*, 490–499. [CrossRef]

22. Mandziej, S.T. Physical simulation of metallurgical processes. *Mater. Technol.* **2010**, *44*, 105–119.

23. Gáspár, M. Effect of Welding Heat Input on Simulated HAZ Areas in S960QL High Strength Steel. *Metals* **2019**, *9*, 1226. [CrossRef]

24. Lukács, J.; Dobosy, Á. Matching effect on fatigue crack growth behaviour of high-strength steels GMA welded joints. *Weld. World* **2019**, *63*, 1315–1327. [CrossRef]

25. Xu, J.; Li, P.; Fan, Y.; Sun, Z. Effect of temperature on fracture toughness in weld thermal. simulated X80 pipeline steels. *Trans. China Weld. Inst.* **2017**, *38*, 22–26.

26. Qiu, H.; Mori, H.; Enoki, M.; Kishi, T. Fracture mechanism and toughness of the welding heat-affected zone in structural steel under static and dynamic loading. *Metall. Mater. Trans.* **2000**, *31*, 2785–2791. [CrossRef]

27. British Standards Institution. *BS-7448-2: Fracture Mechanics Toughness Tests. Part 2: Method for Determination of KIc, Critical CTOD and Critical J Values of Welds in Metallic Materials*; BSI: London, UK, 1997.

28. Det Norske Veritas. Fracture control for pipeline installation methods introducing cyclic plastic strain. In *Recommended Practice DNV-rp-f108*; Det Norske Veritas: Høvik, Norway, 2006.

29. ABAQUS. *ABAQUS User Manual*, version 6.14; Dassault Systemes Simulia Corp.: Providence, RI, USA, 2014.

30. Xu, J.; Zhang, Z.L.; Østby, E.; Nyhus, B.; Sun, D.B. Constraint effect on the ductile crack growth resistance of circumferentially cracked pipes. *Eng. Fract. Mech.* **2010**, *77*, 671–684. [CrossRef]

31. Eikrem, P.A.; Zhang, Z.L.; Nyhus, B. Effect of plastic prestrain on the crack tip constraint of pipeline steels. *Int. J. Press. Vessels Pip.* **2007**, *84*, 708–715. [CrossRef]

32. Xu, J.; Zhang, Z.L.; Østby, E.; Nyhus, B.; Sun, D.B. Effect of crack depth and specimen size on ductile crack growth of SENT and SENB specimens for fracture mechanics evaluation of pipeline steels. *Int. J. Press. Vessels Pip.* **2009**, *86*, 787–797. [CrossRef]

33. Ren, X.B.; Zhang, Z.L.; Nyhus, B. Effect of residual stresses on ductile crack resistance. *Eng. Fract. Mech.* **2010**, *77*, 1325–1337. [CrossRef]

34. Zhang, Z.L.; Thaulow, C.; Ødegård, J. A complete Gurson model approach for ductile fracture. *Eng. Fract. Mech.* **2000**, *67*, 155–168. [CrossRef]

© 2019 by the authors. Licensee MDPI, Basel, Switzerland. This article is an open access article distributed under the terms and conditions of the Creative Commons Attribution (CC BY) license (http://creativecommons.org/licenses/by/4.0/).

Article

On the Behaviour of 316 and 304 Stainless Steel under Multiaxial Fatigue Loading: Application of the Critical Plane Approach

Alejandro S. Cruces [1], Pablo Lopez-Crespo [1,*], Stefano Bressan [2], Takamoto Itoh [3] and Belen Moreno [1]

[1] Department of Civil and Materials Engineering, University of Malaga,
    C/Dr Ortiz Ramos s/n, 29071 Malaga, Spain
[2] Graduate School of Science and Engineering, Ritsumeikan University, Kusatsu–shi, Shiga 525-8577, Japan
[3] Department of Mechanical Engineering, College of Science and Engineering, Ritsumeikan University,
    Kusatsu–shi, Shiga 525-8577, Japan
*   Correspondence: plopezcrespo@uma.es; Tel.: +34-951952308

Received: 23 July 2019; Accepted: 2 September 2019; Published: 3 September 2019

Abstract: In this work, the multiaxial fatigue behaviour of 316 and 304 stainless steel was studied. The study was based on the critical plane approach which is based on observations that cracks tend to nucleate and grow in specific planes. Three different critical plane models were employed to this end, namely Fatemi–Socie (FS), Smith–Watson–Topper (SWT) and the newly proposed Sandip–Kallmeyer– Smith (SKS) model. The study allowed equi-biaxial stress state, mean strain and non–proportional hardening effects to be taken into consideration. Experimental tests including different combinations of tension, torsion and inner pressure were performed and were useful to identify the predominant failure mode for the two materials. The results also showed that the SKS damage parameter returned more conservative results than FS with lower scatter level in both materials, with prediction values between FS and SWT.

Keywords: critical plane model; multiaxial fatigue; non–proportional loading; 316 stainless steel; 304 stainless steel

---

1. Introduction

Fatigue failure is a common problem for a wide range of industries. Since the first reported study, new materials and advanced methods to predict the number of cycles until failure has appeared. Uniaxial or bending rotation cyclic tests are often conducted to characterise the fatigue behaviour of different metals [1]. However, most mechanical applications imply more complex scenarios, real service loads are usually variable and designs include complex profile shapes instead of just flat or cylindrical surfaces. As a consequence, different stresses/strain distributions appear on the real structures subjected to cyclic loads [2–4]. For characterising such complex scenarios there exists more sophisticated methods such as the critical plane approaches as an alternative to the classical models [5]. Critical plane models have been successfully applied for different materials and service loads. For example Chu observed improvements using these methods for AISI 1045 steel under complex loading conditions [6]. Sharifimehr employed critical plane methods to predict the fatigue life of a brittle and a ductile material under variable amplitude loads [7]. Llavori also used the critical plane methods to study the fatigue performance of a welded joint of S275JR, and was able to achieve better predictions as compared to classical methods [8]. One of the main strengths of critical plane methods is that they take into consideration the physical mechanisms involved in the nucleation and growth of the fatigue crack [9,10]. Nevertheless, there exists other alternative approaches that allow more accurate predictions to be achieved. One such approach is the Strip Yield Model (implemented in Nasgro software [11]). Besides the total fatigue life, these cycle by cycle models can also describe with certain accuracy the propagation stage until final failure takes place [12]. Depending on the material and the loading conditions, certain mechanisms will show a dominant presence along the fatigue process. For example, brittle materials tend to show a dominant Mode I crack growth along the fatigue process while ductile materials tend to have a dominant Mode II crack growth [13]. Critical plane methods are based on defining the plane where the highest damage takes place. This also means that they allow the crack growth angle to be predicted. This has been shown by Reis et al. who assessed the crack path initiation and growth for several structural steels [14,15]. The procedure requires evaluating the damage along the cycle from some stress and strain components. In some cases, obtaining such stress and strain components can be difficult and might introduce an additional source of error. Depending on the type of material, different critical plane models have been proposed. Models that include only stress variables are more useful in the high cycle regime but often fail at computing the fatigue damage in the low-cycle regime based on S–N curves. Stress values used in such models are frequently unrealistic and different to the actual stress experienced by the specimen due to the material behaviour above yield stress. Models that include strain variables are more robust in that sense.

To date, there is not a universal critical plane model that is valid for all the types of materials and all loading conditions. A very comprehensive review of different critical plane models can be found in the literature [16].

Well established models such as the Fatemi–Socie [17] or Smith et al. [18] were thoroughly studied, showing good results for ductile and brittle behaviour materials, respectively. Usually these models are chosen as a benchmark to propose new damage parameters [19,20]. In some cases the new model will return better results for the studied material and load condition [21,22], considering it more appropriate for those scenarios.

In this work a newly proposed critical plane damage parameter, called Sandip–Kallmeyer–Smith (SKS) was assessed based on its excellent performance on a low carbon steel [22]. First, the collapse capacity of the newly proposed damage parameter was evaluated. This was done by fitting the model to a set of experimental data with different loading paths both under proportional and non-proportional loads. Then, the fitted curve was used to predict fatigue lives for different multiaxial cases. The study was conducted on two stainless steels, namely 316 stainless steel and 304 stainless steel. The efficacy of the SKS damage parameter was compared with Fatemi–Socie model and Smith–Watson–Topper model.

2. Materials and Methods

The different models were tested on 316 and 304 stainless steels that are widely used in the industry. Previous studies observed better results with Mode II/III dominant critical plane methods for 316 stainless steel and with Mode I for 304 stainless steel [19,23].

All the tests were carried out on hollow cylindrical samples with 8.5 mm gauge length, 14 mm outer diameter and 12 mm inner diameter. The specimens were carefully polished to a surface roughness of approximately 0.3 μm both in the external and the internal surface. An in-house built fatigue machine allowed axial loads as well as inner pressure to be applied, thus allowing a very wide range of loading paths to be applied (see Figures 1 and 2). All tests were conducted in air. More details about the biaxial loading rigs, as well as additional details about the experiments can be found elsewhere [24,25].

The experimental tests employed for fitting the models and to evaluate the collapse capacity of the models on the 316 and the 304 stainless steel are shown in Tables 1 and 2, respectively [26]. Both experimental sets include proportional and non-proportional (out-of-phase between axial and shear strain) fatigue tests, as described in Figure 1. There are no mean stress tests among the experimental tests used for the 316 stainless steel. This is because no significant effect of the means stress on the fatigue life was observed for this material [27]. A comparison of the equivalent tests between 316 and 304 stainless steel (Tables 1 and 2) indicates that 304 stainless steel presents a higher hardening level.

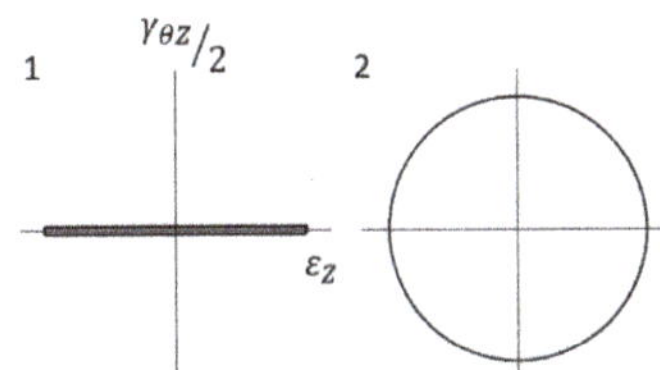

Figure 1. Proportional and non-proportional loading paths.

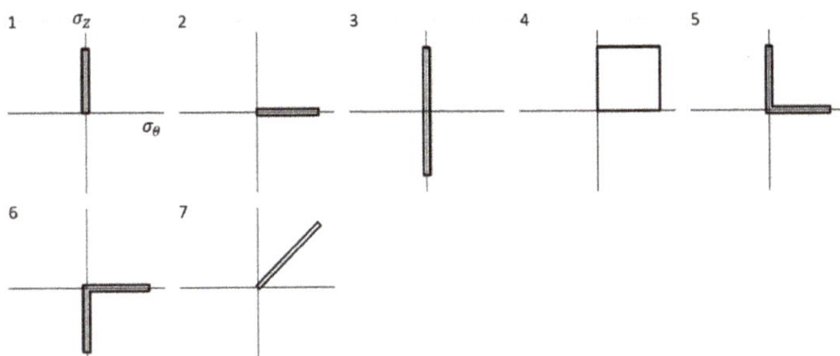

Figure 2. Proportional multiaxial loading paths applied on the 316 stainless steel specimens.

Table 1. Summary of the experimental data used to fit the model parameters for 316 stainless steel.

Path	ID	ε_a	γ_a	σ_a	τ_u	N_f
P	1	0.0025	-	225.5	-	25,100
	2	0.0035	-	252.5	-	8750
	3	0.005	-	278	-	4220
	4	0.0075	-	326	-	2200
NP	1	0.0015	0.0026	219.5	138.5	32,400
	2	0.0025	0.0043	344.5	219.5	4780
	3	0.0035	0.0061	412.5	238.5	3660
	4	0.0050	0.0087	474	299.5	1360
	5	0.0075	0.0130	615	388	410

Table 2. Summary of the experimental data used to fit the model parameters for 304 stainless steel.

Path	ID	ε_a	γ_a	σ_a	τ_a	N_f
P	1	0.0025	-	265	-	49,000
	2	0.0033	-	290	-	23,400
	3	0.004	-	315	-	7100
	4	0.005	-	365	-	1500
NP	1	0.002	0.0035	300	168	50,000
	2	0.002	0.0035	307	176	45,000
	3	0.0035	0.0061	457	256	3730
	4	0.0035	0.0061	477	267	3560

The tests were used to fit the SKS model that was subsequently used to predict fatigue lives for both materials under a range of multiaxial loading conditions. The predictions given by the SKS model are valid for the range of fatigue lives covered in Tables 1 and 2 for 316 and 304 stainless steel, respectively.

Fifteen different tests were conducted on 316 stainless steel to evaluate the different critical plane models, which are described in Table 3. The loading path used for the 316 stainless steel are shown in Figure 2. The load control mode was used to conduct the tests. It was possible to produce a triaxial stress state at the inner surface of the specimen and a biaxial stress state at the outer surface. On the outer surface, cases 1, 2 and 3 produced a uniaxial stress state, cases 4 and 7 a biaxial stress state and cases 5 and 6 an alternating pulsating stress state in perpendicular directions. For all the cases, a high level of ratchetting was observed [27]. The total reverse stress was applied in the case 3 (Figure 1) which promoted a non-zero mean strain probably because of the real stress asymmetry caused by the high load levels. Accordingly, a biaxial stress condition was induced on the outer surface. Since the principal stress directions are constant in time, all loading paths can be considered as proportional, given that the main slip plane does not change along the load cycle. The load ratio for Path 3 in Figure 2 is $R = -1$. The rest of tests had a zero load ratio, $R = 0$. These tests will be used in Section 4 to evaluate the accuracy of the different models, as well as the response of the models under mean stress loads and biaxial conditions.

Table 3. Summary of the 316 stainless steel experimental data used to evaluate the fitted models.

Path	$\Delta\sigma_z$	$\Delta\sigma_\theta$	$\Delta\varepsilon_z$	$\Delta\varepsilon_\theta$	N_f
1	445.15	1.8277	0.0022	0.0007	159,600
2	2.3931	450.32	0.0016	0.0055	29,300
2	1.8885	420.25	0.0013	0.0037	24,800
2	1.6815	366.46	0.0005	0.0033	53,000
3	1,024.3	1.7169	0.0406	0.0059	208
3	884.49	1.4092	0.0245	0.0042	393
4	399.2	424.1	0.0006	0.0051	3560
4	373.95	413.12	0.0033	0.0036	8400
5	6.5677	346.09	0.0023	0.0058	14,486
5	400.92	393.85	0.0057	0.0016	5300
5	376.34	383.25	0.0023	0.0045	14,486
6	399.76	450.21	0.0016	0.0048	25,770
6	442.64	511.01	0.0018	0.0049	13,542
6	375.8	335.22	0.0022	0.0027	31,400
7	389.49	465.73	0.0019	0.0019	24,700

The experimental tests used to evaluate the different models on the 304 stainless steel are shown in Table 4. The loading path used for studying the 304 stainless steel are shown in Figure 3. That is 29 experimental tests, three for path 0, and two for each of the other paths in Figure 3. These tests will allow the different models to be evaluated in terms of their capacity to take into account the fatigue damage produced by the hardening caused by non-proportional loads. For the same range of applied strains, increasing the non-proportionality in the loads requires increasing stresses to conduct the test. Previous results showed a high-hardening level for the 304 stainless steel [26]. Cases 1 and 6 can be considered proportional as the principal stress direction are constant along the cycle. The maximum non-proportionality factor appeared for cases 9, 10, 11, 13 and 14 [28]. Further experimental details are available elsewhere [25,26,28].

The coordinates adopted in this work are shown in Figure 4a. The radial and axial directions are defined as R and Z, respectively. The hoop direction θ is defined as being perpendicular to both other directions. The plane φ is defined by the normal vector $\vec{n}$ Figure 4b. This vector forms an angle α between its projection over the plane $[\theta R]$ and the direction R. It also forms an angle β between $\vec{n}$ and the Z direction. The vector $\vec{p}$, parallel to the intersection between φ and $[\theta R]$ is defined to consider the shear values. In addition, another vector $\vec{s}$ contained on φ and perpendicular to $\vec{p}$ is also defined for handling the shear component.

For the 316 stainless steel loading paths, the stresses and strains are computed at different planes φ. This is done by evaluating α and β angles in 15° increments in the range 0° to 90° [10]. For the 304 stainless steel loading paths, the hoop and radial strain should be the same on the surface (i.e., $\varepsilon_\theta = \varepsilon_R$). The maximum strain values are found on planes perpendicular to the surface ($\alpha = 90°$), with β ranging between 0° and 180°.

Once the strain and stress values are defined on each plane, a cycle counting process was performed using the rainflow method [29]. For dominant Mode II models, the shear strain cycles were counted and for dominant Mode I models, the normal strain cycles were counted. Mean and amplitude values for shear strain and shear stress were obtained using the circumscribed theory proposed by Papadopoulus [30]. Finally the damage was computed following Miner's linear rule [31].

Table 4. Summary of the 304 stainless steel experimental data used to evaluate the models.

Path	$\Delta\varepsilon_z$	$\Delta\gamma_{\theta z}$	$\Delta\sigma_z$	$\Delta\tau_{\theta z}$	N_f
1	0.0113	0	730	730	1700
1	0.012	0	805	805	690
1	0.015	0	825	825	540
2	0.005	0.0087	685	685	9500
2	0.008	0.0139	950	950	1400
3	0.005	0.0087	670	670	20,000
3	0.008	0.0139	860	860	2100
4	0.005	0.0087	670	670	2400
4	0.008	0.0139	975	975	820
5	0.005	0.0087	790	790	3400
5	0.008	0.0139	1010	1010	900
6	0.005	0.0087	485	485	17,500
6	0.008	0.0139	590	590	3200
7	0.005	0.0087	500	500	9700
7	0.008	0.0139	670	670	2600
8	0.005	0.0087	530	530	18,000
8	0.008	0.0139	735	735	1700
9	0.005	0.0087	760	760	2050
9	0.008	0.0139	1055	1055	470
10	0.005	0.0087	780	780	2950
10	0.008	0.0139	1075	1075	660
11	0.005	0.0087	765	765	2600
11	0.008	0.0139	1060	1060	320
12	0.005	0.0087	570	570	14,400
12	0.008	0.0139	850	850	1200
13	0.005	0.0087	660	660	4750
14	0.008	0.0139	940	940	710
14	0.005	0.0087	655	655	3200
	0.008	0.0139	965	965	1000

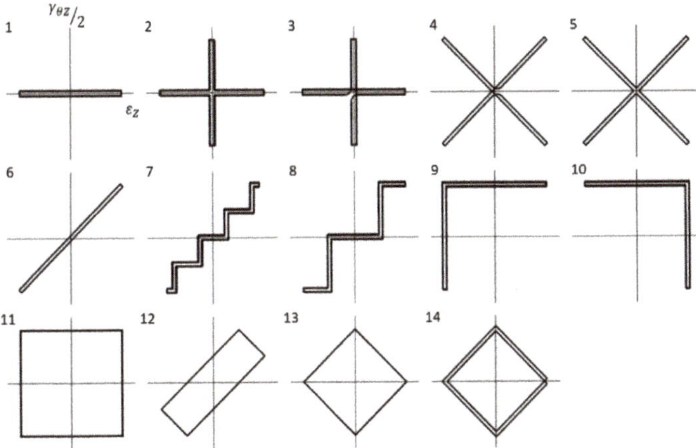

Figure 3. Proportional and non-proportional multiaxial loading paths applied on the 304 stainless steel specimens.

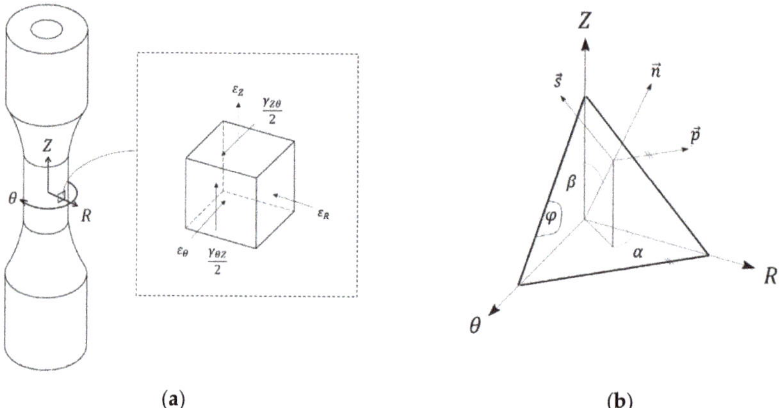

(a)   **(b)**

Figure 4. (**a**) Coordinates adopted in this work and (**b**) definition of the plane φ that was used to evaluate the critical plane.

3. Critical Plane Models

Critical plane models are based on observations of the nucleation and growth of fatigue cracks [10]. They are based on a damage parameter (DP) which incorporates stress and/or strain information that is subsequently used to predict the fatigue life. The plane where the DP is maximised is called the critical plane. The DP is defined for each cycle extracted along the entire loading block. For the sake of computational speed, the damage below 25% of the maximum damage along the loading block was not taken into account in the algorithm. This is because the effect of such low damage values on the fatigue life is negligible. Subsequently, a damage accumulation rule was used to obtain the number of cycles until the failure. In this work, three different critical plane models were used to characterise the multiaxial fatigue behaviour of the 316 and 304 stainless steels. The Fatemi—Socie (FS) critical plane model is normally employed for materials prone to shear failure [17]. The Smith–Watson–Topper (SWT) critical plane model gives accurate predictions for materials with predominant tension failure [21]. In addition, a newly proposed critical plane model by Suman, Kallmeyer and Smith (SKS) was also used to investigate the two materials. By studying the two materials with the FS and SWT models, it will be possible to identify the predominant failure mechanism for each of the materials. In addition, the study will also be useful to assess the predictive capabilities of the newly proposed model via comparison with two widely used critical plane models.

3.1. Fatemi–Socie model (FS)

The Fatemi–Socie model defines a strain type DP (Equation (1)) [17]. The model is based on that proposed by Brown and Miller [1]. They suggested substituting the normal strain component by a normal stress component. The DP is defined on the plane φ^* that maximises the shear strain range, $\Delta\gamma$.

$$DP_{FS} = \frac{\Delta\gamma_{max}}{2}\left(1 + k\frac{\sigma_{n,max}}{\sigma_y}\right) \tag{1}$$

where $\Delta\gamma_{max}/2$ is the maximum shear strain amplitude, $\sigma_{n,max}$ is the maximum tensile stress at φ^*, σ_y is the yield stress and k is a material parameter. The values for the yield stress were set to 260 MPa and 290 MPa for 316 and 304 stainless steel, respectively [26].

The strain hardening effect is considered with the $\Delta\gamma_{max}$ to be 2 times the $\sigma_{n,max}$ product. The mean normal stress effect is also considered via $\sigma_{n,max}$.

The parameter k represents the sensitivity of the material to normal stresses. This parameter can be estimated from the fatigue life N_f [10], through Equation (2).

$$k = \left[\frac{\frac{\tau'_f}{G}(2N_f)^{b_\gamma} + \gamma'_f(2N_f)^{c_\gamma}}{(1+v_e)\frac{\sigma'_f}{E}(2N_f)^{b} + (1+v_p)\varepsilon'_f(2N_f)^{c}} - 1 \right] \frac{\sigma'_y}{\sigma'_f(2N_f)^{b}} \tag{2}$$

where v_e and v_p are the Poisson's ration in the elastic and plastic regimes, respectively, E the Young modulus, σ'_f the fatigue strength coefficient, b the fatigue strength exponent, ε'_f the fatigue ductility coefficient, c the fatigue ductility exponent, σ'_y the cyclic yield stress, G the shear modulus, τ'_f the shear fatigue strength coefficient, b_γ the shear fatigue strength exponent, γ'_f the shear fatigue ductility coefficient and c_γ the shear fatigue ductility exponent.

Figure 5 shows the k values for 316 and 304 stainless steel against fatigue life N_f. For the 316 stainless steel, there is little variation of the k parameter with respect to the fatigue life. In addition the k parameter is very small (around 0.1) throughout the entire life. Figure 5 indicates that k parameter is much more sensitive to the fatigue life for the 304 stainless steel, with values ranging between ~0.5 and ~1.25. It is noted that the sensitivity parameter increases with the fatigue life for both materials, although with a much greater gradient for the 304 steel. For the cases where little information is gathered at either low fatigue lives or high fatigue lives, it is possible to use a constant sensitivity factor [10,19]. Nevertheless, in general this might reduce the accuracy of the fatigue predictions.

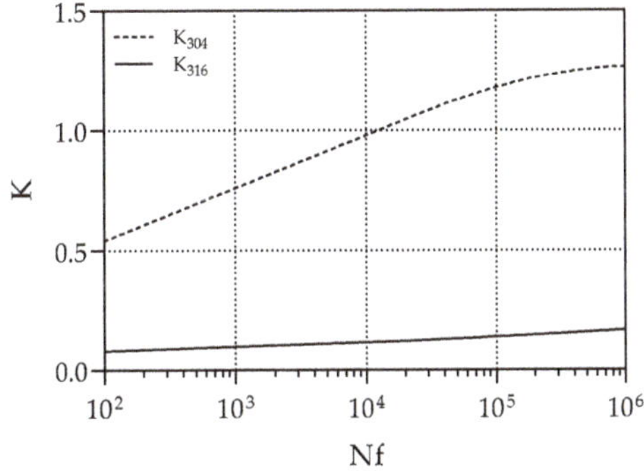

Figure 5. Fatemi–Socie normal stress sensitive factor k for 304 and 316 stainless steel.

3.2. Smith–Watson–Topper Model (SWT)

The Smith, Watson and Topper model [18] defines a strain energy density type DP (Equation (3)). The DP considers the normal strain and stress acting on the critical plane φ^*. The DP is defined on the plane φ^* that maximises the normal strain range, $\Delta\varepsilon$.

$$DP_{SWT} = \frac{\Delta\varepsilon}{2}\sigma_{n,max} \tag{3}$$

where $\Delta\varepsilon/2$ is normal strain amplitude, $\sigma_{n,max}$ is the maximum tensile stress at φ^*.

The strain hardening effect is considered in the SWT model through the $\Delta\varepsilon/2$ and $\sigma_{n,max}$ product. The mean normal stress effect is also taken into account via $\sigma_{n,max}$.

3.3. Sandip–Kallmeyer–Smith Model (SKS)

The multiaxial fatigue behaviour of the two steels is also evaluated with the Suman, Kallmeyer and Smith newly proposed critical plane model [21]. The SKS model defines a stress type DP (Equation (4)). Stress based models, such as Findley [32] and McDiarmid [33] normally give worse predictions for low-cycle fatigue due to lack of real stress information under such conditions. This is overcome with SKS model because it includes a strain component, in a similar way to the FS [17] and SWT [18] models. By using shear strain and shear stress elements, the SKS damage is more suitable for ductile failing materials. Following Sines compilation of ductile failing materials [34], such an effect is more predominant in the low-cycle regime. The DP is defined on the plane φ^* that maximises the shear strain range, $\Delta\gamma$.

$$DP_{SKS} = (G\Delta\gamma)^w \tau_{max}^{(1-w)}\left(1 + k\frac{(\sigma\cdot\tau)_{max}}{\sigma_0^2}\right) \tag{4}$$

where G is the shear modulus, $\Delta\gamma$ is the shear strain range, τ_{max} is the maximum shear stress, $(\sigma\,\tau)_{max}$ is the maximum shear and tensile stress product value, σ_0 is a factor used to maintain unit consistency, w and k are material fitting parameters. The values for the shear modulus were set to 75 GPa to 316 and 304 stainless steel [26]. A value of 500 MPa was set to σ_0, following the suggestions given by the authors [21]. The σ_0 parameter in the SKS damage parameter (Equation (4)) currently does not have a physical meaning, other than consistency of the units. Its value is corrected with the value of k in the fitting.

The strain hardening effect that takes place in the LCF regime is considered by $\Delta\gamma$ and τ_{max} values. The mean shear stress effect in the high cycle fatigue (HCF) regime is also considered by the shear ratio τ_{min}/τ_{max}. The parameter w weights the hardening and mean shear stress effects. The product $(\sigma\,\tau)_{max}$ introduces the detrimental effect over fatigue life observed when sub-cycle load peaks are applied simultaneously. The parameter k gauges the interaction effect between the shear and the normal stresses.

Unlike the FS model, there is not an equation to define the w and k parameters (Equation (4)). w parameter is tuned by fitting the experimental data with a mean shear stress effect. w incorporates the mean shear stress effect and the strain hardening effect. Unlike in the FS model, the w parameter has a constant value for all fatigue lives. By using tests for the fitting of the model with a wide range of lives, the w parameter cross the different fatigue regimes.

3.4. Fitted Models

Models damage parameter DP_{exp} (Equations (1), (3) and (4)), are related to the fatigue life N_f using the same double exponential curve Equation (5). All the material parameters used in the fitting were obtained from previous experimental data, Tables 1 and 2 [26]. The parameters were obtained with an optimisation process based on a least square error minimisation between DP_{exp} and DP_{calc} [9]. As the number of experimental data used to fit the parameter were relatively low, the expected difference between the minimisation of the DP instead of the fatigue life N_f should be negligible over the fitted values.

$$DP_{calc} = AN_f^b + CN_f^d \tag{5}$$

where A, b, C and d are material dependent parameters and N_f is the fatigue life in cycles. When fitting the models, N_f is the experimental value of each test. Thus, utilisation of SWT requires evaluating those four parameters (A, b, C and d). FS requires those four parameters plus the sensitivity factor described in Section 3.1. SKS requires those four parameter plus the two material parameters described in Section 3.2. Fatigue live data are required in order to fit the material parameters for SKS model. Since the SKS model has six fitting parameters, in order to obtain a deterministic (or an over-deterministic) system of equations, six fatigue tests (or more than six tests) were required. These tests should be conducted in conditions as general as possible, to make it as versatile as possible. Accordingly, both proportional and non-proportional tests with a wide range of fatigue lives were employed. In our case we observed an improvement by using an over-deterministic system of equations (Nine and eight tests for 316 and 304 steels, respectively, as shown in Tables 1 and 2).

The collapse capacity of the different models was evaluated by studying the mean and the standard deviation of the error [35]. The error is defined as the difference between the predicted and the experimental life in logarithmic scale (Equation (6)).

$$\text{error} = \log_{10}(N_{mod}) - \log_{10}\left(N_{exp}\right) \tag{6}$$

where N_{mod} is the fatigue life predicted by the fitted model and N_{exp} is the experimental fatigue life.

Table 5 includes the mean and standard deviation of the error values observed in the fitting. Negative mean values are indicative of conservative results and vice versa. A better fitting is obtained with mean values as close as possible to zero. In a similar way, a better fitting is also obtained with the standard deviation value as close as possible to zero. It is observed that the lowest mean values for both materials were obtained with the SKS critical plane model, followed by FS. The mean values in Table 5 indicated that the SWT model appears to produce the least accurate predictions for the type of experiments under study. For both materials, SKS returns the best fit and SWT the worst fit, probably because of the different number of material parameters used in the different models. Materials that normally exhibit a ductile behaviour were more sensitive to damage mechanisms caused by shear stress rather than by normal stress. Materials that normally present a brittle behaviour were more sensitive to damage mechanisms caused by normal stress [16]. Accordingly, the Fatemi–Socie model will be more appropriate for ductile materials failing predominantly under shear mode (Mode II and III); and Smith–Watson–Topper for brittle materials failing predominantly under tension mode (Mode I).

Table 5. Statistical analysis for the comparison of the models collapse capacity.

Statistical Values	FS	SWT	SKS
316 stainless steel	-	-	-
Mean value	0.0091	0.0308	0.0022
Standard deviation	0.0130	0.0370	0.0014
304 stainless steel	-	-	-
Mean value	0.0020	0.0069	0.0016
Standard deviation	0.0026	0.0065	0.0019

4. Results and Discussion

The fatigue life predictions of each fitted model are shown in Figures 6 and 7 for 316 and 304 stainless steels, respectively. The experimental fatigue life N_{exp} is defined in the horizontal axis and the predicted fatigue life N_{mod} in the vertical axis. Logarithmic scale is used in both Figures 6 and 7. The points falling along the solid line present coincidence between N_{mod} and N_{exp}. The values along the dashed lines have a factor 2 deviation between N_{mod} and N_{exp}, that is the prediction given by the model that is twice or half of that measured experimentally [9,36]. The estimations of FS are shown by the blue crosses, SWT by green circles and SKS by purple diamonds on both Figures 6 and 7. It was observed that most of the predictions fall within the factor 2 band deviation.

Figures 6 and 7 show that the predictions returned by SKS are mostly between those of FS and those of SWT. The SKS predictions appear to be overall closer to the FS predictions. For the 316 stainless steel, better results were obtained with SKS and FS models (Figure 6). This is in agreement with previous research that indicated that dominant Mode II critical plane models appear to be more suitable for 316 stainless steel [37]. Figure 6 also shows that both SKS and FS models produced less conservative predictions for the square-shape and equi-biaxial tests (cases 4 and 7 in Figure 2). These cases correspond to the two FS points in Figure 6 where the predictions were beyond the twice fatigue life bound. In these tests, the simultaneous application of the loads in the different directions highly restrict the deformation of the material. As a consequence the range of strains were reduced considerably as compared to the equivalent uniaxial loading test. This in turn reduced the value of the damage parameter thus decreasing the accuracy of the predictions towards the non-conservative side [38,39]. The most conservative prediction by the SKS in Figure 6 has N_{exp} = 159,600 cycles. This is indeed the most conservative prediction given by the SKS model. Since the 316 study was conducted with nearly half the samples of the 304 study, the relative weight of this single prediction is larger on the 316 than on the 304 material. It is not surprising that that point produces the longest fatigue life, since it corresponds to the simplest loading case: Pure uniaxial cyclic tension, as shown in Table 3 and Figure 2. The SWT model appears to yield the most conservative predictions, in agreement with previous research [40] where a different steel with tendency to ductile failure was subjected to proportional loadings.

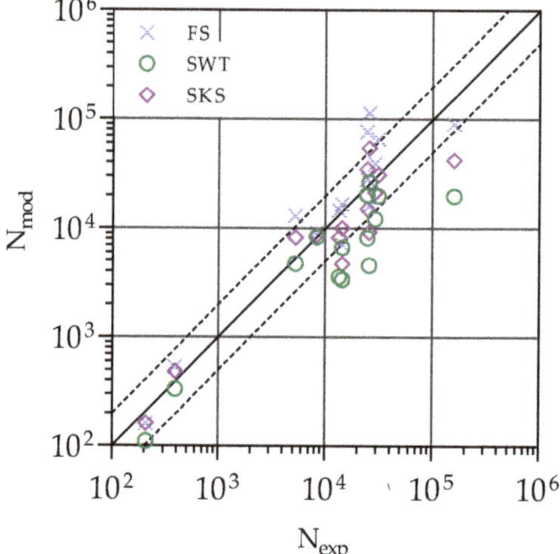

Figure 6. Fatigue life predicted by each model, N_{mod} versus experimental fatigue life, N_{exp} for 316 stainless steel.

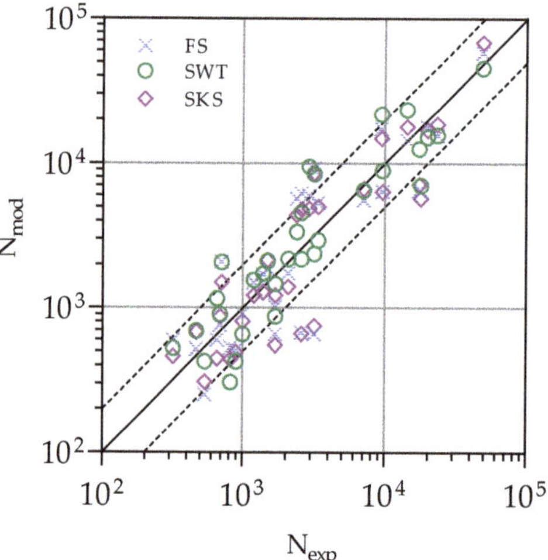

Figure 7. Fatigue life predicted by each model, N_{mod} versus experimental fatigue life, N_{exp} for 304 stainless steel.

Figure 7 shows that most of the predictions given by the different models on the 304 steel are inside the factor of 2 deviation. Comparison between Figures 6 and 7 indicate the best predictions were overall achieved on the 304 steel. In general the best results were obtained by SWT, as it was observed for this material by Socie [23]. Previous analysis showed that only torsion tests promoted predominant Mode II crack growth and only axial loading tests promoted predominant Mode I cracking on 304 stainless steel, for a range of fatigue lives below 10^5 cycles [10]. As mentioned previously, SWT should then produce more accurate predictions for only axial loading tests and FS present better accuracy for purely torsional tests. Loading cases 2 and 3 in Figure 3 pose a challenging problem in this sense because each loading block is formed of alternating cycles of pure Mode I load and pure Mode II load. That is pure axial load and pure torsional load applied but non-simultaneously. Accordingly, the predictions given by FS and by SWT should be similar. This is indeed observed for the loading cases 2 and 3 in Figure 7.

As it can be seen in Figure 5, the sensitivity parameter k in FS model changes from ~0.5 to ~1.25 in the range from 10^2 to 10^5 cycles. If not enough experimental data in the entire range were available, it would be possible to take a constant value of 1 for the FS sensitivity parameter [10]. Nevertheless, the effect in the accuracy of the predictions would be detrimental, producing more conservative predictions in the lower range of the fatigue life (between 10^2 and 10^3 cycles in Figure 5) and non-conservative predictions in the higher range of the fatigue life (between 10^5 and 10^6 cycles in Figure 5).

FS and SWT models allow the additional hardening of the material caused by the non-proportionality of the loads to be taken into account because their damage parameter includes both stress and strain variables. SKS also includes the additional hardening cause by the non-proportionality because its damage parameter has both stress and strain information. This is clear for the loading cases with high non-proportionality (cases 9, 10, 11, 13 and 14 in Figure 3) where the three critical plane models yielded predictions within the factor 2 error bound. Tests with the loads applied proportionally (cases 6, 7 and 8 in Figure 3) were also handled satisfactorily by the three models.

In order to assess numerically the overall performance of the different models, the probability density function (PDF) of the error (Equation (6)) was computed [35]. The results of the PDF for the 316 material are shown in Figure 8a and the results for the 304 material are shown in Figure 8b. In addition, Table 6 summarises the mean value and standard deviation for both materials. The PDF curves closer to a zero mean error and with lower deviation indicated a better accuracy of the model.

For the 316 stainless steel, Table 6 shows a slightly higher mean value for SKS than for FS but a slightly smaller standard deviation for SKS than for FS, thus indicating a similar performance of SKS and FS for the loads analysed on the 316 stainless steel. It was also noted that the mean SKS values were negative while the mean FS values were positive. That is because the SKS predictions are overall more on the conservative side while the FS predictions are more on the non-conservative side for the 316 steel. The larger mean and standard deviation values observed for the SWT indicate overall worst predictions as compared to SKS and FS models. In addition, this is symptomatic of the SKS model being more appropriate for predominantly shear mode failing materials.

On the other hand, for the 304 steel SWT shows the lowest values of both the mean and the standard deviation. This suggests that 304 fails predominantly under tensile mode for the tests described. In addition, the FS the mean value is lower than the SKS, while for the SKS model the standard deviation is lower than that of FS model. The performance of both SKS and FS appears to be similar for the 304 steel but again, the SKS model tends to yield predictions on the conservative side and FS more on the non-conservative side.

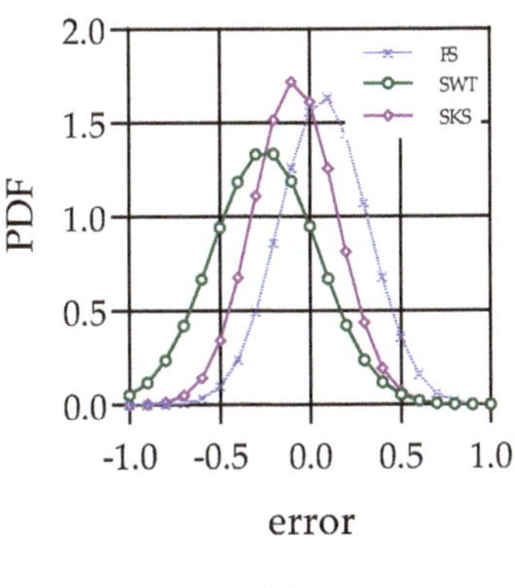
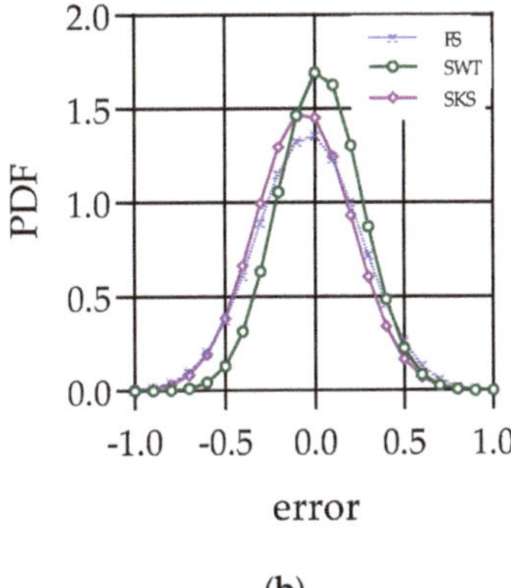

(a)   (b)

Figure 8. Probability density function of prediction error for (**a**) 316 stainless steel and (**b**) 304 stainless steel.

Table 6. Statistical analysis for the comparison of the model prediction errors.

Statistical Values	FS	SWT	SKS
316 stainless steel	-	-	-
Mean value	0.0763	−0.2490	−0.0836
Standard deviation	0.2432	0.2945	0.2315
304 stainless steel	-	-	-
Mean value	−0.0299	0.0291	−0.0595
Standard deviation	0.2919	0.2339	0.2681

Table 6 also indicates that the three mean values are smaller for the 304 than for the 316 material, thus indicating that the predictions obtained for the 304 steel are slightly more accurate than for the 316 steel.

The different performance of the different types of models were useful for identify the predominant failure mode of the two materials. The 316 steel appears to fail predominantly under shear mode because FS produces better estimations. Conversely, the 304 steel appears to fail predominantly under tension mode since SWT generated the best predictions. This appears to hold for the wide range of multiaxial loads analysed. Moreover, the alignment of SKS with FS in terms of predictions suggests that SKS model is most suitable for predominantly shear mode failing materials.

5. Conclusions and Future Works

The multiaxial fatigue behaviour of two widely used materials, 316 and 304 stainless steels, was studied by means of the critical plane approach. The analysis has been performed on a wide range of experimental tests including different combinations of tension, torsion and inner pressure. Three different critical plane models were used, namely FS, SWT and the newly proposed SKS model. First, the collapse capacity of the different models were evaluated. The larger number of material parameters of SKS model appeared to offer a higher flexibility in this sense, thus producing the best fitting. Nevertheless SKS did not offer any expression to define the k and w parameters included in the damage parameter. In addition, the σ_0 parameter also included in the SKS damage parameter did not have any physical meaning. Producing analytical expressions relating the k and w parameters to different fatigue properties of the material remain as challenging prospective research activities. The SKS could also be improved by relating σ_0 to another characteristic material property, to promote a more uniform use of the model. Unlike the FS model, the parameter k and w parameters take the same value across the whole range of fatigue lives. This can be a weakness for design situations where a very wide range of conditions and very different fatigue regimes are studied. In addition, the SKS model should also be applied to other different materials, to evaluate its performance for other types of alloys.

The efficacy of the different models has also been analysed in terms of their accuracy for predicting the fatigue life. To this end, the damage parameter was correlated with the fatigue life using a double exponential curve. The fitted curves for 316 and 304 stainless steel were used to predict fatigue life under different loading path for the same materials. Most of the predictions given by the three models fall in the band defined by the factor of 2 deviation. For cases with a higher level of hardening, the critical plane models have shown to also generate satisfactory predictions. SKS and FS produced the best predictions for the 316 material while SWT generated the best predictions for the 304 material. Such a trend has been useful to identify the predominant failure mode of the two materials: 316 fails predominantly under the shear mode loading and 304 material fails predominantly under the tension mode loading. The results also indicated that the SKS model appears to be most suitable for shear mode failing materials. For the experimental tests described, SKS has proven to generate predictions on the conservative side and FS on the non-conservative side. This suggests that SKS could be more suitable from a design point of view.

Author Contributions: Conceptualization, A.S.C. and P.L.-C.; methodology, A.S.C., P.L.-C., T.I., S.B. and B.M.; software, A.S.C. and B.M.; validation, P.L.-C., B.M. and T.I.; formal analysis, A.S.C., P.L.-C. and B.M.; investigation, A.S.C., P.L.-C., T.I., S.B. and B.M.; resources, A.S.C. and S.B.; data curation, A.S.C. and P.L.-C.; writing—original draft preparation, A.S.C.; writing—review and editing, P.L.-C.; visualization, A.S.C.; supervision, P.L.-C.; project administration, P.L.-C.; funding acquisition, P.L.-C.

Funding: This research was funded by Ministerio de Economia y Competitividad (Spain), grant number MAT2016-76951-C2-2-P. The APC was funded by Ministerio de Economia y Competitividad (Spain).

Acknowledgments: The financial support of Ministerio de Economia y Competitividad (Spain) through grant reference MAT2016-76951-C2-2-P is acknowledged. Fernando V Antunes from University of Coimbra (Portugal) is also greatly acknowledged for his help in conducting the experiments, analysis ideas and interesting discussions. Industrial support from Sandip Suman and UTC Aerospace Systems (CA, USA) is also greatly acknowledged.

Conflicts of Interest: The authors declare no conflict of interest.

References

1. Brown, M.W.; Miller, K.J. A theory for fatigue failure under multiaxial stress-strain conditions. *Proc. Inst. Mech. Eng.* **1973**, *187*, 745–755. [CrossRef]

2. Metcalfe, R.G.; Costanzi, R. Fatigue cracking of dragline boom support strands. *Eng. Fail. Anal.* **2019**, *99*, 46–68. [CrossRef]

3. Gledić, I.; Parunov, J.; Prebeg, P.; Ćorak, M. Low-cycle fatigue of ship hull damaged in collision. *Eng. Fail. Anal.* **2019**, *96*, 436–454.

4. Mamiya, E.N.; Castro, F.C.; Ferreira, G.V.; Nunes Filho, E.L.S.A.; Canut, F.A.; Neves, R.S.; Malcher, L. Fatigue of mooring chain links subjected to out-of-plane bending: Experiments and modeling. *Eng. Fail. Anal.* **2019**, *100*, 206–213. [CrossRef]

5. Chen, X.; Xu, S.; Huang, D. Critical plane-strain energy density criterion for multiaxial low-cycle fatigue life under non-proportional loading. *Fatigue Fract. Eng. Mater. Struct.* **1999**, *22*, 679–686. [CrossRef]

6. Chu, C. Multiaxial fatigue life prediction method in the ground vehicle industry. *Int. J. Fatigue* **1997**, *19*, 325–330. [CrossRef]

7. Sharifimehr, S.; Fatemi, A. Fatigue analysis of ductile and brittle behaving steels under variable amplitude multiaxial loading. *Fatigue Fract. Eng. Mater. Struct.* **2019**, *42*, 1722–1742. [CrossRef]

8. Llavori, I.; Etxeberria, U.; Lopez, A.; Ulacia, I.; Ugarte, D.; Esnaola, J.; Larrañaga, M. A numerical analysis of multiaxial fatigue in a butt weld specimen considering residual stresses. In Proceedings of the 12th International Fatigue Congress (FATIGUE 2018), Poitiers, Futuroscope, France, 27 May 2018; Volume 165, p. 21005.

9. Erickson, M.; Kallmeyer, A.R.; Van Stone, R.H.; Kurath, P. Development of a multiaxial fatigue damage model for high strength alloys using a critical plane methodology. *J. Eng. Mater. Technol.* **2008**, *130*, 0410081–0410089. [CrossRef]

10. Socie, D.F.; Marquis, G.B. *Multiaxial Fatigue*, 1st ed.; Society of Automotive Engineers Inc.: Warrendale, PA, USA, 2000.

11. Ten-Hoeve, H.; De-Koning, A. *Reference Manual of the Strip-Yield Module in NASGRO or ESACRACK Software for the Prediction of Retarded Crack Growth and Residual Strength in Metal Materials*; Report No. TR 97012; National Aerospace Laboratory: Amsterdam, The Netherlands, 1977.

12. Moreno, B.; Martin, A.; Lopez-Crespo, P.; Zapatero, J.; Dominguez, J. Estimations of fatigue life and variability under random loading in aluminum Al-2024T351 using strip yield models from NASGRO. *Int. J. Fatigue* **2016**, *91*, 414–422. [CrossRef]

13. Li, B.; Reis, L.; de Freitas, M. Comparative study of multiaxial fatigue damage models for ductile structural steels and brittle materials. *Int. J. Fatigue* **2009**, *31*, 1895–1906. [CrossRef]

14. Reis, L.; Freitas, M.J. Crack initiation and growth path under multiaxial fatigue loading in structural steels. *Int. J. Fatigue* **2009**, *31*, 1660–1668. [CrossRef]

15. Anes, V.; Reis, L.; Li, B.; Freitas, M. Crack path evaluation on HC and BCC microstructures under multiaxial cyclic loading. *Int. J. Fatigue* **2014**, *58*, 102–113. [CrossRef]

16. Karolczuk, A.; Macha, E. A review of critical plane orientations in multiaxial fatigue failure criteria of metallic materials. *Int. J. Fract.* **2005**, *134*, 267–304. [CrossRef]

17. Fatemi, A.; Socie, D.F. A Critical Plane Approach to Multiaxial Fatigue Damage Including out-of-Phase Loading. *Fatigue Fract. Eng. Mater. Struct.* **1988**, *11*, 149–165. [CrossRef]
18. Smith, K.; Topper, T.H.; Watson, P. A stress-strain function for the fatigue of metals (Stress-strain function for metal fatigue including mean stress effect). *J. Mater.* **1970**, *5*, 767–778.
19. Jin, D.; Tian, D.J.; Li, J.H.; Sakane, M. Low-cycle fatigue of 316 L stainless steel under proportional and nonproportional loadings. *Fatigue Fract. Eng. Mater. Struct.* **2016**, *39*, 850–858. [CrossRef]
20. Anes, V.; Reis, L.; Li, B.; De Freitas, M. New cycle counting method for multiaxial fatigue. *Int. J. Fatigue* **2014**, *67*, 78–94. [CrossRef]
21. Suman, S.; Kallmeyer, A.; Smith, J. Development of a multiaxial fatigue damage parameter and life prediction methodology for non-proportional loading. *Frattura ed Integrità Strutturale* **2016**, *10*, 224–230. [CrossRef]
22. Cruces, A.S.; Lopez-Crespo, P.; Moreno, B.; Antunes, F.V. Multiaxial Fatigue Life Prediction on S355 Structural and Offshore Steel Using the SKS Critical Plane Model. *Metals* **2018**, *8*, 1060. [CrossRef]
23. Socie, D. Multiaxial Fatigue Damage Models. *J. Eng. Mater. Technol.* **1987**, *10*, 293–298. [CrossRef]
24. Ohnami, M.; Hamada, N. Crack Propagation Behavior of Biaxial Low-Cycle Fatigue at Elevated Temperatures (Effects of the Cyclic Principal Stressing in Parallel with the Fatigue Crack and the Rotation of the Principal Stress Axes). *J. Soc. Mater. Sci. Japan* **1981**, *30*, 822–828. [CrossRef]
25. Morishita, T.; Takada, Y.; Ogawa, F.; Hiyoshi, N.; Itoh, T. Multiaxial fatigue properties of stainless steel under seven loading paths consisting of cyclic inner pressure and push-pull loading. *Theor. Appl. Fract. Mech.* **2018**, *96*, 387–397. [CrossRef]
26. Itoh, T.; Yang, T. Material dependence of multiaxial low cycle fatigue lives under non-proportional loading. *Int. J. Fatigue* **2011**, *33*, 1025–1031. [CrossRef]
27. Cruces, A.S.; Lopez-Crespo, P.; Bressan, S.; Itoh, T. Investigation of the multiaxial fatigue behaviour of 316 stainless steel based on critical plane method. *Fatigue Fract. Eng. Mater. Struct.* **2019**, *42*, 1633–1645. [CrossRef]
28. Itoh, T.; Sakane, M.; Ohnami, M.; Socie, D.F. Nonproportional low-cycle fatigue criterion for type 304 stainless steel. *J. Eng. Mater. Technol. ASME* **1995**, *117*, 285–292. [CrossRef]
29. Matsuishi, M.; Endo, T. Fatigue of metals subjected to varying stress. *Japan Soc. Mech. Eng.* **1968**, *68*, 37–40.
30. Papadopoulos, I. A comparative study of multiaxial high-cycle fatigue criteria for metals. *Int. J. Fatigue* **1997**, *19*, 219–235. [CrossRef]
31. Miner, M. Cumulative damage in fatigue. *J. Appl. Mech.* **1945**, *12*, 159–164.
32. Findley, W.N. A theory for the effect of mean stress on fatigue of metals under combined torsion and axial load or bending. *J. Eng. Ind. Trans. ASME* **1959**, *81*, 301–306. [CrossRef]
33. McDiarmid, D.L. A Shear Stress Based Critical-Plane Criterion of Multiaxial Fatigue Failure for Design and Life Prediction. *Fatigue Fract. Eng. Mater. Struct.* **1994**, *17*, 1475–1485. [CrossRef]
34. Sines, G. *Failure of Materials Under Combined Repeated Stresses with Superimposed Static Stresses*; TN3495; NACA: Washington DC, USA, 1955.
35. Ince, A.; Glinka, G. A modification of Morrow and Smith-Watson-Topper mean stress correction models. *Fatigue Fract. Eng. Mater. Struct.* **2011**, *34*, 854–867. [CrossRef]
36. Liu, Y.; Mahadevan, S. Multiaxial high-cycle fatigue criterion and life prediction for metals. *Int. J. Fatigue* **2005**, *27*, 790–800. [CrossRef]
37. Liu, K.C.; Wang, J.A. An energy method for predicting fatigue life, crack orientation, and crack growth under multiaxial loading conditions. *Int. J. Fatigue* **2001**, *23*, 129–134. [CrossRef]
38. Ellison, E.G.; Andrews, J.M.H. Biaxial cyclic high-strain fatigue of aluminum alloy RR58. *J. Strain Anal. Eng. Des.* **1973**, *8*, 209–219. [CrossRef]
39. Ellyin, F.; Gołoś, K.; Xia, Z. In phase and out–of–phase multiaxial fatigue. *Trans. ASME* **1991**, *113*, 112–118. [CrossRef]
40. Lopez-Crespo, P.; Moreno, B.; Lopez-Moreno, A.; Zapatero, J. Study of crack orientation and fatigue life prediction in biaxial fatigue with critical plane models. *Eng. Fract. Mech.* **2015**, *136*, 115–130. [CrossRef]

© 2019 by the authors. Licensee MDPI, Basel, Switzerland. This article is an open access article distributed under the terms and conditions of the Creative Commons Attribution (CC BY) license (http://creativecommons.org/licenses/by/4.0/).

 *metals*

Article

Study on the Corrosion Fatigue Properties of 12Cr1MoV Steel at High Temperature in Different Salt Environments

Jianjun He, Jiangyong Bao, Kailiang Long, Cong Li * and Gang Wang

School of Energy and Power Engineering, Changsha University of Science and Technology, Changsha 410114, China
* Correspondence: liconghntu@csust.edu.cn; Tel.: +86-0731-85258409

Received: 6 June 2019; Accepted: 7 July 2019; Published: 10 July 2019

Abstract: Biomass energy, as a reliable renewable energy source, has gained more and more attention. However, microstructure degradation and corrosion fatigue damage of heat pipes hinder its further application. In this paper, high temperature corrosion fatigue characteristics of 12Cr1MoV steel under a mixed alkali metal chloride salt environment and mixed sulfate salt environment were investigated. Fatigue tests with different total strain amplitudes were performed. Results show that the effect of total strain amplitude on the cyclic stress response of the alloy is approximately the same under three different deformation conditions. With the increase of the cyclic numbers, the alloyed steel mainly exhibited cyclic hardening during loading. The fatigue properties in air environment were the best, which is most obvious when the total strain amplitude is ±0.3%. The fatigue life of samples in mixed alkali metal salts is the shortest. Furthermore, the fatigue fracture morphology of the alloyed steel in different environments were also deeply analyzed. This experimental study attempts to provide a theoretical reference for solving the problem of rapid failure of heat pipes in biomass boilers, and to establish a scientific basis for the material selection and safety operation.

Keywords: high temperature; 12Cr1MoV steel; mixed salt environments; corrosion fatigue; heat pipe failure

---

1. Introduction

As a kind of renewable energy, biomass energy has attracted considerable attention in recent years due to its rich resources and lower pollution. However, most of the biomass fuels generally contain a large number of alkali metals, alkaline earth metals, sulfides, and chlorine elements with high concentration [1]. Alkali metal chlorides and sulfides are the main compounds produced during biomass combustion. They generally exist in the form of KCl, NaCl, K_2SO_4, and Na_2SO_4. Alkali metal compounds in fuels tend to flow with flue gas at high temperatures, then deposit on the wall of the superheater, and form alkali metal salt slagging with complex composition [2]. This phenomenon easily leads to local overheating, which makes the high temperature salt exist in a liquid pool state. On the other hand, when biomass burns, the sulfur content will be released. Part of the released sulfur is combined with oxygen to form SO_2 gas, which is mixed with flue gas and discharged outside of the boiler The other part will deposit in the form of sulfate onto the inside of the boiler. These sulfates will react with the oxide film on the metal surface at high temperatures, which results in corrosion. At the same time, because the heat pipe of the boiler superheater works in high temperature environments [3], the components need to bear more complex loading, which will lead to a local stress concentration, thus resulting in a local plastic deformation and fatigue damage of the material [4]. With the accumulation of deformation, defect structures and cracks are formed [5]. Cracks propagate unstably under the combined action of a hot corrosion environment [6] and high temperature fatigue [7], which leads to bursting and failure of the steel tube.

The molten salt corrosion is a complex process, involving chemical corrosion, electrochemical corrosion, as well as the interface reaction and the dissolution of oxides. The corrosion resistance at high temperatures is largely dependent on the formation of protective oxidation layer. Cl–Cl-containing environments are well known to cause accelerated corrosion, resulting in increased oxidation, metal wastage, void formation, and loose non-adherent scales, hence destroying the protective oxidation layers [8]. During the fatigue process, corrosion caused by alkali chloride and sulfate salt is one of the most important factors that determine the service life of the heat pipe. In terms of research on the corrosion fatigue of alloys in a salt environment, the following papers are listed. Liu et al. found that the NaCl–KCl mixture could accelerate the corrosion process of TP347H stainless steel and C22 alloy in molten chloride [9]. Zhang et al. have reported that Ni–20Cr–18W alloy suffered severe thermal corrosion in mixed molten salt, the main corrosion products are almost the same at different temperatures [10]. Tsaur et al. used a 310 stainless steel with pre-coated NaCl–Na_2SO_4 mixture for hot corrosion tests. The experiments have proved that NaCl is the main corrosion catalyst at high temperature. The presence of the NaCl in deposits inhibits the formation of protective oxidation layers at the initial stage, leading to an acceleration of hot corrosion of 310 stainless steel [11].

However, to our best knowledge, until now most experimental research on corrosion fatigue has been carried out mainly from the aspects of salt corrosion, solution corrosion, and so on. In fact, the actual operating environment of the biomass boiler superheater heat pipe is very complex. In many cases, thermal corrosion and fatigue damage caused by biomass combustion will act on the heat pipe material simultaneously, thus accelerating the degradation and failure of the material. Therefore, it is of great significance to study corrosion fatigue characteristics of the heat pipe materials for biomass boilers from the perspective of an interaction between corrosion and fatigue. As a kind of high strength alloyed steel, 12Cr1MoV steel possesses excellent corrosion resistance and mechanical properties. It is widely used in coal-fired boilers and biomass boiler heat pipes at temperatures below 600 °C. In this study, 12Cr1MoV steel was chosen as the research material, and corrosion fatigue characteristics of 12Cr1MoV steel at a high temperature under different loading conditions and chemical environments were deeply investigated.

2. Materials and Methods

The experimental material used in this study was 12Cr1MoV rolled round steel with a diameter of 20 mm. The chemical composition is shown in Table 1.

Table 1. Chemical composition of 12Cr1MoV steel (wt%).

Element	C	Cr	Mn	Mo	Si	V	Fe
Content	0.1	1.03	0.54	0.3	0.24	0.2	Balance

In this paper, fatigue tests of alloyed steel were carried out under three different environments: an air environment (reference test), a mixed alkali metal salt environment, and a mixed sulfate environment. The alkali metal salts used in this experiment are KCl and NaCl, which are mixed according to the mass ratio of 70% NaCl + 30% KCl (mixed chloride salt). The composition of mixed sulfates is 70% Na_2SO_4 + 30% K_2SO_4. Most biomass boilers set their rated steam temperatures at 450 °C, 540 °C, and 580 °C. Considering that there is a heat storage process on the surface of the steam pipe when the boiler is in operation, the temperature of the heat pipe surface is actually higher than that of the rated steam temperature, therefore in our experiments, the temperature was set to be 600 °C. The total strain amplitudes of each group were selected to be +0.5%, +0.4%, +0.3%, and +0.2% for low-cycle fatigue tests, the frequency was 0.2 Hz, and the strain rate was 10^{-3}/s. Specific experimental schemes are shown in Table 2.

Table 2. Fatigue experimental scheme of 12Cr1MoV alloy.

Temperature (°C)	Strain Amplitude	Waveform	Frequency (HZ)	Environment
600	±0.4% ±0.5% ±0.3% ±0.2%	Cosine wave	0.2	Air/Mixed Alkali metal chloride salt (Type A)/Mixed sulfate salt (Type B)

The fatigue specimens with a dimension of 40 mm standard distance, 8 mm diameter, and M14 threads at both sides were prepared, as shown in Figure 1a. Specimens were polished smoothly with different sandpapers. Before any further experiments, samples were cleaned with alcohol for 15 min and repeated twice in order to remove impurities and grease on the surface of the samples. The mechanical property tests were conducted on a RDL05 electronic creep-fatigue testing machine. Extensometers and thermocouples were used to measure the strain and temperature respectively. The installation is schematically shown in Figure 1b. The tensile tests were performed at room temperature and 600 °C in the air with a strain rate of 10^{-3} s^{-1}. In order to assure the repeatability of the results, each test was performed three times.

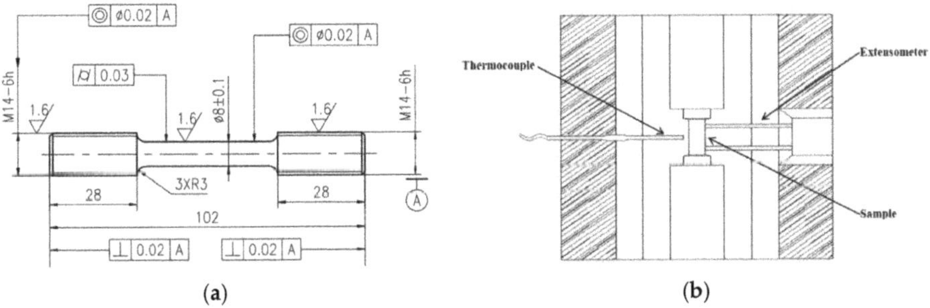

Figure 1. Experimental sample and device, (**a**) shape and dimensions of fatigue test specimens (dimensions in mm), and (**b**) installation of extensometers and thermocouples.

Saturated salt solutions were prepared according to the set mass ratio evenly coated on the sample surface using a small brush. In order to volatilize the water, samples were preheated in a furnace at 100 °C. After several minutes, a salt layer of a certain thickness was adhered to the surface of the samples. This process was repeated until a certain thickness of salt layer was uniformly attached. The fracture surfaces of the fatigue samples were observed by Quanta FEG 250 SEM using a voltage of 20 kV.

3. Results and Discussion

3.1. Effect of Total Strain Amplitudes

In order to identify the effect of total strain amplitudes on the fatigue property of 12Cr1MoV steel, low-cycle fatigue experiments with different total strain amplitudes under a mixed chloride salt environment, a mixed sulfate salt environment, and an air condition were carried out at 600 °C. Figure 2 shows the cyclic stress as a function of cyclic numbers. As seen in Figure 2, the effect of total strain amplitude on the cyclic stress response of the alloy is approximately the same under three different deformation conditions. For all samples, with the increase of total strain amplitude, the cyclic stress values increased and the fatigue life decreased significantly. This is a very normal and reasonable result. On the other hand, with the increase of the cyclic numbers, all of the samples presented an increment of the cyclic stress, especially at the strain amplitudes of ±0.3% and ±0.2%. Therefore, it can be concluded that the alloyed steel mainly exhibited cyclic hardening during the loading. This phenomenon is consistent with the data calculated from the uni-axial tensile test, as shown in Table 3. Generally, cyclic hardening and softening of material can be determined by the data of the uni-axial tensile test. When $\sigma_b/\sigma_{0.2} > 1.4$, the material exhibits cyclic hardening; when $\sigma_b/\sigma_{0.2} < 1.2$, the material exhibits cyclic softening; and when $1.4 > \sigma_b/\sigma_{0.2} > 1.2$, it is impossible to determine whether the material is cyclic hardening or cyclic softening [12]. Based on the data of the tensile test at 600 °C, it can be calculated that: $\sigma_b/\sigma_{0.2} = 544/311 = 1.749 > 1.4$. According to the calculation, cyclic hardening should occur during the fatigue test at 600 °C for the 12Cr1MoV steel, which is consistent with the curves drawn from the experiments.

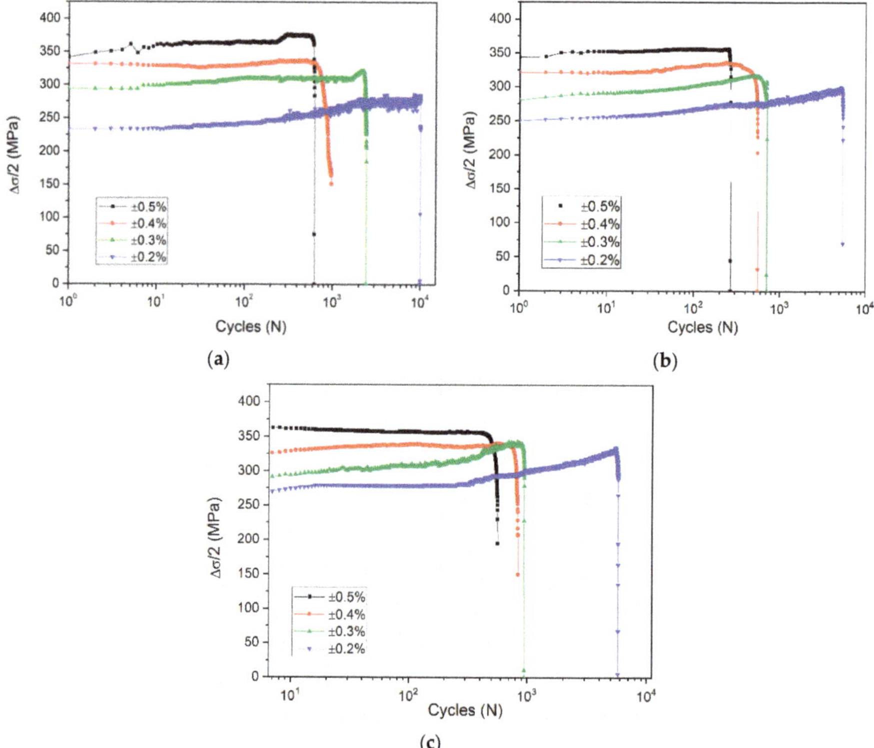

Figure 2. Cyclic stress response of the steel at 600 °C (**a**) air environment, (**b**) mixed chloride salt environment, and (**c**) mixed sulfate salt environment.

Table 3. Tensile mechanical properties of 12Cr1MoV steel.

Temperature (°C)	σ_b (MPa)	$\sigma_{0.2}$ (MPa)	A (%)
Room temperature	715 ± 32	521 ± 24	21.5 ± 1.1
600	544 ± 22	311 ± 17	42.7 ± 2.3

The Holomon expression is used to depict the relationship between the magnitude of the stress and the amplitude of the plastic strain [13], as shown in Equation (1):

$$\frac{\Delta\sigma}{2} = K'\left(\frac{\Delta\varepsilon_p}{2}\right)^{n'} \tag{1}$$

where K' is the cyclic strength coefficient, and n' is the strain hardening exponent. After taking logarithms on both sides, the relationship between the total strain amplitude and stress of the samples under different deformation environments was plotted in Figure 3. In this figure, Type A represents mixed alkali metal salt samples, and Type B represents mixed sulfate samples. It can be seen that for all of the samples, the three fitting lines are almost parallel to each other. The n' value is the slope of a straight line, and the K' value is the intercept of a straight line on the longitudinal axis. It can be concluded the values of the cyclic strength coefficient K' and the strain hardening exponent n' in the three environments are almost the same. After extensive research on high-strength materials commonly used in industry, Landgra [14] proposed that the strain hardening exponent n' can be used to evaluate the effect of cyclic strain on material properties. When $n' < 0.1$, the material behaves as cyclic softening, when $n' > 0.1$, the material exhibits cyclic hardening or cycle stability. After fitting, the strain hardening exponent is obtained, as shown in Table 4, with $n'_{air} = 0.1083 > 0.1$, $n'_{mcs} = 0.1102 > 0.1$, $n'_{mss} = 0.1107 > 0.1$. According to this standard, 12Cr1MoV steel mainly exhibits cyclic hardening or cycle stability, which is consistent with the results of the monotonic tensile determination and the cyclic stress response determination. In addition, all of the samples show a cyclic softening before the final fracture, with the cyclic stress decreasing rapidly during the last several cycles. The reason is that the fatigue crack becomes unstable and propagates rapidly and fractures eventually after the nucleation and coalescence.

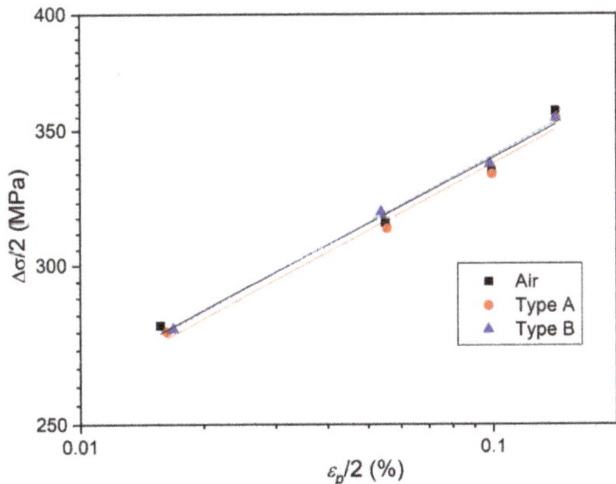

Figure 3. Cyclic stress–strain relationships of samples under different deformation environments.

Table 4. Low-cycle fatigue characteristics parameters K' and n'.

Parameter	Air	Mixed Chloride Salt	Mixed Sulfate Salt
K'	435	434	438
n'	0.1083	0.1102	0.1107

3.2. Effect of Deformation Environments

The cyclic stress response of alloyed steel under different loading environments is shown in Figure 4. It can be seen that different deformation environments have a great influence on the mechanical properties of the specimens. Generally speaking, the fatigue properties of samples in the air environment are the best, which is most obvious when the total strain amplitude is ±0.3%. The fatigue life of samples in mixed alkali metal salts is the shortest. The fatigue life (loading cycles) of all samples are listed in Table 5. It can be clearly seen that the differences in fatigue life caused by environments are different under various total strain amplitudes. When the total strain amplitude is ±0.2%, the fatigue life of the alloy under a mixed alkali metal salt environment is almost the same as that under a mixed sulfate environment, but it is lower than that of the sample under an air environment. When the total strain amplitude is ±0.3%, the fatigue life of the samples under the three environments has the greatest distinction. When the total strain amplitude is ±0.4% and ±0.5%, the effect of mixed sulfate on the fatigue properties of samples is very limited.

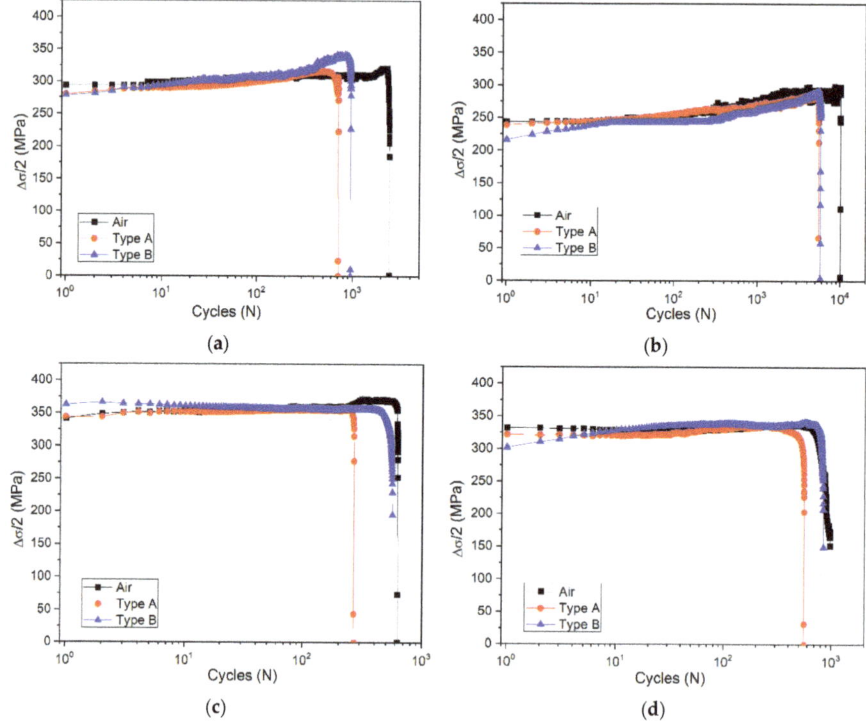

Figure 4. Cyclic stress response of alloyed steel in different environments (**a**) ±0.2%, (**b**) ±0.3%, (**c**) ±0.4%, and (**d**) ±0.5%.

Table 5. Fatigue life of specimens in the three environments (cycles).

Strain Amplitudes	Air	Mixed Alkali Metal Chloride Salt	Mixed Sulfate Salt
±0.5%	627 ± 33	271 ± 15	562 ± 22
±0.4%	978 ± 46	559 ± 21	840 ± 39
±0.3%	2444 ± 128	713 ± 37	951 ± 47
±0.2%	10,053 ± 484	5524 ± 278	5778 ± 282

In low-cycle fatigue tests, the plastic fatigue strain is relatively high, therefore the life–stress curve (N–S) cannot be used to describe the fatigue property of the material, and the strain fatigue curve is applied. Coffin and Manson proposed a fatigue life description method in which plastic strain amplitude was taken as a key parameter. The formula is given below:

$$\Delta\varepsilon_p/2 = \varepsilon_f'(2N_f)^c \tag{2}$$

where $\Delta\varepsilon_p/2$ is the plastic strain, ε_f' is the fatigue ductility coefficient, and c is the fatigue ductility exponent. According to Equation (2), the plastic strain fatigue life of three different steel samples is fitted by using the logarithmically processed Coffin–Manson relationship, as shown in Figure 5. It can be clearly seen that there is a linear relationship between the plastic deformation and the service life, in both air and mixed salt environments. The c value is the slope of the fitting line, and the value ε_f' is the intercept of the line on the longitudinal axis. After fitting, ε_f' and c values are obtained as listed in Table 6. The fatigue ductility coefficient ε_f' represents the fatigue resistance of materials, with a higher value of ε_f' indicating a better fatigue resistance of materials. Additionally, according to the Coffin–Manson equation, the lower the absolute value of fatigue ductility index c, the longer the fatigue life. The values of fatigue ductility coefficient and fatigue ductility exponent obtained by the data fitting, show that the fatigue resistance of 12Cr1MoV steel in two mixed salt environments are lower than that of an air environment.

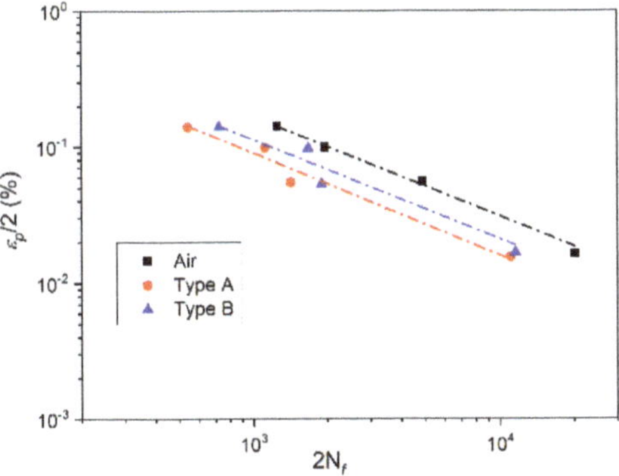

Figure 5. Relationship between plastic deformation and fatigue life of steel tested in different environments.

Table 6. Low-cycle fatigue characteristics parameters ε_f' and c.

Parameter	Air	Mixed Chloride Salt	Mixed Sulfate Salt
ε_f'	26.72	16.43	17.76
c	−0.73413	−0.75219	−0.73448

3.3. Fracture Surface Observation

Figure 6 shows the fatigue fracture morphology of the alloyed steel in different environments with the total strain amplitude of +0.2% and +0.5% respectively. All fracture surfaces can be clearly divided into either a fatigue crack initiation zone, a fatigue crack propagation zone, or an instantaneous fracture zone, which are marked by A, B, and C in figures respectively. It can be seen that in region A, there are many crack sources located near the surface. Additionally, the number of crack sources is the least in the air environment (Figure 6a,b), and the number of crack sources is the most in the mixed alkali metal chloride environment (Figure 6c,d). This indicates that under a mixed salt environment, cracks are easier to form due to the combined action of corrosion and cyclic stress. In the same environment, the higher the total strain amplitude, the more the crack sources. In the fatigue crack propagation area (B zone), there are obvious cowrie patterns, especially under low load conditions (Figure 6b,d,f). These cowrie patterns look like a group of arcs centered on the fatigue source, with the concave side pointing to the fatigue source area, and the convex side pointing to the direction of crack propagation. In places near the fatigue source, the crack propagation is slow and the cowrie patterns are dense. In places far away from the fatigue source area, the cowrie patterns are sparse and the fatigue crack propagation is fast. Under the same conditions, the higher the total strain amplitude, the smaller the fatigue crack growth zone and the less cowrie patterns. With an increase of loading cycles, the cracks increase continuously. When the critical length is reached, the stress intensity factor is higher than the fracture toughness of the material, and the cracks expand rapidly, leading to an instantaneous fracture. Furthermore, it can be seen that shear lips exists in the instantaneous fracture zone, and the shear lip surface is at a certain angle with the loading direction, indicating that the instantaneous fracture is mainly caused by shear stress.

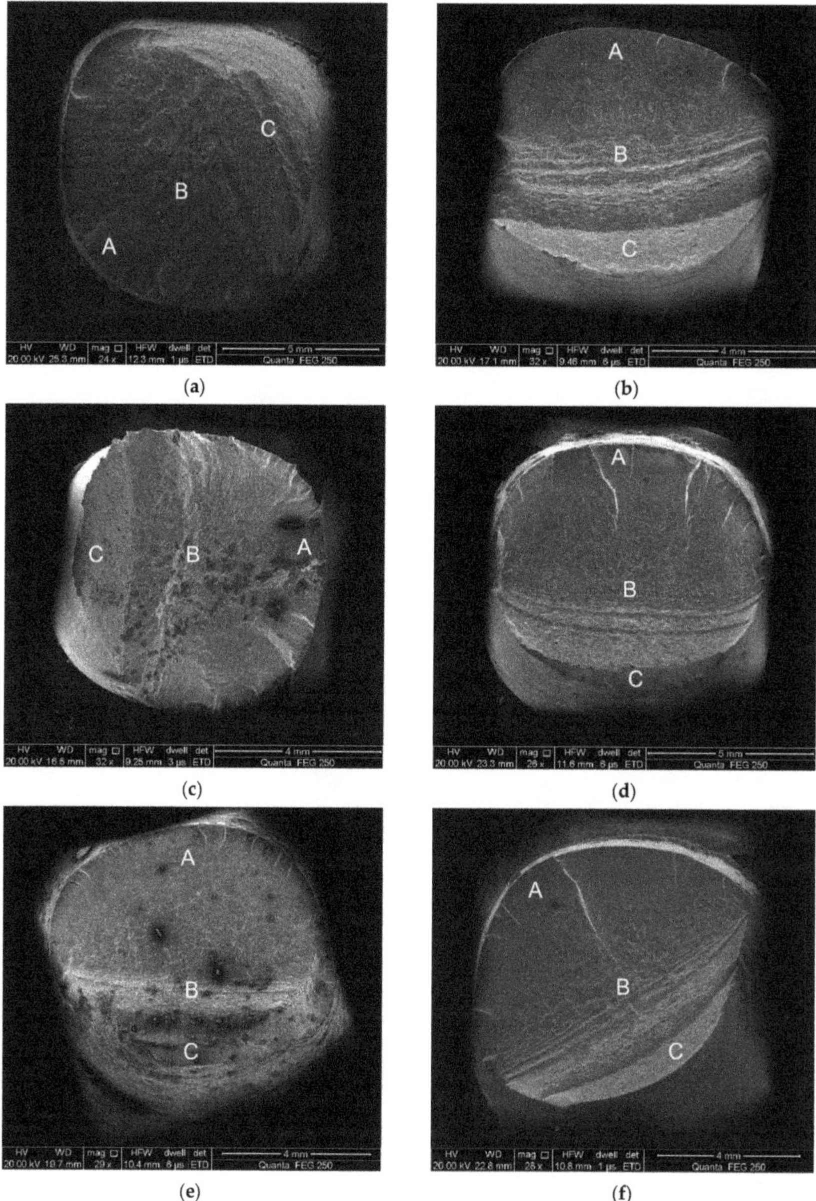

Figure 6. Overview of the fatigue fracture surface (**a**) ±0.5% (air), (**b**) ±0.2% (air), (**c**) ±0.5% (mixed chloride salt), (**d**) ±0.2% (mixed chloride salt), (**e**) ±0.5% (mixed sulfate salt), and (**f**) ±0.2% (mixed sulfate salt).

Figure 7 shows the detailed morphology of the fatigue crack propagation region. It can be seen that the spacing distance between fatigue striations increases with an increase of strain amplitude. This is because the higher the strain amplitude is, the faster the crack propagation is, subsequently resulting in a shortening of the macroscopic fatigue life. The crack growth region are not as flat as the crack nucleation area, instead the surface is rough and there are many parallel striations on it.

During this stage, the stress peaks and valleys are relatively stable, and the cracks propagate at a certain rate. Usually the area of the crack propagation region is very large, which consumes most of the fatigue life of the specimens. When comparing the morphology of crack propagation area between an air environment and a mixed chloride salt environment under the same strain amplitude, it can be seen that the spacing distance of fatigue striation in a mixed chloride salt environment is larger than that in an air environment, and that the depth of cracks in the mixed chloride salt environment are deeper, as shown in Figure 7a–d. This result indicates that the crack growth rate in a mixed chloride salt environment is greater than that in an air environment. The observation of fracture surface shows that a mixed alkali metal chloride environment can be harmful to the fatigue crack resistance of this alloy. Figure 7e,f are images of the fatigue crack propagation region of alloyed steel in a mixed sulfate environment. Its characteristics lie between the air environment and the mixed alkali metal salt environment.

Figure 8 is the typical dimple morphology of the instantaneous fracture region. It can be seen that the surface of the instantaneous fracture zone is rough, and there are many holes and voids with some inclusions. When the interface of inclusions is plastically deformed, it is easy to cause stress concentration, and then cracks are generated and propagate continuously. Subsequently, the matrix between the inclusions forms an "internal plastic neck", which is torn or sheared when the internal plastic neck reaches a certain extent, thus connecting the voids. Finally, forming the observed dimple fracture morphology.

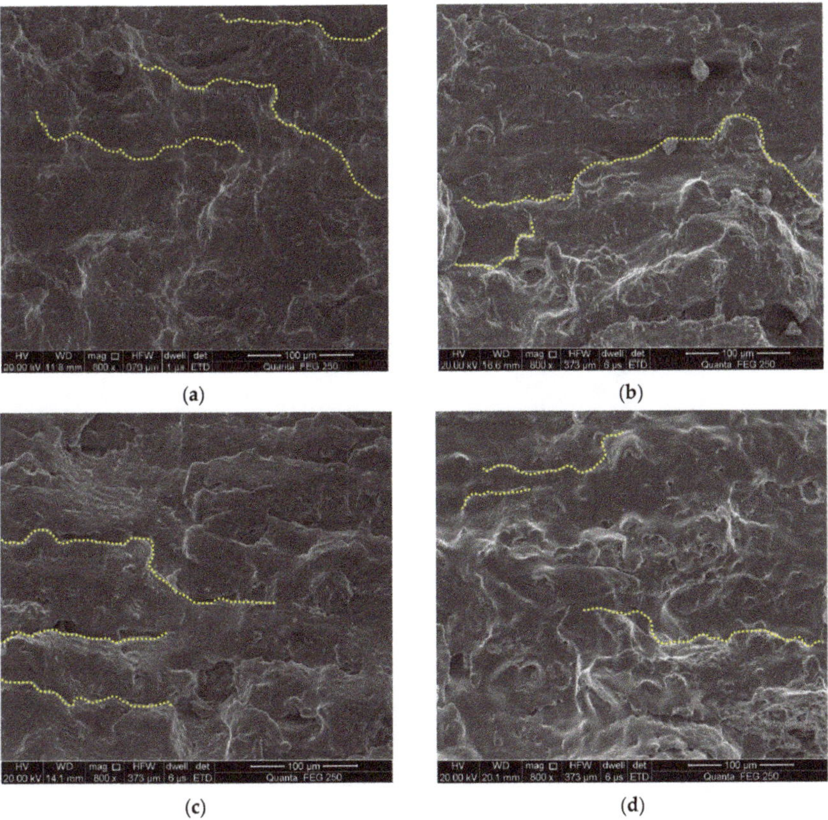

(a)

(b)

(c)

(d)

Figure 7. *Cont.*

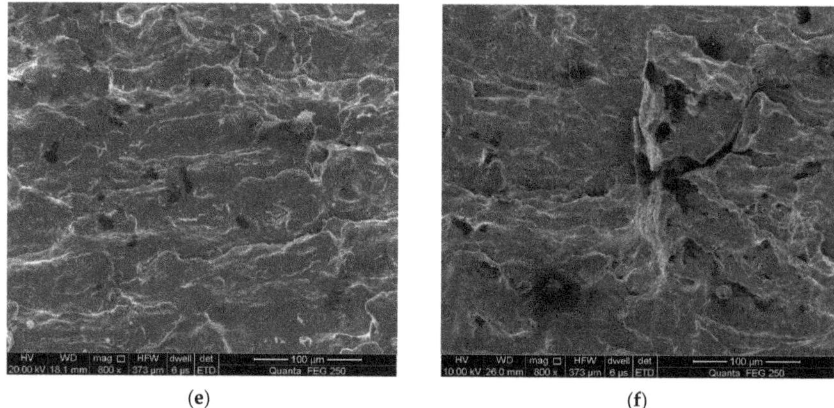

(e) (f)

Figure 7. The fatigue crack propagation region (**a**) ±0.5% (air), (**b**) ±0.2% (air), (**c**) ±0.5% (mixed chloride salt), (**d**) ±0.2% (mixed chloride salt), (**e**) ±0.5% (mixed sulfate salt), and (**f**) ±0.2% (mixed sulfate salt).

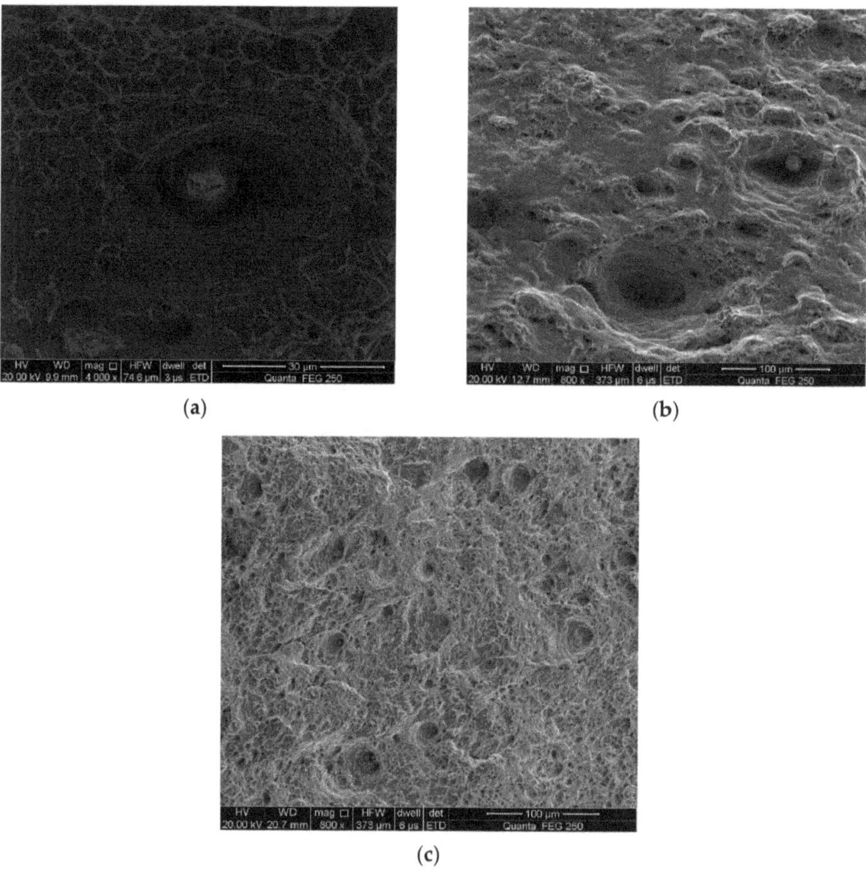

(a) (b)

(c)

Figure 8. The fatigue fracture region of the alloy with strain amplitude of ±0.5% (**a**) air, (**b**) mixed chloride salt, and (**c**) mixed sulfate salt.

From the above analysis, it can be concluded that the surface and matrix of 12Cr1MoV steel will be corroded seriously under a high temperature mixed molten chloride environment. With the increase of KCl content in alkali metal molten salt, the structure of corrosion products on the surface of 12Cr1MoV steel will change from round granular to flocculent state. The corrosion rate and thickness of corrosion products increase obviously. The corrosion mechanism of test steel in molten salt is the activation and oxidation behavior of chlorine element. The corrosion products of a section are mostly in porous distribution, and the pore size is larger when Cl element content is higher. The thickness of the corrosion inner layer increases further with an increase of temperature. When corrosion reaction continues, it is easy to change the material quality from weight gain to weight loss by separating some corrosion products from matrix. Under the action of high temperature thermal stress, stress concentration easily occurs near the hole. This results in the initiation and instability propagation of corrosive cracks, further affecting the strength and properties of materials. In a sulfate environment, such corrosion rate is lower, therefore in a fatigue test we can observe that the fatigue property of the samples in chloride environment is the worst.

4. Conclusions

The corrosion fatigue properties of 12Cr1MoV steel under different amplitudes and different environments were studied. The following conclusions can be drawn:

1. The effect of total strain amplitude on the cyclic stress response of the alloy is approximately the same under three different deformation conditions. For all samples, with an increase of total strain amplitude, the cyclic stress values of alloyed steel increased and the fatigue life decreased significantly.
2. With the increase of the cyclic number, all of the samples presented an increment of the cyclic stress, especially at the strain amplitudes of ±0.3% and ±0.2%. The alloyed steel mainly exhibited cyclic hardening during the loading.
3. The fatigue properties of samples in an air environment are the best, which is most obvious when the total strain amplitude is ±0.3%. The fatigue life of samples in a mixed alkali metal salts is the shortest.
4. Total strain amplitudes and loading environments have great influence on the fatigue fracture morphology of the alloyed steel.

Author Contributions: Data curation, J.B.; investigation, G.W.; methodology, K.L.; writing—original draft, J.H.; writing—review and editing, C.L.

Funding: This research was funded by National Natural Science Foundation of China, grant number 51275058.

Conflicts of Interest: The authors declare no conflict of interest.

References

1. Zhang, H.; Zhang, F.; Su, W.; Jiang, Y.; Li, Y. Research and evaluation of T91 superheater material for high temperature corrosion in biomass power plants. *Anti-Corros. Methods Mater.* **2015**, *62*, 133–135. [CrossRef]
2. He, J.; Xiong, W.; Zhang, W.; Li, W.; Long, K. Study on the high temperature corrosion behavior of superheater steels of biomass fired boiler in molten alkali salts' mixtures. *Adv. Mech. Eng.* **2016**, *8*. [CrossRef]
3. Retschitzegger, S.; Gruber, T.; Brunner, T.; Obernberger, I. Short term online corrosion measurements in biomass fired boilers. Part 2: Investigation of the corrosion behavior of three selected superheater steels for two biomass fuels. *Fuel Process. Technol.* **2016**, *142*, 59–70. [CrossRef]
4. He, J.; Xiong, W. Effect of high temperature hot corrosion on the compression ceep behabvior of 12Cr1MoV alloys. *High Temp. Mater. Process.* **2017**, *36*, 1011–1023. [CrossRef]
5. Zhu, M.L.; Xuan, F.Z.; Chen, J. Influence of microstructure and microdefects on long term fatigue behavior of a Cr-Mo-V steel. *Mater. Sci. Eng. A* **2012**, *546*, 90–96. [CrossRef]
6. Zhu, Z.; Jiao, X.; Tang, X.; Lu, H. Effect of SO_4^{2-} concentration on corrosion behavior of carbon steel. *Anti-Corros. Methods Mater.* **2015**, *62*, 322–326. [CrossRef]

7.   Chauhan, A.; Hoffmann, J.; Litvinov, D.; Aktaa, J. Hight-temperatuer low cycle fatigue behavior of a 9Cr-ODS steel: Part 2-hold time influence, microstructural evolution and damage characteristics. *Mater. Sci. Eng. A* **2018**, *730*, 197–206. [CrossRef]

8.   Xiong, S.; Zhu, Z.; Jing, L. Influence of Cl-ions on the pitting corrosion of bolier water wall tube and its principle. *Anti-Corros. Methods Mater.* **2012**, *59*, 3–9. [CrossRef]

9.   Liu, S.; Liu, Z.; Wang, Y.; Tang, J. A comparative study on the high temperature corrosion of TP347H stainless steel, C22 alloy and laser-cladding C22 coating in molten chloride salts. *Corros. Sci.* **2014**, *83*, 396. [CrossRef]

10.   Zhang, T.B.; Dong, R.F.; Rui, H.; Kou, H.C.; Li, J.S. Hot corrosion characteristics of Ni-20Cr-18W superalloy in molten salt. *Trans. Nonferr. Met. Soc. China* **2015**, *25*, 3840–3846. [CrossRef]

11.   Tsaur, C.C.; Rock, J.C.; Wang, C.J.; Su, Y.H. The hot corrosion of 310 stainless steel with precoated $NaCl/Na_2SO_4$ mixtures at 750 °C. *Mater. Chem. Phys.* **2005**, *89*, 445–453. [CrossRef]

12.   Halford, G.; Hirschberg, M.; Manson, S. *Creep-Fatigue Analysis by Strain Range Partitioning*; NASA Lewis Research Center: Cleveland, OH, USA, 1971.

13.   Zenr, C.; Holloman, J. Effect of strain rate upon the plastic flow of steel. *J. Appl. Phys.* **1944**, *15*, 22. [CrossRef]

14.   Seifert, T.; Riedel, H. Mechanism-based thermomechanical fatigue life prediction of cast iron. Part 1: Models. *Int. J. Fatigue* **2010**, *32*, 1358–1367. [CrossRef]

 © 2019 by the authors. Licensee MDPI, Basel, Switzerland. This article is an open access article distributed under the terms and conditions of the Creative Commons Attribution (CC BY) license (http://creativecommons.org/licenses/by/4.0/).

Article

Influence of Residual Stress on Fatigue Weak Areas and Simulation Analysis on Fatigue Properties Based on Continuous Performance of FSW Joints

Guoqin Sun [1],*, Xinhai Wei [1], Jiangpei Niu [2], Deguang Shang [1] and Shujun Chen [1]

[1] College of Mechanical Engineering and Applied Electronics Technology, Beijing University of Technology, Beijing 100124, China; 18810816630@163.com (X.W.); shangdg@bjut.edu.cn (D.S.); sjchen@bjut.edu.cn (S.C.)
[2] Production R&D center, CRRC Tangshan Co. Ltd., No.3 Changqian Rd, Tangshan 064000, China; sjc-niujiangpei@tangche.com
* Correspondence: sguoq@bjut.edu.cn; Tel.: +86-010-6739-6888

Received: 1 February 2019; Accepted: 26 February 2019; Published: 2 March 2019

Abstract: The fatigue weak area of aluminum alloy for a friction stir-welded joint is investigated based on the hardness profile, the residual stress measurement and the simulation analysis of fatigue property. The maximum residual stresses appeared at the heat-affected zone of the joint in the fatigue damage process, which was consistent with the fracture location of the fatigue specimen. The fatigue joint model of continuous performance is established ignoring the original negative residual stress; considering that it will be relaxed soon when the joint is under tension-tension cyclic loading. The fatigue parameters of joint model is based on the static mechanical properties of the joint that obtained from the micro-tensile tests and four-point correlation method. The predicted results for the fatigue weak locations and fatigue lives based on the continuous performance joint model are closer to the fatigue experimental results by comparison with the simulation results of the partitioned performance joint model.

Keywords: friction stir welding; residual stress; weak area; finite element simulation; life prediction

---

1. Introduction

Friction stir welding (FSW) of aluminum alloy has been widely used in the automotive, aerospace and ship industries. The anti-fatigue design is worthy of attention for the load-carrying FSW components. The influencing factors on the failure location of the joints are needed to be focused on. The hardness distribution of the friction stir-welded joints is related to the materials and welding process. It is considered that it has a relation to the fatigue fracture position of alloy joints [1–6]. But some studies showed that fatigue cracks were independent from hardness distributions in the weld seams and the fatigue cracks initiated at the inhomogeneous microstructure [7]. Fatigue fracture of the large plate happened at the lowest hardness location in the heat affect zone (HAZ) [8]. But the tensile strength of the FSW joint has a linear relationship with the weld nugget hardness [9]. The fatigue properties of aluminum alloy FSW joints are also affected by the residual stress [2–4]. The residual stresses of aluminum alloy welded joints are dependent on the welding parameters and their distribution has a typical "M" profile [10–13]. The locations of the maximum residual tensile stress are different for different alloy joints. The compressive residual stress delays the crack growth and increases the fatigue life of the welded joint and the tensile residual stress accelerates the propagation of the crack [14–17]. Fratini et al. [4] found that the residual stress had an influence on the crack propagation of base metal area and had no obvious effect on the welding area. The effect of residual stress on the joint would weaken with the increase of fatigue cycle or the crack length [18,19].

Finite element numerical simulation has been applied on the evaluation of mechanical properties for FSW joints. In addition, the residual stress can be preloaded into the joint model as a prestress if it is non-negligible [20]. Rao and Simar [21,22] established the finite element model of the FSW joint through the partition method to obtain the tensile properties. The friction stir welding joints were divided into different regions to simulate the mechanical properties of the weak regions [23,24].

The simulation method with the partitioned performance joint model has the stress concentration at the regional boundary positions because of the discontinuity of material properties. The continuous performance joint model does not have the effect of the area partition and obtains more accurate stress–strain data. Therefore, the fatigue finite element model of continuous performance for the FSW joint is established to simulate the weak areas and the stress–strain response. The life prediction is proceeded based on the simulation results and a comparison between the continuous performance joint model and the partitioned performance joint model is carried out. Also, the relationship between the fatigue weak areas of 2219-T6 aluminum alloy friction stir-welded joints and mechanical properties is analyzed with experimental.

2. Experimental

2.1. Materials and Specimens

The test specimens is cut from 2219-T6 aluminum alloy FSW butt plate of 800 mm × 300 mm × 6 mm size. The chemical compositions and mechanical properties of base metal are given in Tables 1 and 2. The alloy plates were welded perpendicular to the rolling direction with a rotation speed of 800 rpm and an advancing speed of 180 mm/min. The FSW tool consists of a shoulder with a diameter of 18 mm and a tool pin of 5.7 mm in length and 6 mm in diameter. The tool axis is tilted by 2° with respect to the vertical axis of the plate surface. The welding was finished in the China Academy of Launch Vehicle Technology. The fatigue samples were wire cut from the plate, which its axial direction was perpendicular to the weld line. The surfaces and sides of the specimens were ground with silicon carbide sandpaper from 150 to 2000 grit and the surfaces were polished with diamond polishing pastes from 4000 to 10,000 grit. The thickness of the fatigue specimen is 6 mm and the dimensions is shown in Figure 1.

Table 1. Chemical composition of 2219 aluminum alloy.

Chemical Composition (wt%)								
Cu	Mn	Fe	Zn	Si	Zr	Ti	V	Al
6.48	0.32	0.23	0.04	0.49	0.2	0.06	0.08	balance

Table 2. Mechanical properties of 2219-T6 aluminum alloy.

Material	Elasticity Modulus (GPa)	Ultimate Strength σ_u (MPa)	Yield Strength $\sigma_{p0.2}$ (MPa)
2219-T6	72	416	315

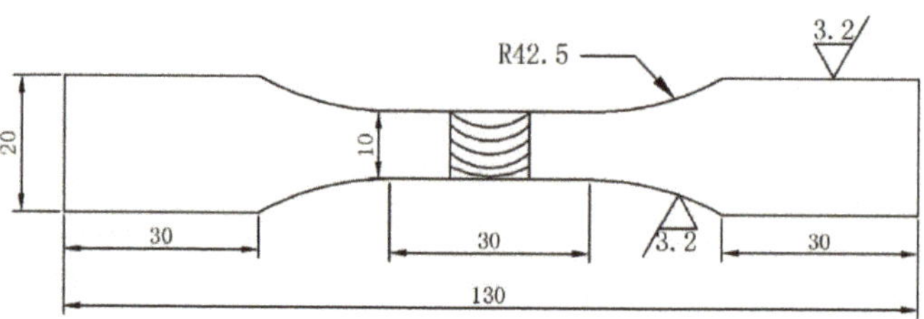

Figure 1. Shape and size of fatigue specimen.

2.2. Metallographic Morphology

The cross-sectional metallographic morphology of the specimen was observed with the optical microscope (Shanghai optical instrument factory, Shanghai, China) after the specimen was etched with keller's reagent consisting of 2.5 mL HNO_3, 1.5 mL HCL, 1 mL HF and 95 mL H_2O for about 15 s. According to the metallographic morphology of the welded joint in Figure 2 and the sizes of the grain, the welded joint is divided into different regions: the weld nugget zone (WNZ), the thermo-mechanically affected zone (TMAZ), the HAZ and the base material (BM). The HAZ is further divided into high-hardness heat affect zone (HHAZ) and low-hardness heat affect zone (LHAZ) based on the hardness distribution and the phase sizes. The phase in the HHAZ has the similar size as that in the BM, while the hardness in HHAZ is lower than that of BM. In addition, according to the rotation direction of the welding tool, the joint is divided into the advancing side (AS) and the retreating side (RS). The dimensions of the different zones that marked in the hardness profiles can be obtained by combining the metallographic morphology, phase size and the hardness profile.

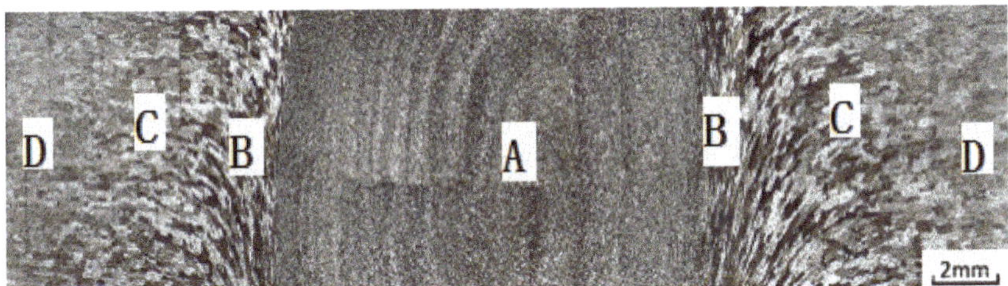

Figure 2. Metallograph of the aluminum alloy FSW joints: (**A**) WNZ, (**B**) TMAZ, (**C**) HAZ, (**D**) BM.

2.3. Hardness Measurements

The microhardness distributions were measured on the upper and the lower surfaces of the welded joint with a load of 1.95 N for 15 s with a SCTMC DHV-1000Z Vickers hardness tester (Shangcai tester machine Co, Ltd, Shanghai, China). The measurement spacing between each two adjacent points was 0.5 mm.

2.4. Fatigue Experiments

All fatigue tests were carried out using MTS858 hydraulic servo system (MTS Systems Corporation, Minneapolis, MN, USA) under the axial stress range from 216 to 261 MPa with a stress ratio of 0.1 and a frequency of 10 Hz until the specimens fractured. The fatigue loading parameters and total fatigue lives are available from Ref. [25]. The curve of cyclic stress range versus fatigue lives for the FSW specimens is shown in Figure 3.

2.5. Residual Stress Measurements

The residual stresses of the original and the fatigue-damaged specimens were measured by D8 Discover X-ray diffraction meter (Bruker Corporation, Karlsruhe, Germany). The specimen was electrochemically polished to remove the effect of mechanical polishing before the residual stress measurement. The original residual stress distributions of the upper and the lower surfaces for the joint were measured before fatigue test. Then, the cyclic stress range of 216 MPa with a stress ratio of 0.1 was loaded on the joint and the fatigue test was stopped when the fatigue life was 30,000 cycles. Then, the surface residual stresses of the fatigue-damaged specimen were measured again to analyze the effect of the residual stress distribution in the fatigue process on fatigue damage.

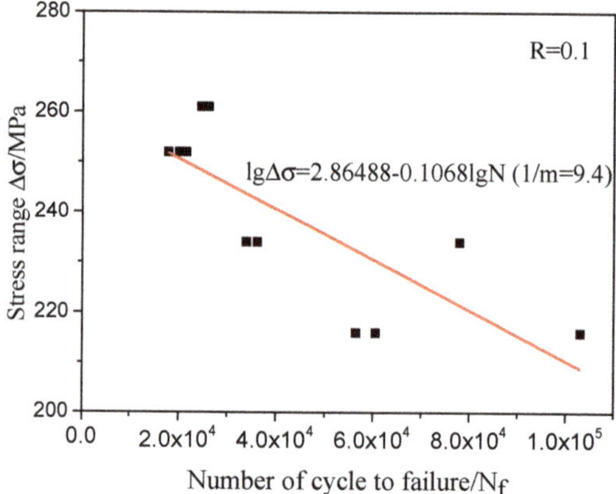

Figure 3. S-N curve of FSW specimens.

3. Experimental Results

3.1. Hardness Distribution

The upper and the lower surfaces hardness profiles of the welded joint are shown in Figure 4. The hardness distributions of the joint upper and lower surfaces present the approximate "W" shape. The hardness in the welded region is obviously lower than the base material. The minimum hardness values of the upper and the lower surfaces are both in the LHAZ according to the hardness distribution of the joints, which is not in coincidence with the failure position of HHAZ for most fatigue specimens. Hardness has a relation with the material strength but the fatigue weak area of the joint is not necessarily corresponding to the location of minimum hardness [7]. It has a relation with the variation of the hardness gradient, which corresponds to the variation of the mechanical property and heterogeneous microstructure.

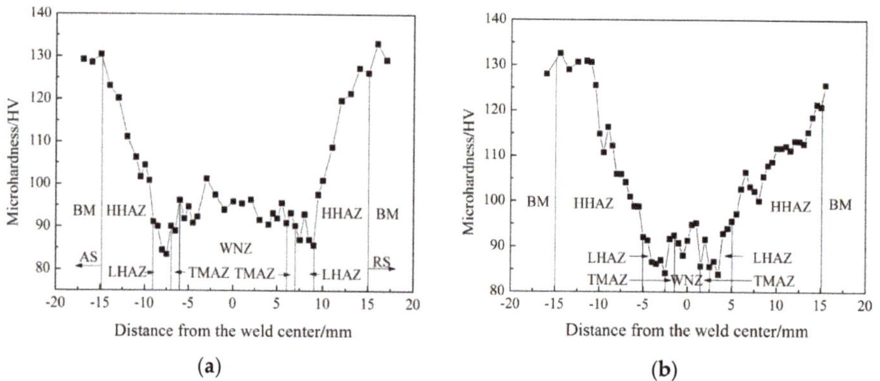

Figure 4. Hardness distributions of (a) upper and (b) lower surfaces in the joint.

3.2. Fatigue Experimental Results

The failure locations in the joints were observed with the optical microscope after the specimens fractured. First, the locations of the fatigue crack sources are needed to be found and observed. Then, the sides or the surfaces of the sample near the location of fatigue crack sources are etched with Keller's reagent. The failure locations can be affirmed through the metallographic morphologies of the samples.

The statistics of failure locations for 11 specimens are listed as follow: 10 specimens fractured in HHAZ, 1 specimen broke in WNZ.

3.3. Residual Stress Distribution

The residual stress distributions are shown in Figures 5 and 6. The transverse residual stresses are the stresses perpendicular to the weld and the longitudinal residual stresses are the stresses parallel to the weld direction. The center of the WNZ was taken as the zero coordinate and the test points were selected on both sides along the axial direction of the specimen. The measured residual stresses before fatigue test are called as the original residual stresses of the specimen in this paper. The residual stresses of the fatigue-damaged specimens were measured after the fatigue cycle reached 30,000 cycles under the cyclic stress range of 216 MPa with a stress ratio of 0.1.

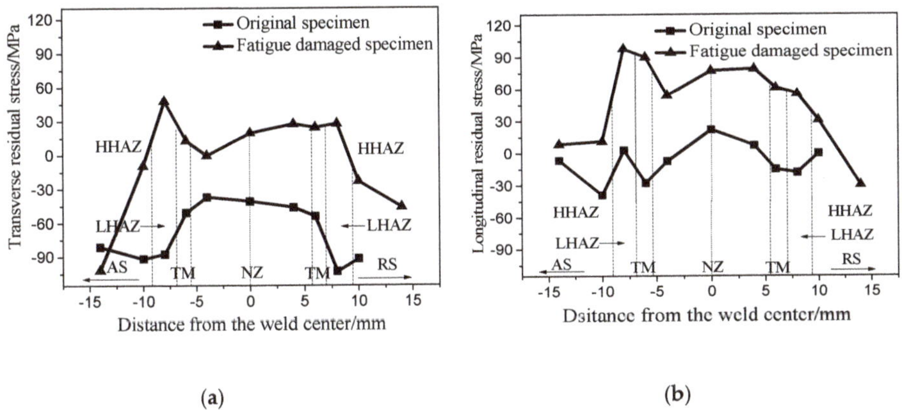

Figure 5. (**a**) Transverse and (**b**) longitudinal residual stress distributions of upper surface in the joint.

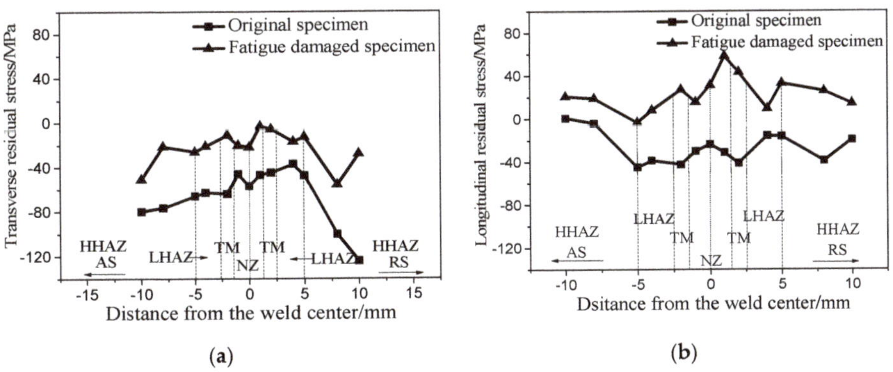

Figure 6. (**a**) Transverse and (**b**) longitudinal residual stress distributions of lower surface in the joint.

Figure 5 shows the variation of the transverse and longitudinal residual stress distribution curves in the upper surface of the welded joint. It can be seen that the maximums of the transverse and longitudinal residual stresses in the original welded joint appear in the WNZ and the values are basically negative or near zero. The original negative residual stresses in the welded joint is beneficial to the fatigue life. However, the original compress residual stress will be relaxed and the positive stress will appear soon once the tension-tension stress is loaded. The positive transverse and longitudinal residual stresses arose in the fatigue process and the maximum residual stress appeared at the LHAZ during the cyclic loading.

Figure 6 shows the variation of the transverse and longitudinal residual stress distribution curves in the lower surface of the welded joint. The residual stresses is negative in the original joint. The maximum stresses appeared near the boundary of NZ and TMAZ in the fatigue process and the values were far lower than the residual stresses in the upper surface. The residual stress of the upper surface in the joint that produced in the fatigue process should have more effect on the fatigue damage.

4. Fatigue Numerical Simulation

4.1. Fatigue Parameters of the Welded Joint

The fatigue finite element numerical analysis is further proceeded in order to obtain the stress–strain responses and evaluate the weak area of the joint with the simulation method. Since the original residual stress was negative and would be relaxed once the tension-tension stress was loaded. It did not considered to be added on the joint as a prestress in the numerical analysis under the tension-tension cyclic loading.

The fatigue parameters are different in different zones since the materials of different zones in the joint have different microstructures and mechanical properties. They can be used in some fatigue life prediction models to calculate fatigue lives or as the intermediate variables to calculate other cyclic strength parameters. The fatigue parameters of different positions in the welded joint were obtained by four-point correlation method proposed by Manson [26].

The four-point correlation method is based on the elastic and plastic strain-life lines. The respective two points on elastic and plastic lines are definite. On the elastic line, a point is at $1/4$ cycle with a strain range of $2.5\,\sigma_f/E$, where σ_f is the fracture strength, E is the elastic modulus. The other point is at 10^5 cycles with a strain range of $0.9\,\sigma_u/E$, where σ_u is the ultimate tensile strength. On the plastic line, a point locates at 10 cycles with a strain range of $1/4D^{3/4}$, where D is the logarithmic ductility of the material. Another point is at 10^4 cycles with a strain range of $\frac{0.0132-\Delta\varepsilon_e^*}{1.91}$, $\Delta\varepsilon_e^*$ indicates the value of elastic strain range at 10^4 cycles on the elastic line, as shown in Figure 7.

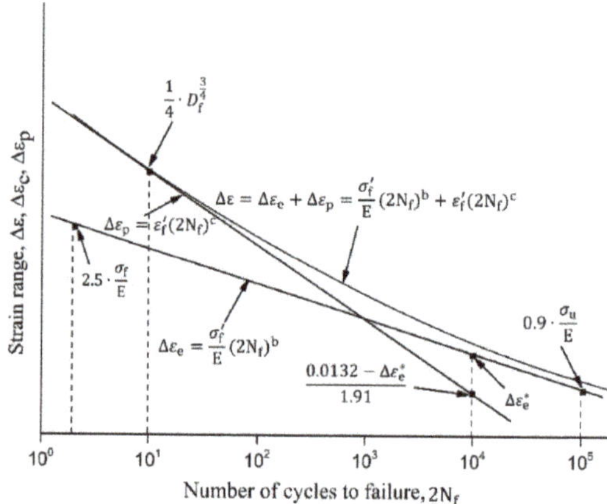

Figure 7. Four-point correlation method by Manson.

The static mechanical properties of different positions in the joint were obtained from the micro-tensile tests. Microspecimens from different regions were cut parallel to the weld direction of welded joints. The static mechanical properties of each zone were measured by using Instron 5948 Microtester (Instron Corporation, Norwood, MA, USA). The calculated fatigue parameters of different positions in the welded joint are listed in Table 3.

Table 3. Fatigue parameters of the welded joint.

Cutting Cites of the Micro Specimens X (mm)	Elastic Modulus E (GPa)	Fatigue Strength Coefficient σ'_f (MPa)	Fatigue Strength Exponent b	Fatigue Ductility Exponent c	Fatigue Ductility Coefficient ε'_f	Cyclic Strength Coefficient K' (MPa)	Cyclic Strain Hardening Exponent n'
−11	64	364	−0.078	−0.33	0.041	781	0.24
−7	53	275	−0.079	−0.31	0.037	635	0.25
−5	58	324	−0.077	−0.37	0.062	578	0.21
0	59	334	−0.078	−0.37	0.059	610	0.21
6	61	313	−0.076	−0.35	0.055	581	0.21
8	62	263	−0.058	−0.33	0.043	458	0.18
11	63	328	−0.076	−0.34	0.047	656	0.23

4.2. Continuous Performance Joint Model

The continuous performance joint model is established by inputting different elastic moduli and fatigue stress–strain data at different areas of the joint with ABAQUS software. The WNZ center of the joint is the zero coordinate of the X axis in the model. The different coordinates correspond to the different locations of the joint along the axial loading direction. The elastic moduli of different positions with the coordinate X that listed in Table 3 were obtained from the micro tension tests. The elastic moduli of the other positions in the joint can be obtained by interpolation according to the data in Table 3. The mechanical properties of the joint also vary with change of the coordinate. The materials in different zones of the joint have different stress–strain responses. The cyclic stress–strain at different locations of the joint were obtained using Ramberg–Osgood equation and applied to the finite element simulation. The Ramberg–Osgood equation is shown below.

$$\varepsilon_a = \frac{\sigma_a}{E} + \left(\frac{\sigma_a}{K'}\right)^{\frac{1}{n'}} \tag{1}$$

where ε_a is the strain amplitude, σ_a is the stress amplitude, K' is cyclic strength coefficient, n' is the cyclic strain hardening exponent. They can be obtained from the following equations.

$$K' = \frac{\sigma'_f}{\varepsilon'^{n'}_f} \tag{2}$$

$$n' = b/c \tag{3}$$

The cyclic yield strength was measured by the stress value of 0.2% plastic strain in the cyclic stress–strain curve. The nonlinear kinematic hardening model of the material attribute was adopted for the joint material.

User subroutine USDFLD in the ABAQUS software can be used to describe material properties. The material properties is defined as a function of field variables and the variables can be solved with USDFLD subroutine. The variation of elastic modulus, yield stress and plastic strain with the coordinate X were computed with the interpolation method using the known data of the specific coordinate through the subroutine.

The joint model is constrained at one end and loaded the cyclic stress range of 216 MPa with a stress ratio of 0.1 at another end of the joint in the X-axial direction. The loaded cyclic stress is schematically drawn in Figure 8. The hexahedron mesh with eight nodes element type of C3D8R is selected to solve the structure. The size of each element is approximately 1.3 mm × 0.8 mm × 1.2 mm. 3600 elements and 4758 nodes are obtained to simulate the FSW joints, as shown in Figure 9. The model does not include the clamping part and the cyclic stress as a surface load is directly applied to the section of the right end.

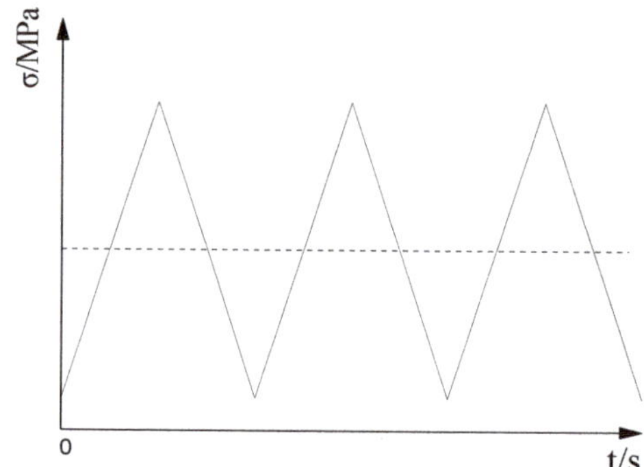

Figure 8. Schematic of loaded cyclic stress on the joint.

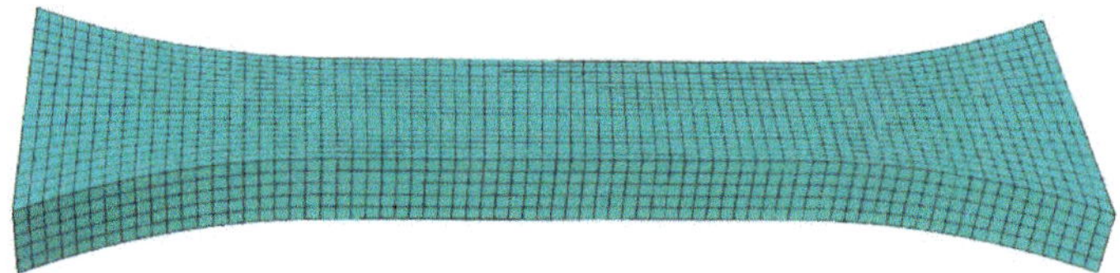

Figure 9. Meshed joint.

4.3. Partitioned Performance Joint Model

To compare the effectiveness of the continuous performance joint model for the fatigue performance simulation of the FSW joints, the partitioned performance joint model was established according to the metallurgical morphologies, hardness profile of the joints and the material attributes in different regions of the joint [25]. The joint is composed of the WNZ, TMAZ, HHAZ, LHAZ and BM, as shown in Figure 10. The fatigue parameters of different positions of the FSW joints were calculated according to the four-point correlation methods. The stress–strain data in different zones of the joint were obtained with Ramberg–Osgood equation. The material property does not have a change within each region.

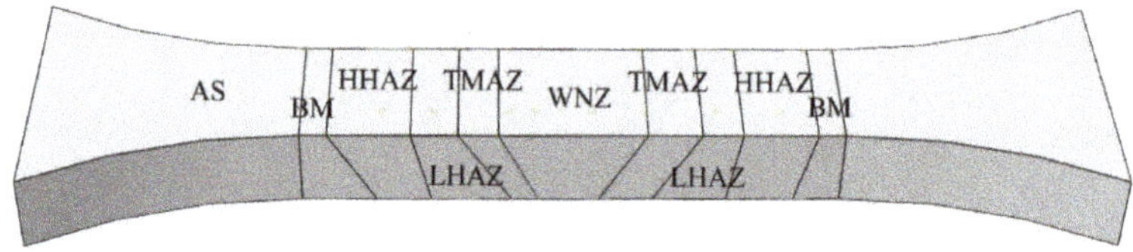

Figure 10. Partitioned performance joint model of the FSW joint.

5. Simulation Results

5.1. Stress Distribution

Von Mises stress distributions of continuous and partitioned performance joints are shown in Figure 11a,b. The stress distribution of the partitioned performance joint shows a significant mutation at the junction of different regions, especially at the junction of TMAZ and HAZ. The simulation results of continuous performance joints show the changes of stress and strain are smoother.

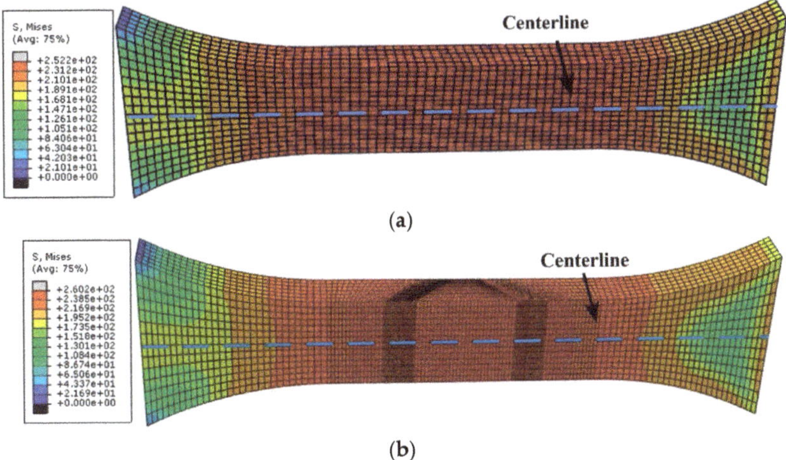

Figure 11. Von Mises stress contour plots of (**a**) continuous joint model and (**b**) partitioned joint model.

To observe the fatigue performance of the joint, the cyclic stress range of 216 MPa with the stress ratio of 0.1 was loaded on the joint. The maximum von Mises equivalent stress appeared in the upper surface of the joint in the partitioned performance joint model. The stress distribution trend of the upper and the lower surfaces in the continuous performance joint model is the same. The von Mises stress data of upper surface centerline were extracted from the simulation results of the continuous and the partitioned performance joint models. The maximum von Mises stresses both appeared in the HHAZ for the two models, as shown in Figure 12.

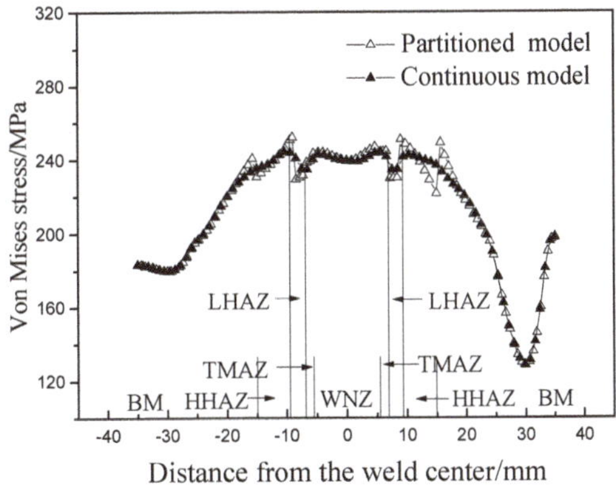

Figure 12. Von Mises stresses distributions of upper surface in the joint model.

The sudden change of stress in the partitioned performance joint model is obvious at the junction of adjacent regions in the joint. The distribution of von Mises equivalent stress obtained from the continuous performance joint model is relatively continuous for the joint, which is closer to the practical condition. The maximum stress appeared at the boundary of LHAZ and HHAZ for the partitioned performance joint model and it occurred at HHAZ near the LHAZ for the continuous performance joint model. The location of the maximum stress has a little difference for the two models.

The stress components of continuous and partitioned joint models were analyzed. There is no stress in the Z direction. The shear stresses are so small that their effects on the fatigue performance of the welded joint are ignored. It can be considered that the stresses in the X and the Y directions determine the total stress distribution of the FSW joints. The stresses in the X and the Y directions of upper surface in the models were extracted for analysis as an example, as shown in Figure 13. The stresses in the X direction are larger in the two models since the tension-tension stress is loaded in the X direction. The stress fluctuation appeared in the X and the Y directions of the partitioned performance joint model. The stress distribution of the continuous performance joint model has no obvious sudden change.

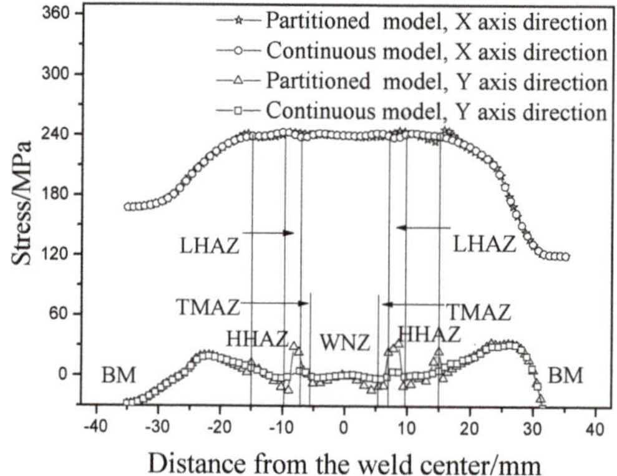

Figure 13. Stress distributions in the X and Y directions in upper surface of the joint model.

5.2. Strain Distribution

The weak area of joint fatigue performance can be determined by the stress and strain response of different zones in the joint from the simulation results of the fatigue property. The maximum principal strains at the centerline of upper surface, which extracted from the continuous and the partitioned performance joint models under the cyclic stress range of 216 MPa with a stress ratio of 0.1, are presented in Figure 14. Due to the abrupt change of material properties, the partitioned performance joint model caused a corresponding stress concentration at the junction of different regions and resulted in larger maximum principal strains than that of the continuous performance joint model. The larger the loaded cyclic stress in the model, the more obvious the abrupt change of strain distribution. The maximum principal strain is in the LHAZ near the TMAZ for the partitioned performance joint model and it appears in the HHAZ close to the LHAZ for the continuous performance joint model, which is approximately consistent with the fatigue failure position. The simulated fatigue weak area with the continuous performance joint model is more close to the fatigue experimental results.

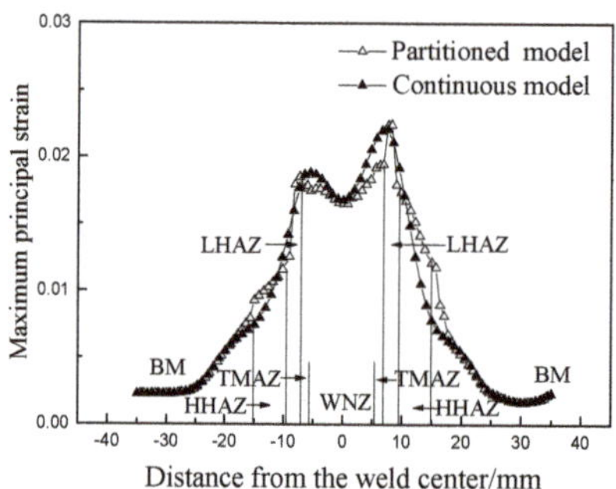

Figure 14. Maximum principal strain distributions of upper surface in the joint model.

By comparison with the partitioned performance joint model, the continuous performance joint model has a higher accuracy on predicting the fatigue weak areas of the FSW joint according to the stress and strain distributions. It eliminates the large stress concentration caused by the abrupt change of material properties in the adjacent area and more accurately simulates the fatigue performance of the joint.

6. Fatigue Life Prediction

The joint models established in this paper not only evaluate the fatigue weak areas and obtain the stress and strain values of the joint, but they also predict the fatigue lives of the joints. In this section, the stress and strain values of the weak areas obtained from the continuous and the partitioned performance joint models are used to predict the fatigue lives of the FSW joints. The validity of the continuous performance joint model is further verified by comparison with the experimental results.

Crack initiation lives were estimated with Smith–Watson–Topper (SWT) method, which considered the effect of average stress [27]. The maximum principal stresses and corresponding maximum principal strain ranges of the weak area of the joint were extracted from the simulation results of the joint models. Fatigue parameters of weak areas are used to predict the fatigue crack initiation lives of joints. The SWT formula is shown below.

$$\sigma_{\max}\frac{\Delta\varepsilon}{2} = \frac{\sigma_f'^2}{E}(2N)^{2b} + \sigma_f'\varepsilon_f'(2N)^{b+c} \tag{4}$$

where $\sigma_{\max}$ is the maximum principal stress, $\Delta\varepsilon$ is the maximum principal strain range, N is the fatigue life.

The experimental results in Ref. [28] showed that the size of fatigue crack initiation was defined as 1 mm and the crack initiation lives for the HAZ and the WNZ of aluminum alloy welded joints were 40.21% and 60.67% of the total fatigue lives, respectively. The research in Ref. [29] showed that the crack initiation lives accounted for 40–50% of total fatigue lives. Therefore, the crack initiation life is selected as 50% of the total fatigue life in this paper.

The life prediction results of FSW joints are shown in Figure 15. The predicted errors of the fatigue lives based on the continuous and the partitioned performance joint models are basically within the factor of two by comparison with the experimental results. The fatigue life prediction results with the simulation data obtained from the continuous performance joint model are closer to the experimental lives because the continuous performance joint model has not the stress and strain concentrations caused by the area partition. It is shown that the continuous performance joint model is suitable for life prediction of the FSW joints.

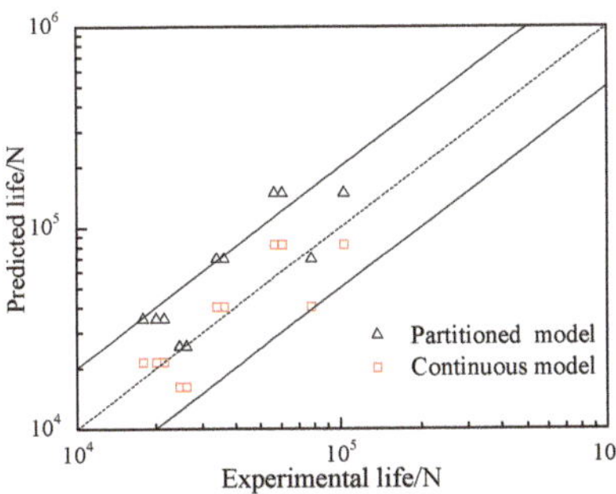

Figure 15. Life prediction of joint based on continuous and partitioned performance joint models.

7. Conclusions

(1) The original transverse and longitudinal residual stresses in the welded joint before fatigue are negative and the maximum tensile residual stress occurs in the HAZ during the tension–tension cyclic loading process.

(2) The fatigue parameters of different areas are obtained with four-point correlation method and the static mechanical property parameters of micro-tensile specimen at different locations of the joint. The continuous performance joint model is established by inputting different elastic moduli and fatigue stress–strain data at different locations of the FSW joint with user subroutine USDFLD. The stress components in the X and the Y directions determine the stress distribution of the FSW joints. The continuous performance joint model eliminates the stress and strain concentration caused by the area partition and more accurately simulates the fatigue performance of the joint by comparison with the partitioned performance joint model. The simulated fatigue weak area with the continuous performance joint model is more close to the fatigue experimental results.

(3) The fatigue life prediction of the FSW joints is proceeded with SWT method based on the simulation results of the continuous and the partitioned performance joint models. The results show that the predicted lives with the continuous performance joint model are closer to the experimental results and the life prediction error is within the factor of two.

Author Contributions: G.S. designed and performed the experiments; G.S., J.N. and X.W. analyzed the experimental data and simulated the joint model; D.S. and S.C. gave some advices for the research; G.S. and X.W. wrote the paper.

Funding: This research was founded by the National Natural Science Foundation of China (Grant No. 11672010, 51535001 and 51575012).

Conflicts of Interest: The authors declare no conflicst of interest.

Abbreviations

FSW	Friction stir welding
WNZ	Weld nugget zone
TMAZ	Thermo-mechanically affected zone
HAZ	Heat affect zone
BM	Base material
HHAZ	High-hardness heat affect zone
LHAZ	Low-hardness heat affect zone
AS	Advancing side
RS	Retreating side
SWT	Smith-Watson-Topper

References

1. Charitidis, C.A.; Dragatogiannis, D.A.; Koumoulos, E.P.; Kartsonakis, I.A. Residual stress and deformation mechanism of friction stir welded aluminum alloys by nanoindentation. *Mater. Sci. Eng. A* **2012**, *540*, 226–234. [CrossRef]
2. Abdulstaar, M.A.; Al-Fadhalah, K.J.; Wagner, L. Microstructural variation through weld thickness and mechanical properties of peened friction stir welded 6061 aluminum alloy joints. *Mater. Charact.* **2017**, *126*, 64–73. [CrossRef]
3. Tan, Y.; Wang, X.; Ma, M.; Zhang, J.; Liu, W.C.; Fu, R.D.; Xiang, S. A study on microstructure and mechanical properties of AA 3003 aluminum alloy joints by underwater friction stir welding. *Mater. Charact.* **2017**, *127*, 41–52. [CrossRef]
4. Fratini, L.; Pasta, S.; Reynolds, A.P. Fatigue crack growth in 2024-T351 friction stir welded joints: Longitudinal residual stress and microstructural effects. *Int. J. Fatigue* **2009**, *31*, 495–500. [CrossRef]
5. Eslami, N.; Hischer, Y.; Harms, A.; Lauterbach, D.; Böhm, S. Optimization of process parameters for friction stir welding of aluminum and copper using the taguchi method. *Metals* **2019**, *9*, 63. [CrossRef]
6. Metz, D.F.; Weishaupt, E.R.; Barkey, M.E.; Fairbee, B.S. A microstructure and microhardness characterization of a friction plug weld in friction stir welded 2195 Al-Li. *J. Eng. Mater. Technol.* **2012**, *134*, 021005. [CrossRef]
7. Besel, M.; Besel, Y.; Alfaro Mercado, U.; Kakiuchi, T.; Uematsu, Y. Fatigue behavior of friction stir welded Al–Mg–Sc alloy. *Int. J. Fatigue* **2015**, *77*, 1–11. [CrossRef]

8. Sillapasa, K.; Surapunt, S.; Miyashita, Y.; Mutoh, Y.; Seo, N. Tensile and fatigue behavior of SZ, HAZ and BM in friction stir welded joint of rolled 6N01 aluminum alloy plate. *Int. J. Fatigue* **2014**, *63*, 162–170. [CrossRef]

9. Rajakumar, S.; Balasubramanian, V. Correlation between weld nugget grain size, weld nugget hardness and tensile strength of friction stir welded commercial grade aluminium alloy joints. *Mater. Des.* **2012**, *34*, 242–251. [CrossRef]

10. Sun, T.; Reynolds, A.P.; Roy, M.J.; Withers, P.J.; Prangnell, P.B. The effect of shoulder coupling on the residual stress and hardness distribution in AA7050 friction stir butt welds. *Mater. Sci. Eng. A* **2018**, *735*, 218–227. [CrossRef]

11. Lemos, G.V.B.; Cunha, P.H.C.P.; Nunes, R.M.; Bergmann, L.; Dos Santos, J.F.; Clarke, T. Residual stress and microstructural features of friction-stir-welded GL E36 shipbuilding steel. *Mater. Sci. Technol.* **2018**, *34*, 95–103. [CrossRef]

12. Xu, W.; Liu, J.; Zhu, H. Analysis of residual stresses in thick aluminum friction stir welded butt joints. *Mater. Des.* **2011**, *32*, 2000–2005. [CrossRef]

13. Aval, H.J. Microstructure and residual stress distributions in friction stir welding of dissimilar aluminium alloys. *Mater. Des.* **2015**, *87*, 405–413. [CrossRef]

14. Prime, M.B.; Thomas, G.H.; Baumann, J.A.; Lederich, R.J.; Bowden, D.M. Residual stress measurements in a thick, dissimilar aluminum alloy friction stir weld. *Acta Mater.* **2006**, *54*, 4013–4021. [CrossRef]

15. Toribio, J.; Matos, J.C.; González, B.; Escuadra, J. Fatigue crack growth in round bars for rock anchorages: The role of residual stresses. *Procedia Struct. Integr.* **2016**, *2*, 2734–2741. [CrossRef]

16. Benachoura, M.; Benachoura, N.; Benguediab, M. Effect of compressive residual stress generated by plastic preload on fatigue initiation of 6061 Al-alloy. *Procedia Struct. Integr.* **2016**, *2*, 3090–3097. [CrossRef]

17. Citarella, R.; Carlone, P.; Lepore, M.; Palazzo, G.S. Numerical-experimental crack growth analysis in AA2024-T3 FSWed butt joints. *Adv. Eng. Softw.* **2015**, *80*, 47–57. [CrossRef]

18. Shen, F.; Zhao, B.; Li, L.; Chua, C.K.; Zhou, K. Fatigue damage evolution and lifetime prediction of welded joints with the consideration of residual stresses and porosity. *Int. J. Fatigue* **2017**, *103*, 272–279. [CrossRef]

19. Sowards, J.W.; Gnäupel-Herold, T.; McColskey, J.D.; Pereira, V.F.; Ramirez, A.J. Characterization of mechanical properties, fatigue-crack propagation, and residual stresses in a microalloyed pipeline-steel friction-stir weld. *Mater. Des.* **2015**, *88*, 632–642. [CrossRef]

20. Servetti, G.; Zhang, X. Predicting fatigue crack growth rate in a welded butt joint: The role of effective R ratio in accounting for residual stress effect. *Eng. Fract. Mech.* **2009**, *76*, 1589–1602. [CrossRef]

21. Rao, D.; Huber, K.; Heerens, J.; Dos Santos, J.F.; Huber, N. Asymmetric mechanical properties and tensile behaviour prediction of aluminum alloy 5083 friction stir welding joints. *Mater. Sci. Eng. A* **2013**, *565*, 44–50. [CrossRef]

22. Simar, A.; Bréchet, Y.; Meester, B.D.; Denquin, A.; Pardoen, T. Microstructure, local and global mechanical properties of friction stir welds in aluminum alloy 6005A-T6. *Mater. Sci. Eng. A* **2008**, *486*, 85–95. [CrossRef]

23. Nielsen, K.L. Ductile damage development in friction stir welded aluminum (AA2024) joints. *Eng. Fract. Mech.* **2008**, *75*, 2795–2811. [CrossRef]

24. Nielsen, K.L.; Pardoen, T.; Tvergaard, V.; Meester, B.D.; Simar, A. Modelling of plastic flow localisation and damage development in friction stir welded 6005A aluminium alloy using physics based strain hardening law. *Int. J. Solids Struct.* **2010**, *47*, 2359–2370. [CrossRef]

25. Sun, G.; Niu, J.; Wang, D.; Chen, S. Fatigue experimental analysis and numerical simulation of FSW joints for 2219 Al-Cu alloy. *Fatigue Fract. Eng. Mater. Struct.* **2015**, *38*, 445–455. [CrossRef]

26. Manson, S.S. A complex subject-some simple approximations. *Exp. Mech.* **1965**, *5*, 193–226. [CrossRef]

27. Smith, K.N.; Watson, P.; Topper, T.H. A stress-strain function for the fatigue of metals. *J. Mater.* **1970**, *5*, 767–778.

28. Liu, X.; Zhang, L.; Wang, L.; Wu, S.; Fang, H. Fatigue behavior and life prediction of A7N01 aluminum alloy welded join. *Trans. Nonferr. Met. Soc. China* **2012**, *22*, 2930–2936. [CrossRef]

29. Zhang, Y.; Maddox, S. Fatigue life prediction for toe ground welded joints. *Int. J. Fatigue* **2009**, *31*, 1124–1136. [CrossRef]

© 2019 by the authors. Licensee MDPI, Basel, Switzerland. This article is an open access article distributed under the terms and conditions of the Creative Commons Attribution (CC BY) license (http://creativecommons.org/licenses/by/4.0/).

Article

Analytical and Numerical Crack Growth Analysis of 1:3 Scaled Railway Axle Specimens

David Simunek [1,*], Martin Leitner [1], Jürgen Maierhofer [2], Hans-Peter Gänser [2] and Reinhard Pippan [3]

[1]   Chair of Mechanical Engineering, Department Product Engineering, Montanuniversität Leoben, 8700 Leoben, Austria; martin.leitner@unileoben.ac.at
[2]   Materials Center Leoben Forschung GmbH, 8700 Leoben, Austria; juergen.maierhofer@mcl.at (J.M.); hp.gaenser@mcl.at (H.-P.G.)
[3]   Erich Schmid Institute of Materials Science, 8700 Leoben, Austria; reinhard.pippan@oeaw.ac.at
*   Correspondence: david.simunek@stud.unileoben.ac.at; Tel.: +43-3842-402-1452

Received: 28 December 2018; Accepted: 29 January 2019; Published: 3 February 2019

Abstract: This paper deals with experimental fatigue crack propagation in rotating bending loaded round bar specimens as well as an analytical and numerical analysis of the residual lifetime. Constant amplitude (CA) load tests are performed with the surface crack length being evaluated using an optical measurement system. Fracture surfaces are microscopically analyzed to determine crack growth in depth as well as the crack shape. In spite of identical testing conditions, the experimental results show some scatter in residual lifetime, which is mainly caused by different residual stress states. Although X-ray residual stress measurements reveal only minor values, a superposition of the residual stress state with the load-induced stress leads to a significant impact on the residual lifetime calculations, which explains the experimental scatter. Numerical analyses are conducted to consider the residual stress state and their effect on crack propagation by different options. Considering the residual stress distribution in depth within the residual lifetime assessment, the deviation to the most conservative experiment is reduced from +48% to +2%. In conclusion, the results in this paper highlight that it is of utmost importance to consider local residual stress conditions in the course of a crack propagation analysis in order to properly assess the residual lifetime.

Keywords: fatigue crack growth; railway axle; semi-elliptical crack; residual stresses

---

1. Introduction

Fatigue crack propagation is generally influenced by a multitude of different effects. With common crack propagation material parameters, determined using laboratory specimens, crack growth in full-scale components can be additionally affected by the manufacturing process and the operational loads. Load sequences, stress concentration due to notches and press fits as well as manufacturing-induced residual stress states majorly affect the crack propagation during service. Thus, the residual lifetime estimation of railway axles is still a demanding task and may also lead to a non-conservative assessment in the case of inadequate information. Gänser et al. [1] describe the issue of transferability from small-scale laboratory specimens to full-scale components. Many papers deal with fatigue crack growth behavior and assessment methods in railway axles. Numerically based stress intensity factor solutions for fatigue cracks in rotating bending loaded railway axles are given by Beretta et al. [2], Madia et al. [3,4] and Luke et al. [5,6], where various sections of a railway axle (T- and V-notches and the axle body) are analyzed as well as the influence of press fits on stress intensity factor (SIF) solutions. In [4] a comprehensive collection of different stress intensity factor solutions from several authors is presented. An overview on safe life and damage tolerance methods for railway axles, failure scenarios and causes is provided by Zerbst et al. in [7,8]. Special focus on fatigue and crack growth in railway axles under corrosion is presented in [9,10].

Contributing to the ongoing research on this topic, this paper deals with the fatigue crack propagation behavior in round bar specimens on a scale of 1:3 extracted from railway axle blanks. The investigations are performed within the framework of the international project 'Probabilistic fracture mechanics concept for the assessment of railway wheelsets' (Eisenbahnfahrwerke 3, EBFW3) aiming of the transferability of crack propagation parameters determined on standard laboratory specimens to full-scale test axles [11,12]. In the framework of this project, extensive fatigue crack propagation experiments in 1:1 axles, 1:3 axle specimens and single edge notch bending (SENB) specimens have been performed. This paper focuses on the comparison of 1:3 and SENB results. Special attention is denoted on the effect of residual stresses because they affect the crack propagation rate as well as the crack shape. Besides the stress intensity factor range ΔK, the crack propagation rate is mainly influenced by the stress intensity factor ratio R, see Equation (1).

$$R = \frac{K_{min}}{K_{max}} \tag{1}$$

As shown, the ratio R depends on the minimum and maximum stress intensity factor K_{min} and K_{max} respectively. In practice, local conditions may influence the local stress and strain fields. One issue is the proper determination of the local stresses and stress intensity factors. In the case of real components such as railway axles, one challenge is the fact that residual stresses and press fit induced stresses, which lead to a residual stress intensity factor K_{res}. This factor superimposes with the external load, thereby influencing the local minimum and maximum effective stress intensity factors $K_{min,eff}$ and $K_{max,eff}$ leading to a change of the load-dependent ratio R to an local effective ratio R_{eff} at the crack tip, see Equation (2). Although ΔK is well known, the superposition leads to a local change of K_{min} and K_{max}, which may significantly affect the residual lifetime due to acceleration or delay of the crack growth rate.

$$R_{eff} = \frac{K_{min,eff}}{K_{max,eff}} = \frac{K_{min} + K_{res}}{K_{max} + K_{res}} \tag{2}$$

Other crucial issues are the determination of the material properties, load sequence effects as well as the crack propagation models, used for the estimation of the residual lifetime. A review of several crack propagation models under constant and variable amplitude loading is given by Beden et al. in [13]. The original Paris/Erdogan model [14] is a commonly applied method, which is comparatively easy to handle for a simple crack growth estimation; however, one disadvantage is that it does not cover the effect of the load ratio. Hence, a high deviation of the estimated residual lifetime may result. Walker [15] modified the Paris/Erdogan equation to account for the influence by the load ratio. Crack propagation models from Erdogan/Ratwani [16] or the NASGRO equation according to Forman/Mettu [17] also consider the load ratio at residual lifetime estimations and may lead to more accurate assessments. Additionally, the NASGRO equation, see Equation (3), considers crack closure mechanisms similar to Newman [18], which is included in the factor F_{lc} additionally depending on the effective ratio R_{eff}, see Appendix B.

$$\frac{da}{dN} = C \cdot F_{lc} \cdot \Delta K^{m} \cdot \frac{\left(1 - \frac{\Delta K_{th}}{\Delta K}\right)^{p}}{\left(1 - \frac{K_{max}}{K_c}\right)^{q}} \tag{3}$$

Maierhofer et al. [19] modified the NASGRO equation for physically short cracks according to Equation (4) considering that short cracks can grow even though the stress intensity factor range is below the long crack threshold value $\Delta K_{th,LC}$, see Equation B1 for $\Delta K_{th}(R_{eff}, \Delta a)$ in Appendix B. Note that in Equation Error! the transition region III of the fatigue crack curve is not considered due to neglecting the parameter q ($q = 0$). In the remainder of this article, the short crack model (SCM) according to Maierhofer is used for analytical residual lifetime estimation. Detailed information about crack closure mechanisms and short crack behavior is provided in [19–22].

$$\frac{da}{dN} = C \cdot F\left(R_{eff}, \Delta a\right) \cdot \Delta K^{m-p} \cdot \left(\Delta K - \Delta K_{th}\left(R_{eff}, \Delta a\right)\right)^{p} \tag{4}$$

The differences between the NASGRO equation and its modification according to Maierhofer are in the determination of the crack velocity factor F and F_{lc} respectively as well as the fatigue crack propagation threshold ΔK_{th}. For detailed information see [19].

In general, round bars under pure rotating bending exhibit a semi elliptical crack front [3,23,24]. In Figure 1, representative fractographies of semi elliptical cracks in 1:1 railway axle specimens are illustrated.

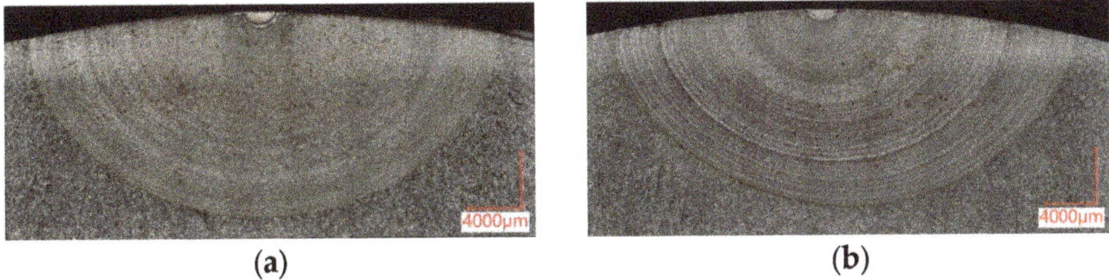

(a) **(b)**

Figure 1. Two typical fractographies of semi elliptical cracks in rotating bending loaded 1:1 railway axle specimens: (**a**) Moderate visibility of beach marks; (**b**) Improved visibility of beach marks.

Furthermore, in some cases deviations from typical reported fracture surfaces are noticed. Figure 2 exhibits two extreme examples of such discrepancies due to local residual stress fields. While in Figure 2a, two-thirds of the crack growth period exhibit semi elliptical crack extension (as can be seen from beach marks), the final fracture surface shows minor deviation of the semi elliptical shape. It seems that crack growth in depth is retarded compared to surface. In Figure 2b, one sided near surface crack growth over a wide range of the fracture surface is shown. The effective stress intensity factor range is not exceeding the threshold on one side of the initial notch, hence unsymmetrical and non-semi-elliptical crack propagation can be observed.

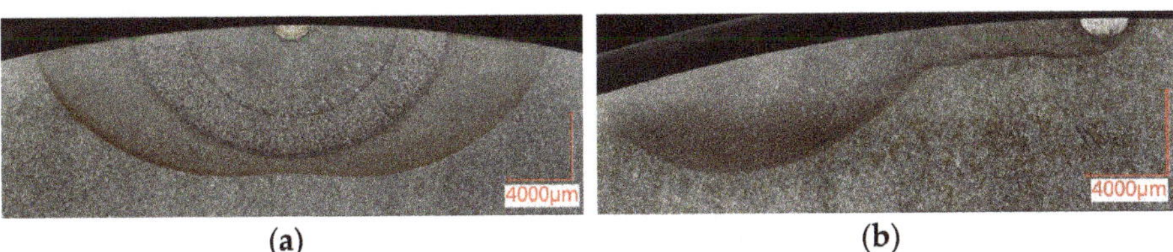

(a) **(b)**

Figure 2. Examples for the deviation from semi elliptical crack front in rotating bending loaded 1:1 railway axle specimens: (**a**) Symmetrical crack growth; (**b**) Unsymmetrical crack propagation.

While the depicted examples of Figure 2 are curiosities, the influence of local residual stress fields on crack propagation is clearly visible. Note that the illustrated fracture surfaces in Figure 1 do not give information about the quantity of the residual stress field, but an indication of the homogeneity. In this paper, only semi-elliptical crack propagation, as illustrated in Figure 1, is considered, being the focus of this article. In the analytical approach, semi-elliptical crack growth is assumed and hence divergences of a semi-elliptical shape are not provided. Numerical methods in this case are more flexible and offer an opportunity for non-semi-elliptical crack growth estimation. In summary, this paper scientifically contributes to the following research topics:

- Transferability of small scale SENB crack growth test results to round 1:3 scaled railway axle specimens incorporating influences of different crack front geometries and size effects focusing on varying residual stress conditions.
- Comparison of analytical and numerical crack growth assessment methods involving both short and long crack growth regime.
- Detailed investigation regarding the effect of different residual stress states on crack shape evolution and residual lifetime.

2. Materials and Methods

The material used for all experimental investigations is the railway axle steel EA1N, which is a normalized 0.35% carbon steel with a minimum yield stress $R_{eH} \geq 320$ MPa, see [25]. Figure 3 shows the investigated specimen geometries. Single edge notch bending specimen (SENB) tests are performed to determine crack propagation parameters at different stress ratios. These parameters are input values for the analytical and numerical fatigue crack propagation assessment in order to estimate the crack growth behavior in round bar specimens with a semi-elliptical crack front. These specimens are extracted from railway axle blanks with a scale of one-third; hence, denoted as 1:3-scale round bar specimens. The model predictions are finally compared to fatigue crack growth experiments under rotating bending. Details about manufacturing of the specimens are depicted in Appendix A.

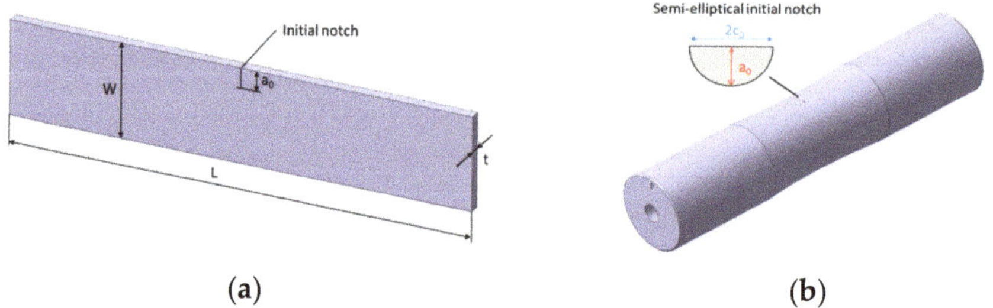

(a)   **(b)**

Figure 3. (a) Illustration of the investigated single edge notch bending (SENB) specimen with straight initial notch and (b) 1:3-scaled axle specimen with semi-elliptical initial notch.

2.1. Single Edge Notch Bending (SENB) Specimens

As introduced, experimental fatigue crack growth tests with SENB specimens have been performed to determine fatigue crack growth parameters. The specimens exhibit a thickness $t = 6$ mm, the width $W = 50$ mm, and a length $L = 250$ mm (see Figure 3a). All specimens are tested under four-point bending at different load ratios under laboratory conditions at room temperature. Crack growth was measured by using the direct current potential drop (DCPD) technique [26–28]. An initial notch with a depth of $a_0 = 10$ mm was spark eroded and sharpened by polishing the notch root with a razor blade and a diamond paste. All specimens were fatigue pre-cracked under compression at a load ratio of $R = 20$ [29,30]. The stress intensity factor solution for the SENB specimen was used based on ISO 12108, see [31], and determined according to Equations (5) and (6).

$$K_I\left(\frac{a}{W}\right) = \frac{F_B}{t \cdot W^{1/2}} \cdot g\left(\frac{a}{W}\right) \tag{5}$$

$$g\left(\frac{a}{W}\right) = 3 \cdot (2 \cdot \tan\theta)^{\frac{1}{2}} \cdot \left[\frac{0.923 + 0.199 \cdot (1 - \sin\theta)^4}{\cos\theta}\right] \quad with\, \theta = \frac{\pi \cdot a}{2 \cdot W} \tag{6}$$

Figure 4a illustrates the results of the crack growth experiments at a load ratio $R = -1$, in Figure 4b results and fitted data using Equation (4) at load ratios from $R = -1$ to $R = 0.7$ are depicted. As shown in Figure 4a, the short crack effect is clearly visible at the beginning of the experiment. Further details are provided in [19].

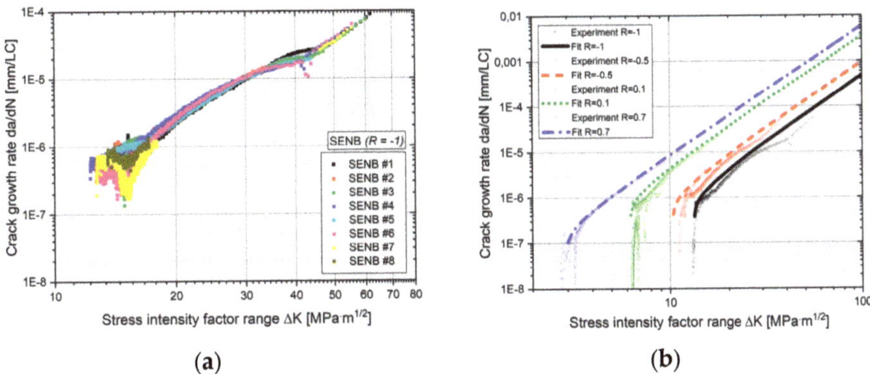

(a)   (b)

Figure 4. Experimental crack growth diagrams of SENB samples; (**a**) test data of eight specimens at a stress ratio $R = -1$; (**b**) Comparison of results and fit for different stress ratios from $R = -1$ to $R = 0.7$.

The material crack propagation parameters of the fitted data according to Figure 4b are shown in Table 1. The Coefficients for the crack opening function f according to Newman [18] are determined with $\sigma_{max}/\sigma_F = 0.3$ and the constraint factor $\alpha = 2.5$. Detailed description about the parameters and influence on the crack growth curve are given in Appendix B.

Table 1. Crack growth material parameters of fitted SENB experiments.

C [mm/MPa$\sqrt{m}$]	m [-]	p [-]	v_1 [-]	v_2 [-]	l_1 [mm]	l_2 [mm]	$\Delta K_{th,eff}$ [MPa$\sqrt{m}$]	$\Delta K_{th,0}$ [MPa$\sqrt{m}$]	C_{th}^{+} [-]
1.72×10^{-8}	2.8	0.21	0.43	0.57	2.09×10^{-3}	1.27	2.0	7.12	3.09

2.2. Round Bar Specimens (1:3 Scaled Railway Axle Specimens)

Round bar specimens with a testing diameter $d = 55$ mm (scale of 1:3 to a real railway axle) have been tested in a rotating bending test rig. The geometry of the samples is depicted in Figure 3b and detailed information on the manufacturing procedure is given in [32]. The nomenclature of the semi-elliptical crack front is illustrated in Figure 5. The crack length on the surface of the specimen is observed by an in-situ optical measurement system. Crack propagation tests are performed starting with surface crack lengths between $2s = 4$ to 5 mm up to a final value of about $2s = 18$ mm, which corresponds to the limit of the optical measurement system. Detailed information about the optical crack length acquisition system and calibration is given in [33]. A geometrical recalculation of the projected surface crack length (shortest distance between S_1 and S_2) to the real surface arc crack length $2s$ is performed in order to properly evaluate the crack growth behavior.

After testing, all samples are cooled down in liquid nitrogen atmosphere and fractured. Optical microscopical analyses of the fracture surfaces are conducted to evaluate the crack shape evolution of the semi-elliptical crack front. To this purpose, a self-written software code is established to measure the semi-elliptical crack front evolution at the fracture surfaces. Starting from the initial notch, beach marks and the final fracture surface are evaluated by non-linear least-squares curve fitting so that the axes of the semi-elliptical crack front can be determined. The data points of the investigated specimens provide information of crack depth evolution depending on the surface crack length ($a = f(2s)$ and $a = f(2c)$ respectively). In Figure 6, the evolution of the crack shape $a = f(2c)$ for 10 tested round bar specimens is depicted. The variation of the shape between the different specimens is small, hence a single fitting function has been determined.

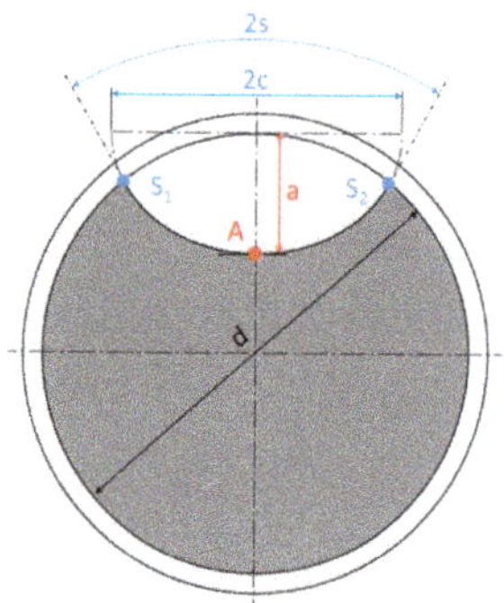

Figure 5. Nomenclature of the semi-elliptical crack.

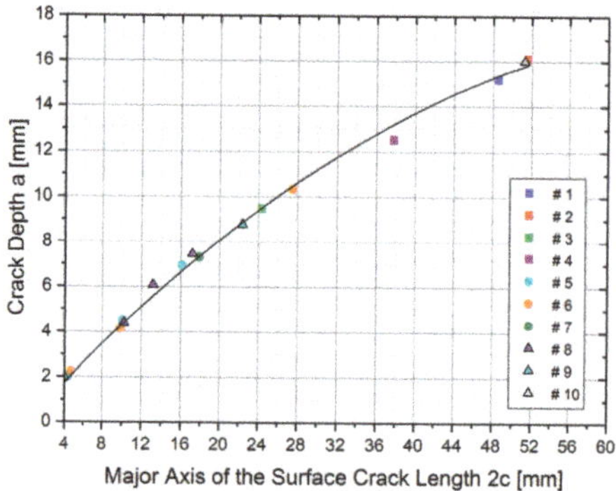

Figure 6. Illustration of crack depth as a function of surface crack length 2*c*.

Based on the fitted crack shape evolution, three dimensional (3D) finite element (FE) models of the specimen with different crack shapes in Abaqus are built up to investigate the stress intensity factor K_I along the crack front under rotating bending. Note that a fatigue crack in a shaft under rotating bending is a Mode I case and the crack grows perpendicular to the maximum principal stress. The other opening modes are negligible. In literature, many papers deal with numerical investigations of semi-elliptical surface cracks in round bars and their crack shape evolution, see [34–41]. In this study, the surface crack length was extended with an increment of $\Delta c = 0.25$ mm and the associated crack depth of the semi-elliptical crack front was adapted by the function of the fitted data points from fractography, see Figure 6. For each increment (crack length), a new numerical model is built up. In the FE model, the four-point rotating bending load was realized by an alternating bending moment. There are different numerical methods to evaluate the stress intensity factor (SIF) [42]. Courtin et al. [43] compared different numerical techniques for estimating the SIF and show the advantages of the J-Integral approach [44] using Abaqus (Version 6.14, Dassault Systemes Simulia Corp., Providence, RI, USA) [45]. The J-Integral can also be adopted for 3D crack problems. In this case, rings of elements surrounding the crack line need to be defined. Abaqus automatically identifies the contours around the crack line, whereas the first contour includes all nodes on the crack line. The first few contours are not recommended for SIF evaluation and may lead to inaccurate results [45]. Hence, a tube-shaped partition along the crack line was generated and the rings of elements have been meshed so that nine contours for evaluation were available, see Figure 7. The model is built up with hexahedral elements exhibiting reduced integration scheme and an element length of approximately 0.15 mm along the crack front. In radial direction the elements exhibit a size of 0.05 mm to ensure an accurate computation result.

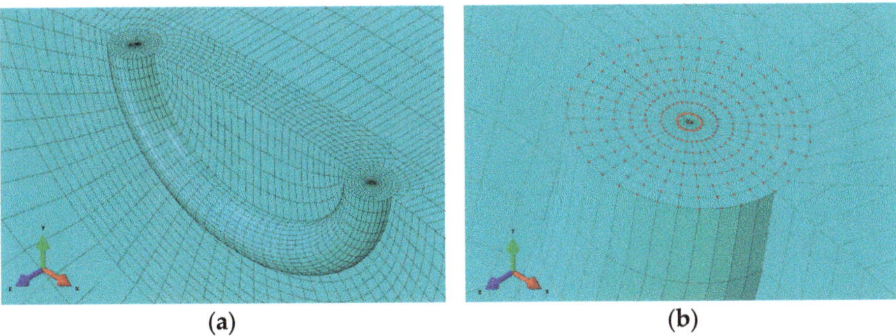

(a)   (b)

Figure 7. Mesh around crack front for contour integral evaluation; (**a**) Tubular mesh around crack line; (**b**) Detail view of contours at surface

The J-Integral leads to inaccurate results for contour integral evaluation on free surfaces (at the end of the crack front) due to the boundary layer effect [45,46]. This means that the $1/\sqrt{r}$-singularity of the stress field on the surface domain is not fulfilled. To this purpose, the SIF-values from contour integrals below the surface are fitted and extrapolated to the surface in order to properly evaluate the stress intensity factor ΔK_S at the surface points S_1 and S_2, which majorly influence the crack growth behavior. In the case of linear elastic fracture mechanics, the J-Integral is equivalent to the strain energy release rate G and the stress intensity factor K_I can be determined according to Equation (7),

$$K_I = \sqrt{J \cdot E'} \tag{7}$$

whereas E' is the Young's modulus and related to $E' = E$ at plane stress or $E' = E/(1-v2)$ at plane strain condition. In Figure 8, the results of the stress intensity factor range ΔK for the surface points S_1, S_2 and the crack depth point A as a function of the surface crack length $2s$ are depicted. Based on the measured surface crack length, the SIF-range for the surface ΔK_S and depth ΔK_A is evaluated and used to generate ds/dN-ΔK diagrams for comparing the crack growth behavior of the different specimens.

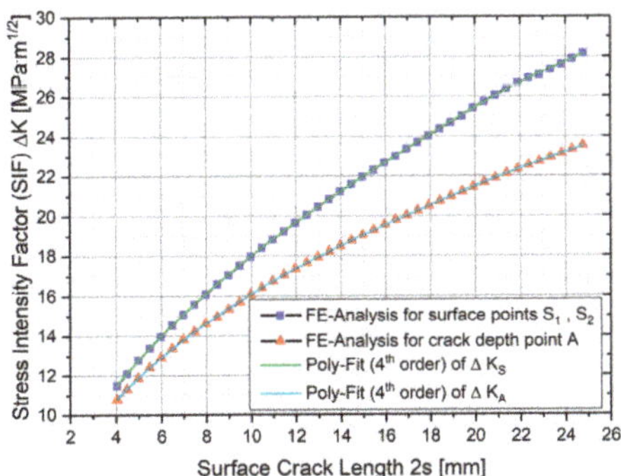

Figure 8. Stress intensity factor of the surface points (S_1 and S_2) and the depth (point A) as a function of the surface crack length $2s$.

Although all experiments show similar slopes and minor shift of the different curves in the Paris region (see Figure 9), a maximum deviation of 1.65 at residual lifetime between Experiment #3 and #4 was observed (evaluated at a surface crack length $2s = 15$ mm).

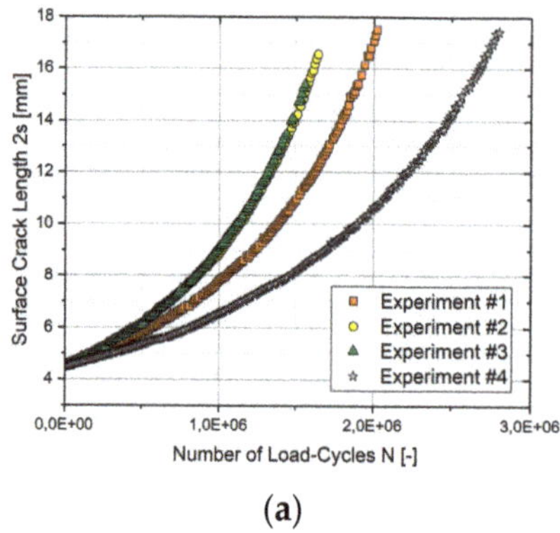
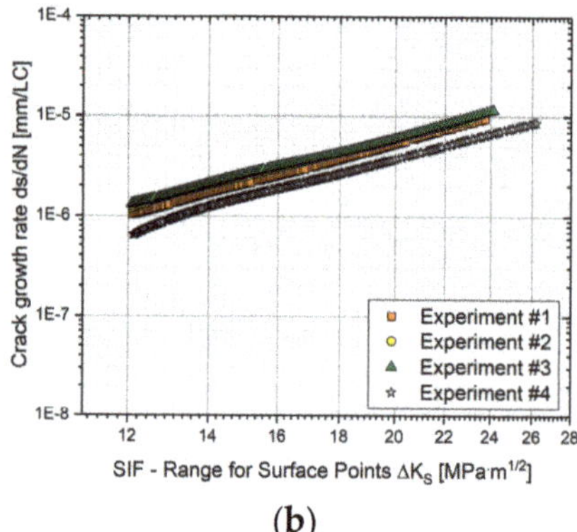

(a)

(b)

Figure 9. Results of rotating bending experiments: (**a**) Surface crack length in dependence of load-cycles and (**b**) Crack growth rate vs. ΔK_S.

The comparison of fitted data points from SENB specimens and round bar specimens is depicted in Figure 10 and shows that the crack growth curves of the round bar samples are in the range of SENB specimens between load ratios $R = -1$ and $R = -0.5$. Although all round bar specimens are tested at $R = -1$, a shift of the crack growth curves can be observed.

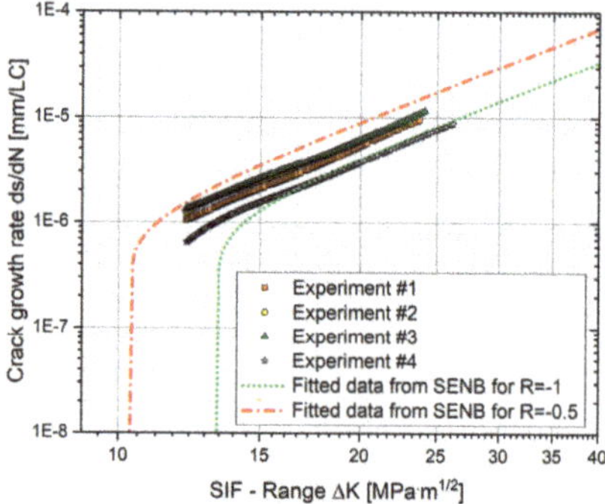

Figure 10. Comparison da/dN vs. ΔK_S of round bar specimens to SENB.

Figure 10 shows that experiments #1–#3 of the round bar samples tend to a load ratio higher than the nominal one; thus, leading to higher crack growth rates compared to SENB specimens at $R = -1$ and to the round bar sample of experiment #4. The shift of the load ratio can be influenced by internal or external superimposed mean stresses. Internal residual stresses act like mean stresses and thus may lead to a shift in load ratio. The samples are extracted from railway axle blanks with a diameter of 190 mm, for further details see also [32]. Whereas 1:1 railway axles usually exhibit compressive axial residual stresses 10–20 mm below the surface and minor tensile stresses below (see [1,47,48]), the residual stress distribution is significantly influenced due to cutting and machining small scale specimens out of these blanks. Hence, X-ray diffraction (XRD) residual stress measurements are performed on round bar specimens up to a maximum depth of 2.5 mm at the position of the notch. The measurements reveal high compressive residual stresses up to −250 MPa on the surface in axial direction due to machining, but almost immediately reducing to zero at a depth of approximately 100 μm. At depths below 100 μm, comparably minor axial tensile residual stresses are measured. Schindler [48] neglected the compressive residual stress peak directly on the surface in calculations due to the fact that right below the effect is not present and problems in calculating the crack growth rate occur. The averaged XRD residual stress values σ_{res} of three different measurements are depicted in Table 2. The measurements are performed from 0.1 mm to a maximum depth of 2.5 mm at three different specimens. Each measurement point exhibits some scatter, hence the minimum (best case) and maximum values (worst case) are specified additionally.

Table 2. Averaged residual stresses from XRD measurements in three different specimens.

Averaged σ_{res} [MPa]	Best Case	Mean Value	Worst Case
Measurement #1	6.1	13.3	20.5
Measurement #2	−1.8	6.0	13.8
Measurement #3	8.0	15.4	22.7

These measurement results are considered for the analytical and numerical assessment within the subsequent chapters. Additionally, performed XRD measurements of the SENB specimens showed negligible residual stresses, see [29]. Hence, the defined load ratio of the SENB results can be taken as reference without further modification of the local residual stress state.

3. Results

As shown in the previous chapters, parameters for different crack growth models are generated based on experimental fatigue crack growth tests with single edge notched bending (SENB) specimens. Analytical and numerical tools for residual lifetime estimation are used to compare calculations to experimental investigations of the round bar specimens.

3.1. Analytical Residual Lifetime Estimation

The analytical assessments are performed with INtegrity Assessment for Railway Axles (INARA) (Version 19-3-2018_13-47, Materials Center Leoben Forschungs GmbH, Leoben, Austria), a software tool to analyze semi-elliptical crack propagation in railway axles within the scope of the research project "Eisenbahnfahrwerke 3". The crack propagation model, used for all analytical calculations, is the short crack model according to Maierhofer et al. [19] and the parameters are determined from SENB specimen results, see Section 2.1. The stress intensity factor solutions for the semi-elliptical crack front in the solid round bar is based on numerical calculations according to Varfolomeev [49] and also reported in [4,50]. The finite element software Abaqus [45] was used for stress intensity factor determination along the crack front for different crack aspect ratios, crack depths and position of the shaft. Based on these results, polynomial influence functions were generated and implemented in the software tool. According to Equation (8), the stress intensity factors are determined [4,49,50]:

$$K_I\left(\tfrac{a}{c}, \tfrac{a}{R_s}, \phi\right) = \sqrt{\pi \cdot a} \sum_{m=0}^{4} \sum_{n=0}^{4} \left[D_{mn}^{(1)} \cdot f_{mn}^{(1)}\left(\tfrac{a}{c}, \tfrac{a}{R_s}, \phi\right) + D_{mn}^{(2)} \cdot f_{mn}^{(2)}\left(\tfrac{a}{c}, \tfrac{a}{R_s}, \phi\right) \right] \cdot \left(\tfrac{a}{R_s}\right)^{m+n} \cdot \left(\tfrac{a}{c}\right)^{-n} \tag{8}$$

First, crack growth calculations without any consideration of residual stresses have been performed. The results highlight that the estimation is satisfying for experiment #4, but non-conservative for the results of experiments #1–#3. Consequently, the averaged results of residual stress measurement #1, see Table 2, are considered as a mean stress state in the analysis. The consideration of such mean stresses leads to a variation of the local load ratio and thus changes the crack growth rate. The results without and with consideration of residual stresses are depicted in Figure 11 in comparison to the experiments.

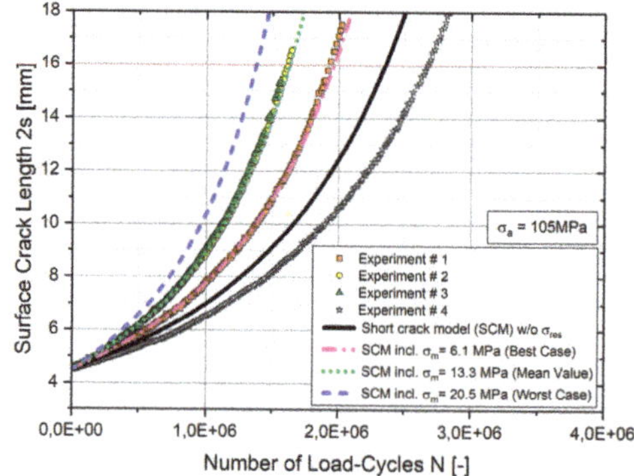

Figure 11. Comparison of analytical assessments and rotating bending experiments #1–#4.

Although the residual stresses considered are quite small, the influence on the residual lifetime is shown to be significant. Here, a mean stress of $\sigma_m = 6.1$ MPa for the best case situation shows a reduction of 16% estimated residual lifetime at a final surface crack length of $2s = 18$ mm. For comparison, in the worst case with $\sigma_m = 20.5$ MPa mean stress, the residual lifetime is even 41% lower. Although the assumption of average constant residual stresses as a mean stress state is an approximation, the results of the evaluated residual lifetime in Figure 11 show sound accordance with the experiments. The analytically estimated crack shape evolution of the semi-elliptical crack front compared to the fitted data points of the fracture surface analysis is depicted in Figure 12. The comparison exhibits some deviation of the crack shape evolution, whereas the experiments show a pronounced parabolic shape evolution after crack initiation, the calculations exhibit a less pronounced development. However, a maximum deviation of only 6% between the fracture surface analysis and the assessment without any residual stress consideration is observed at an a/R_S-ratio of about 0.37.

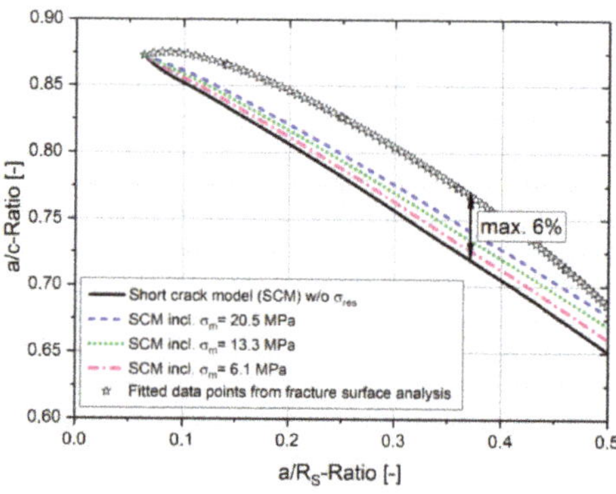

Figure 12. Crack shape evolution compared to fracture surface analysis.

Note that residual stresses are considered as constant mean stresses over the cross section for simplification, which does not describe the real residual stress distribution. Notwithstanding, satisfying results are achieved in case of the investigated round bar specimens.

As mentioned, residual stresses are measured up to a maximum depth of 2.5 mm by X-ray diffraction (XRD). For depths below, no further information is available due to measurement limitations. Based on the measured data points, residual stress distributions up to a depth of 12 mm are extrapolated to achieve an improved crack shape evolution in the course of the crack growth calculations.

3.2. Improved Analytical Assessment Based on Residual Stress Distribution

The influence of minor residual stresses considered as averaged overlapping constant mean stresses is shown in the preceding section. In this section, residual stress depth profiles are generated based on XRD measurements and their influence on the residual lifetime and crack shape evolution is analyzed. To this purpose, a multitude of different radial symmetrical stress distributions are investigated. Figure 13 illustrates two residual stress distributions up to a depth of 8 mm, which are estimated based on the measured XRD data points up to a depth of 2.5 mm. Additionally, the mean values of the data points from XRD measurements are depicted.

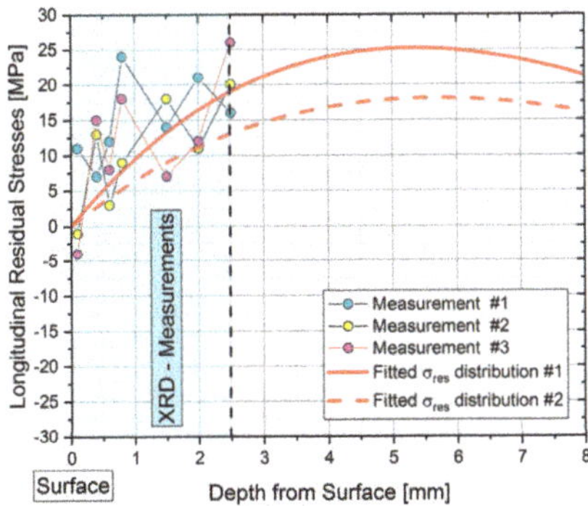

Figure 13. Illustration of measured and fitted residual stress distributions for assessments.

These two residual stress distributions are subsequently considered within the analytical approaches. The analysis reveals an improved crack shape evolution compared to the preceding calculations assuming a constant mean stress. Figure 14a depicts the two distributions compared to the fracture surface analysis and the calculations with constant mean stresses. The comparison highlights only a minor overestimation of the a/c-ratio using the two residual stress distributions compared to the experiments, which leads to conservative results for a crack growth assessment. A maximum deviation of 2.5% was observed at an a/R_S-ratio of 0.19 for the evaluation including residual stress distribution #1.

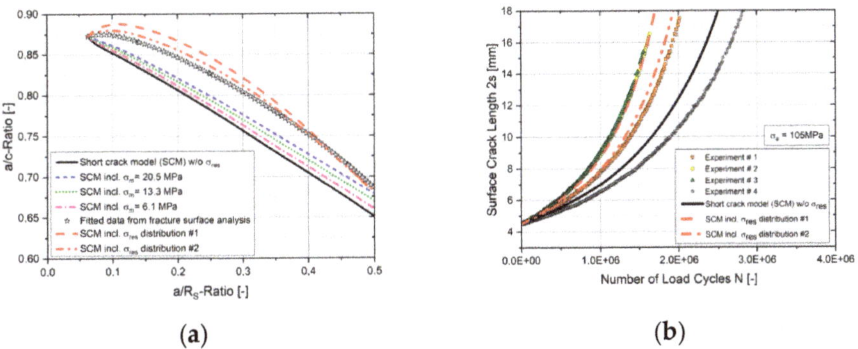

Figure 14. (a) Influence of residual stress distributions on crack shape evolution; (b) and residual lifetime.

In summary, the results of the residual lifetime assessments show sound accordance with the experimental investigations if the estimated residual stress depth profiles are included. Furthermore, it is highlighted that even comparably minor residual stresses in depth may significantly affect the crack growth rate and the residual lifetime estimation. Hence, it is of utmost importance to incorporate the exact stress conditions, such as local residual stress states, in the crack propagation analysis to ensure a proper fatigue assessment and to avoid non-conservative results.

3.3. Numerical Residual Lifetime Estimation

The numerical analyses are conducted with Franc3D (FRacture ANalysis Code 3D Version 7.1.0.2, Fracture Analysis Consultants, Inc., Ithaca, NY, USA), which is a 3D finite element fracture analysis software to simulate crack growth [51]. It is used in combination with a general Finite Element program, such as Abaqus in this case. The crack free model is built up and meshed in Abaqus including boundary and loading conditions. The input file is imported to Franc3D and a sub-model technique is used for the round bar specimen. A semi-elliptical crack is inserted and re-meshed automatically by Franc3D considering the singularity at the crack tip by 3D quarter point singular elements, for detailed information see [52]. Based on this model, crack growth is simulated with the stress and strain analyses conducted by Abaqus and crack extension and re-meshing being done by Franc3D. Different consideration of crack face traction and surface residual stresses and their influence on the residual lifetime are analyzed. In Franc3D, different crack growth models are deposited. In the case of the investigated round bar specimens, all analyses are based on the NASGRO model [17]. Similar to the previously described analytical calculations with INARA, the influence of the residual stress state on the residual lifetime is observed. A comparison of the numerical results based on the NASGRO model with and without consideration of the residual stress condition is provided in Figure 15. The results of the assessment for the surface crack length excluding any residual stress influence are similar to the analytical evaluation with INARA.

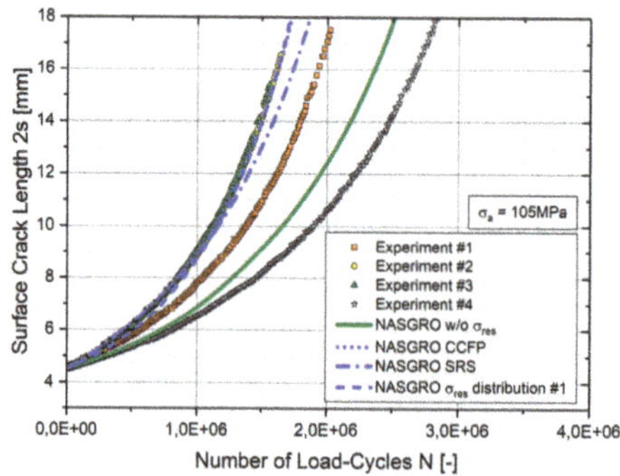

Figure 15. Results of numerical assessments compared to experiments.

Franc3D provides different possibilities for considering residual stresses in crack growth simulations. In the case of the round bar specimen, three options are used for crack propagation analyses. First the mean value (σ_{res} = 13.3 MPa) from XRD measurements is considered as a constant crack face pressure (CCFP), which allows to apply a uniform pressure or tensile load on the crack face. The results show sound accordance with the experimental investigations. Another option is to respect only surface residual stresses. To this purpose, only XRD measurement points up to a depth of 2.5 mm are taken into account (NASGRO SRS). The residual lifetime at a surface crack length $2s$ = 18 mm was slightly non-conservative for experiment #2 and #3; however, it is shown to work well in the case of experiment #1. Finally, the residual stress distribution #1 is considered within the numerical analysis as a 1D radial symmetrical stress distribution (NASGRO σ_{res} distribution #1). This computation leads to almost the same results as for the assessments with constant crack face pressure (NASGRO SRS). Similar to the analytical assessments, minor residual stresses, considered in calculations, significantly reduce the residual lifetime.

Finally, the crack shape evolution in the course of the different numerical analyses has also been investigated. The results are depicted in Figure 16a and significant different evolutions of the a/c-ratio for the different simulations are observed. The decrement of a/c-ratio is generally higher in the first crack growth steps, compared to the analytical assessments by INARA. An improved shape evolution is observed for constant crack face pressure (CCFP) as well as in the case of considering the residual stress distribution. On the contrary, the assessment including only surface residual stresses exhibits higher deviation. This seems logical due to the fact that in this case the crack propagation is only influenced at the surface, where tensile residual stresses accelerate the crack growth, whereas in depth the load ratio is basically not affected.

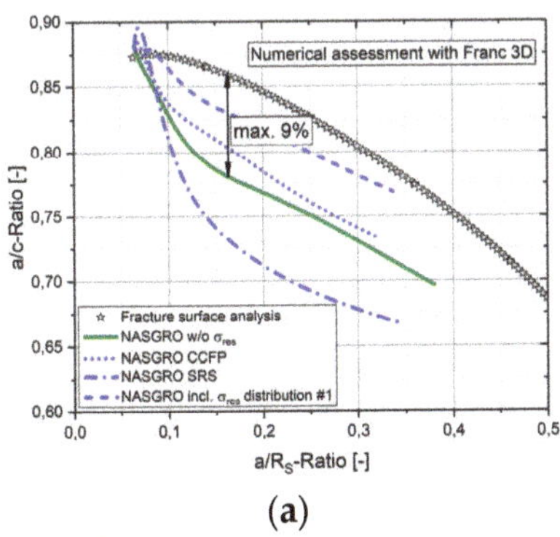

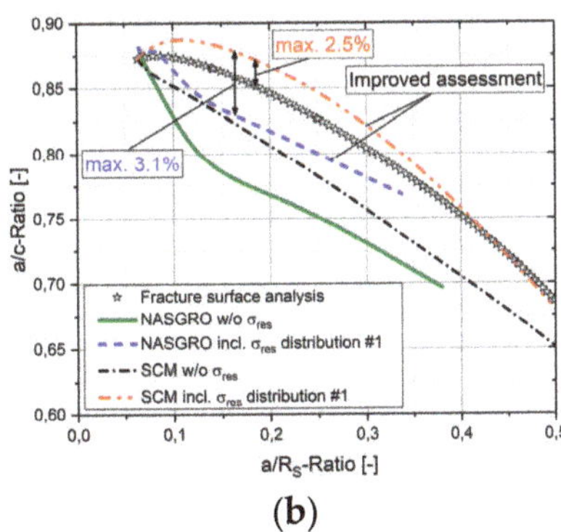

(a)   (b)

Figure 16. (**a**) Crack shape evolution of numerical computations with Franc3D; (**b**) Comparison of SCM model (analytically by INARA) and NASGRO model (numerically by Franc3D) to the experimental fit.

The deviations compared to the experimental investigations are slightly higher than the crack shape evolution according to INARA. Especially in the first few crack growth steps between $a/R_S = 0.06$–0.18 a steep decrease can be observed. Anyway, except the assessment with surface residual stresses (SRS), a maximum deviation of 9% was noticed at $a/R_S \approx 0.17$ for the analysis without any residual stresses, see Figure 16a. As depicted in Figure 16b, the consideration of residual stress distribution #1 leads to a maximum deviation of 3.1%.

Figure 16b illustrates the crack shape evolutions of the analytical and numerical assessments with and without the consideration of the residual stress distribution compared to the experimental investigations. For both assessment methods, an improved estimation of the crack shape evolution is achieved.

4. Discussion

Although minor residual stresses in depth at round bar specimens are measured, the influence can be significant. Table 3 shows the results of the experimental investigated residual lifetime compared to calculated residual lifetime estimations evaluated at a surface crack length of $2s = 15$ mm. For that purpose, the lifetime of the most conservative experiment and the mean lifetime value of all experiments are depicted.

Table 3. Comparison of residual lifetime for calculations and experiments at $2s = 15$ mm.

Residual Stress Condition	SCM (INARA)	NASGRO (Franc3D)	Experiments (Mean of All Tests)	Experiments (Most Conservative Test)
$W/o\ \sigma_{res}$	2.26×10^6	2.28×10^6	1.88×10^6	1.53×10^6
σ_{res} distribution #1	1.56×10^6	1.56×10^6		

A comparison of the experimental mean value shows that both calculation methods (SCM and NASGRO) with residual stress distribution #1 lead to a conservative assessment and exhibit a deviation of −17%. On the contrary, estimation without considering residual stresses results in a non-conservative assessment with a deviation of +20% compared to the mean value of the experiments.

The residual lifetime estimation considering residual stress distribution #1 is in sound accordance with the most conservative experiment with a minor difference of only +2%. Again, neglecting residual stresses within the lifetime assessment leads to a significant overestimation of +48% in residual lifetime, which proves the importance of considering local residual stress states in the crack propagation analysis.

In general, X-ray diffraction measurements are limited in depth. Other methods, such as the cut-compliance method, exhibit the potential for residual stress distribution over the total cross section. To that purpose, cut-compliance measurements are planned to determine the residual stress distributions at real railway axles. Furthermore, the applicability of the numerical crack growth approach presented in this paper will be validated for real components exhibiting varying residual stress conditions, which will lead to non-semi-elliptical crack fronts as shown Figure 2.

5. Conclusions

Analytical and numerical assessment of rotating round bar specimens was conducted and the influence of residual stresses on crack propagation was analyzed. Based on X-ray diffraction measurements, residual stresses are determined and included in calculations. The consideration showed an improved assessment of residual lifetime for the investigated specimens. Based on fracture surface analysis, the crack shape evolution of the semi-elliptical crack front was observed and compared to calculations. In addition, the consideration of residual stress distributions showed an enhanced crack shape evolution. Based on the investigated assessment and experimental analyses, the following conclusions can be drawn:

- The transferability of material parameters evaluated on the basis of small-scale SENB specimens to real components exhibiting different geometries, crack shapes and residual stress conditions is a challenging task in residual lifetime estimations.
- Even minor residual stress states may lead to uncertainties within the assessment and can result in a non-conservative estimation of the residual lifetime.
- The results within this study reveal that a consideration of the residual stress distribution in depth reduces the deviation from the most conservative experiment from +48% down to +2%. This highlights the importance of including effective residual stress conditions in crack propagation analyses to properly estimate the residual lifetime.
- Real railway axles generally exhibit compressive residual stresses up to a depth of 20 mm. Considering real residual stresses can improve the accuracy of the lifetime assessment as well as the definition of inspection intervals.
- The influence of residual stresses on lifetime can be significant. Especially at low loading conditions, the fraction of residual stresses compared to the external load is high and thus, the influence of residual stresses on effective crack growth rate is more pronounced.

Author Contributions: Conceptualization, D.S. and M.L.; investigation, D.S.; J.M.; methodology, D.S.; software, D.S., M.L., J.M. and H.-P.G.; validation, D.S. and M.L.; data curation, D.S. and M.L.; writing—original draft preparation, D.S.; writing—review and editing, M.L., J.M., H.-P.G. and R.P.

Funding: Scientific support was given within the framework of the COMET K2-Programme, whereby the Austrian Federal Government represented by Österreichische Forschungsförderungsgesellschaft mbH and the Styrian and the Tyrolean Provincial Government, represented by Steirische Wirtschaftsförderungsgesellschaft mbH and Standortagentur Tirol, is gratefully acknowledged.

Conflicts of Interest: The authors declare no conflict of interest.

Appendix A

Both SENB and 1:3 scaled specimens are extracted out of railway axle blanks. The distance of the initial notch for both types of specimen to the primary surface of the blank is equal (20 mm) to guarantee similar microstructure, as schematically illustrated for 1:3 scaled specimens in Figure A1. Although the microstructure should be comparable for both SENB and 1:3 scaled specimens, the residual stress state is influenced by the truncated size of the final sample.

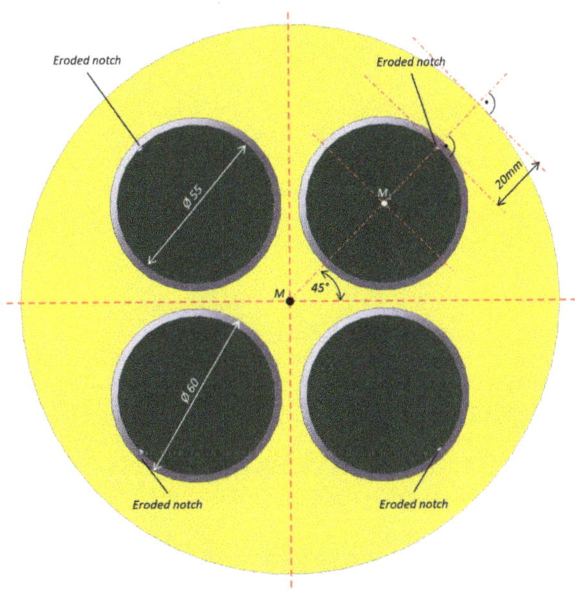

Figure A1. Extraction of 1:3 scaled specimens out of a railway axle blank.

Appendix B

The short crack behavior can be considered according Maierhofer's modification [19] of the NASGRO-model. Based on Equation (4) in Section 1 and the crack growth material parameters of Section 2.1 (Table 1), the crack growth rate can be determined by accounting for the built-up of the crack growth threshold value in dependence of the crack extension Δa. The threshold value ΔK_{th} can be determined according Equation (A1), which describes the transition from the effective threshold $\Delta K_{th,eff}$ to the long crack growth threshold $\Delta K_{th,lc}$ based on the fictitious length scales l_1 and l_2. The constraint factors are denoted as ν_1 and ν_2.

$$\Delta K_{th} = \Delta K_{th,eff} + \left(\Delta K_{th,lc} - \Delta K_{th,eff}\right) \cdot \left[1 - \left(\nu_1 \cdot \exp\left(-\frac{\Delta a}{l_1}\right) + \nu_2 \cdot \exp\left(-\frac{\Delta a}{l_2}\right)\right)\right] \qquad (A1)$$

The long crack growth threshold can be determined with Equation (A2). Newman's crack opening function f [18] and the Newman coefficient A_0 are used. $\Delta K_{th,0}$ is the long crack growth threshold at a load ratio $R = 0$. The curve control coefficient C_{th} depends on the load ratio. In the case that the investigated material C_{th} for positive R-ratios is specified in Table 1, for negative R-values ($R < 0$) $C_{th} = 0$.

$$\Delta K_{th,lc} = \frac{\Delta K_{th,0}}{\left[\frac{1-f}{(1-A_0)\cdot(1-R)}\right]^{(1+C_{th}\cdot R)}} \qquad (A2)$$

The crack velocity factor F is determined according to Equations (A3) and (A4),

$$F = 1 - (1 - F_{lc}) \cdot \left[1 - \left(\nu_1 \cdot \exp\left(-\frac{\Delta a}{l_1} \right) + \nu_2 \cdot \exp\left(-\frac{\Delta a}{l_2} \right) \right) \right] \tag{A3}$$

whereas F_{lc} describes the behavior for long cracks.

$$F_{lc} = \left(\frac{1-f}{1-R} \right)^m \tag{A4}$$

References

1. Gänser, H.-P.; Maierhofer, J.; Tichy, R.; Zivkovic, I.; Pippan, R.; Luke, M.; Varfolomeev, I. Damage tolerance of railway axles—The issue of transferability revisited. *Int. J. Fatigue* **2016**, *86*, 52–57. [CrossRef]
2. Beretta, S.; Madia, M.; Schode, M.; Zerbst, U. SIF Solutions for Cracks in Railway Axles Under Rotating Bending. *Fract. Nano Eng. Mater. Struct.* **2006**, 263–264. [CrossRef]
3. Madia, M.; Beretta, S.; Zerbst, U. An investigation on the influence of rotary bending and press fitting on stress intensity factors and fatigue crack growth in railway axles. *Eng. Fract. Mech.* **2008**, *75*, 1906–1920. [CrossRef]
4. Madia, M.; Beretta, S.; Schödel, M.; Zerbst, U.; Luke, M.; Varfolomeev, I. Stress intensity factor solutions for cracks in railway axles. *Eng. Fract. Mech.* **2011**, *78*, 764–792. [CrossRef]
5. Luke, M.; Varfolomeev, I.; Lütkepohl, K.; Esderts, A. Fracture mechanics assessment of railway axles: Experimental characterization and computation. *Eng. Fail. Anal.* **2010**, *17*, 617–623. [CrossRef]
6. Luke, M.; Varfolomeev, I.; Lütkepohl, K.; Esderts, A. Fatigue crack growth in railway axles: Assessment concept and validation tests. *Eng. Fract. Mech.* **2011**, *78*, 714–730. [CrossRef]
7. Zerbst, U.; Beretta, S.; Köhler, G.; Lawton, A.; Vormwald, M.; Beier, H.T.; Klinger, C.; Černý, I.; Rudlin, J.; Heckel, T.; et al. Safe life and damage tolerance aspects of railway axles—A review. *Eng. Fract. Mech.* **2013**, *98*, 214–271. [CrossRef]
8. Zerbst, U.; Klinger, C.; Klingbeil, D. Structural assessment of railway axles—A critical review. *Eng. Fail. Anal.* **2013**, *35*, 54–65. [CrossRef]
9. Beretta, S.; Carboni, M.; Fiore, G.; Lo Conte, A. Corrosion–fatigue of A1N railway axle steel exposed to rainwater. *Int. J. Fatigue* **2010**, *32*, 952–961. [CrossRef]
10. Beretta, S.; Carboni, M.; Lo Conte, A.; Regazzi, D.; Trasatti, S.; Rizzi, M. Crack Growth Studies in Railway Axles under Corrosion Fatigue: Full-scale Experiments and Model Validation. *Procedia Eng.* **2011**, *10*, 3650–3655. [CrossRef]
11. Deisl, A.; Gänser, H.-P.; Jenne, S.; Pippan, R. *Eisenbahnfahrwerke 3—EBFW3 Description and Aims of the New Project. Railway Axles: Advances in Durability Analysis and Maintenance*; ESIS TC24: Milano, Italy, 2014.
12. Weber, F.-J.; Jenne, S.; Gänser, H.-P.; Kunter, K.; Maierhofer, J.; Deisl, A. *Zwischenbericht Forschungsprojekt Eisenbahnfahrwerke 3*; 43. Tagung "Moderne Schienenfahrzeuge": Graz, Austria, 2016.
13. Beden, S.M.; Abdullah, S.; Ariffin, A.K. Review of Fatigue Crack Propagation Models for Metallic Components. *Eur. J. Sci. Res.* **2009**, *28*, 364–397.
14. Paris, P.C.; Erdogan, F. A critical analysis of crack propagation laws. *J. Basic Eng.* **1963**, *85*, 528–534. [CrossRef]
15. Walker, K. The effect of stress ratio during crack propagation and fatigue for 2024-T3 and 7075-T6 aluminum. In *Effects of Environment and Complex Load History on Fatigue Life*; ASTM International: West Conshohocken, PA, USA, 1970.
16. Erdogan, F.; Ratwani, M. Fatigue and fracture of cylindrical shells containing a circumferential crack. *Int. J. Fract.* **1970**, *6*. [CrossRef]
17. Forman, R.G.; Mettu, S.R. *Behavior of Surface and Corner Cracks Subjected to Tensile and Bending Loads in Ti-6A1-4V Alloy*; NASA-TM-102165; NASA: Houston, TX, USA, 1990.
18. Newman, J.C. A crack opening stress equation for fatigue crack growth. *Int. J. Fract.* **1984**, *24*, R131–R135. [CrossRef]
19. Maierhofer, J.; Pippan, R.; Gänser, H.-P. Modified NASGRO equation for physically short cracks. *Int. J. Fatigue* **2014**, *59*, 200–207. [CrossRef]

20. Maierhofer, J.; Kolitsch, S.; Pippan, R.; Gänser, H.-P.; Madia, M.; Zerbst, U. The cyclic R-curve—Determination, problems, limitations and application. *Eng. Fract. Mech.* **2018**, *198*, 45–64. [CrossRef]
21. Madia, M.; Zerbst, U. Application of the cyclic R-curve method to notch fatigue analysis. *Int. J. Fatigue* **2016**, *82*, 71–79. [CrossRef]
22. Zerbst, U.; Vormwald, M.; Pippan, R.; Gänser, H.-P.; Sarrazin-Baudoux, C.; Madia, M. About the fatigue crack propagation threshold of metals as a design criterion—A review. *Eng. Fract. Mech.* **2016**, *153*, 190–243. [CrossRef]
23. Zerbst, U.; Vormwald, M.; Andersch, C.; Mädler, K.; Pfuff, M. The development of a damage tolerance concept for railway components and its demonstration for a railway axle. *Eng. Fract. Mech.* **2005**, *72*, 209–239. [CrossRef]
24. Beretta, S.; Ghidini, A.; Lombardo, F. Fracture mechanics and scale effects in the fatigue of railway axles. *Eng. Fract. Mech.* **2005**, *72*, 195–208. [CrossRef]
25. European Standard. *Railway Applications—Wheelsets and Bogies—Axles—Product Requirements*; European Committee for Standardization: Brussels, Belgium, 2003.
26. Riemelmoser, F. Möglichkeiten und Grenzen der Potentialmethode zur Risslängenbestimmung. Diploma Thesis, Montanuniversität, Leoben, Austria, 1993.
27. Lieb, K.C.; Horstman, R.; Peters, K.A.; Enright, C.F.; Meltzer, R.L.; Bruce Vieth, M.; Schwalbe, K.-H.; Hellmann, D. Application of the Electrical Potential Method to Crack Length Measurements Using Johnson's Formula. *J. Test. Eval.* **1981**, *9*, 218. [CrossRef]
28. Ke, Y.; Ståhle, P. Crack length measurements with a potential drop method: A finite element simulation. *Int. J. Numer. Meth. Eng.* **1993**, *36*, 3205–3220. [CrossRef]
29. Maierhofer, J.; Gänser, H.-P.; Pippan, R. Crack closure and retardation effects—Experiments and modelling. *Procedia Struct. Integr.* **2017**, *4*, 19–26. [CrossRef]
30. Pippan, R.; Plöchl, L.; Klanner, F.; Stuwe, H.P. Use of fatigue specimens precracked in compression for measuring threshold values and crack growth. *J. Test. Eval.* **1994**, *22*, 98–103.
31. International Organization for Standardization. *Metallic Materials. Fatigue Testing: Fatigue Crack Growth Method. ISO 12108*, 1st ed.; ISO: Geneva, Switzerland, 2012.
32. Simunek, D.; Leitner, M.; Maierhofer, J.; Gänser, H.-P. Crack growth under constant amplitude loading and overload effects in 1: 3 scale specimens. *Procedia Struct. Integr.* **2017**, *4*, 27–34. [CrossRef]
33. Simunek, D.; Leitner, M.; Grün, F. In-situ crack propagation measurement of high-strength steels including overload effects. *Procedia Eng.* **2018**, *213*, 335–345. [CrossRef]
34. Carpinteri, A. Elliptical surface cracks in round bars. *Fatigue Fract. Eng. Mater. Struct.* **1992**, *15*, 1141–1153. [CrossRef]
35. Carpinteri, A. Shape change of surface cracks in round bars under cyclic axial loading. *Int. J. Fatigue* **1993**, *15*, 21–26. [CrossRef]
36. Carpinteri, A. Part-through cracks in round bars under cyclic combined axial and bending loading. *Int. J. Fatigue* **1996**, *18*, 33–39. [CrossRef]
37. Carpinteri, A. Surface flaws in cylindrical shafts under rotary bending. *Fatigue Fract. Eng. Mater. Struct.* **1998**, *21*, 1027–1035. [CrossRef]
38. Shin, C.S.; Cai, C.Q. Experimental and finite element analyses on stress intensity factors of an elliptical surface crack in a circular shaft under tension and bending. *Int. J. Fract.* **2004**, *129*, 239–264. [CrossRef]
39. Shih, Y.-S.; Chen, J.-J. The stress intensity factor study of an elliptical cracked shaft. *Nucl. Eng. Des.* **2002**, *214*, 137–145. [CrossRef]
40. Rubio, P.; Rubio, L.; Muñoz-Abella, B.; Montero, L. Determination of the Stress Intensity Factor of an elliptical breathing crack in a rotating shaft. *Int. J. Fatigue* **2015**, *77*, 216–231. [CrossRef]
41. Shih, Y.-S.; Chen, J.-J. Analysis of fatigue crack growth on a cracked shaft. *Int. J. Fatigue* **1997**, *19*, 477–485.
42. Kuna, M. *Numerische Beanspruchungsanalsye von Rissen. Finite Elemente in der Bruchmechanik*, 1st ed.; Vieweg+Teubner Verlag/GWV Fachverlage: Wiesbaden, Germany, 2008.
43. Courtin, S.; Gardin, C.; Bézine, G.; Ben Hadj Hamouda, H. Advantages of the J-integral approach for calculating stress intensity factors when using the commercial finite element software ABAQUS. *Eng. Fract. Mech.* **2005**, *72*, 2174–2185. [CrossRef]
44. Rice, J.R. A Path Independent Integral and the Approximate Analysis of Strain Concentration by Notches and Cracks. *J. Appl. Mech.* **1968**, *35*, 379. [CrossRef]

45. Dassault Systemes Simulia Corp. *Abaqus Analysis User's Guide Version 6.14*; Dassault Systemes Simulia Corp.: Providence, RI, USA, 2014.

46. Lebahn, J.; Heyer, H.; Sander, M. Numerical stress intensity factor calculation in flawed round bars validated by crack propagation tests. *Eng. Fract. Mech.* **2013**, *108*, 37–49. [CrossRef]

47. Hutař, P.; Pokorný, P.; Poduška, J.; Fajkoš, R.; Náhlík, L. Effect of residual stresses on the fatigue lifetime of railway axle. *Procedia Struct. Integr.* **2017**, *4*, 42–47. [CrossRef]

48. Schindler, H.-J. Effect of Residual Stresses on Safe Life Prediction of Railway Axles. *Procedia Struct. Integr.* **2017**, *4*, 48–55. [CrossRef]

49. Varfolomeev, I.; Burdack, M.; Luke, M. *Fracture Mechanics as a Tool for Specifying Inspection Intervals of Railway Axles*; 39. Tagung des DVM Arbeitskreises: Paderborn, Germany, 2007.

50. Lütkepohl, K. *Sicherer und Wirtschaftlicher Betrieb von Eisenbahnfahrwerken, Abschlussbericht Band I*; Bundesministerium fкr Wirtschaft und Technologie: Clausthal, Germany, 2009.

51. Fracture Analysis Consultants, Inc. FRANC3D Documentation. Available online: http://www.fracanalysis.com/software.html (accessed on 6 September 2018).

52. Wawrzynek, P.A.; Carter, B.; Ingraffea, A. Advances in simulation of arbitrary 3D crack growth using FRANC3D/NG. *J. Comput. Struct. Eng. Inst. Korea* **2010**, *23*, 607–613.

© 2019 by the authors. Licensee MDPI, Basel, Switzerland. This article is an open access article distributed under the terms and conditions of the Creative Commons Attribution (CC BY) license (http://creativecommons.org/licenses/by/4.0/).

 *metals* 

Article

A Fatigue Life Prediction Model Based on Modified Resolved Shear Stress for Nickel-Based Single Crystal Superalloys

Jialiang Wang [1,2]**, Dasheng Wei** [1,2]**, Yanrong Wang** [1,2,]*** and Xianghua Jiang** [1,2]

[1] School of Energy and Power Engineering, Beihang University, Beijing 100191, China; naz.wjl.150@buaa.edu.cn (J.W.); dasheng.w@163.com (D.W.); jxh@buaa.edu.cn (X.J.)

[2] Collaborative Innovation Center for Advanced Aero-Engine, Beijing 100191, China

*   Correspondence: yrwang@buaa.edu.cn; Tel.: +86-010-8233-9300

Received: 4 January 2019; Accepted: 29 January 2019; Published: 2 February 2019

Abstract: In this paper, the viewpoint that maximum resolved shear stress corresponding to the two slip systems in a nickel-based single crystal high-temperature fatigue experiment works together was put forward. A nickel-based single crystal fatigue life prediction model based on modified resolved shear stress amplitude was proposed. For the four groups of fatigue data, eight classical fatigue life prediction models were compared with the model proposed in this paper. Strain parameter is poor in fatigue life prediction as a damage parameter. The life prediction results of the fatigue life prediction model with stress amplitude as the damage parameter, the fatigue life prediction model with maximum resolved shear stress in 30 slip directions as the damage parameter, and the McDiarmid (McD) model, are better. The model proposed in this paper has higher life prediction accuracy.

Keywords: fatigue; nickel-based single crystal superalloy; life modeling; resolved shear stress

---

1. Introduction

Nickel-based single crystal superalloy materials are mainly used in engine turbine blades, and their working environment is very harsh, which is why they are one of the components with the most structural failures. Many authors have conducted a number of experimental and theoretical studies on turbine blades. Some authors have focused on experiments and microscopic observations to investigate the fatigue behavior of nickel-based single crystals with different orientations, and some nickel-based single crystal fatigue life prediction models have been proposed. However, these have mostly been improved models of isotropic material life prediction and are often phenomenological life prediction models. Based on experiments, other scholars have conducted in-depth studies on the deformation behavior of nickel-based single crystals and have established some nickel-based single crystal constitutive models to simulate deformation, finally realizing prediction of fatigue life.

A large amount of literature has described the effect of resolved shear stress on fatigue. Ref. [1] mentions that plastic deformation mainly concentrates on the octahedral slip plane at low and medium temperatures, but both octahedral slip systems and cubic slip systems are activated at high temperatures (>600 °C). According to [2], for directional solidification superalloys, it is found that the crack mainly lies in the primary octahedral slip systems when the temperature is between 500 °C and 600 °C. The fracture surface presents crystallographic features. When the temperature is higher than 700 °C, the I type crack is dominant. With temperatures between 600 °C and 700 °C, the fracture surface has the two aforementioned fracture characteristics. Ref. [3] mentions that at lower temperatures, octahedral slip systems will cause crystallographic fracture. At higher temperatures the 'wavy' slip will cause the fracture to be perpendicular to the loading direction and crack initiation sites will be mostly located on micro-holes near the subsurface. Most crack initiations comprise multiple sources.

Ref. [4] considers that at room temperature and at 300 °C the cracks exhibit a non-crystallographic expansion mode and the crack initiation sites are mostly located on the persistent slip band. At 600 °C, cracks tend to expand along the crystallographic slip plane, and cracks propagating along slip lines on surfaces of specimens have been observed. Sliding surfaces and surface slip lines correspond to the primary octahedral slip system, so resolved shear stress plays an important role in crack initiation and propagation.

A large amount of literature describes the activation of nickel-based single crystal slip systems. For the SC16 nickel-based single crystal superalloy, [5] considers that the primary octahedral slip system is activated in the [001] loading direction at a temperature of 950 °C. Plastic deformation was found on the corresponding crystallographic plane. Ref. [6] mentions that inhomogeneous planar dislocations distributed in strip form have been observed from room temperature to 800 °C, and that the dislocation structure became gradually homogeneous as the temperature was further increased. For the nickel-based single crystal superalloys, it is believed that with increases in experimental temperature, the cubic slip system is gradually activated when loaded in the [001] direction [7]. As the temperature increases further, the quantity of the cubic slip also increases further, and the quantity of the octahedral slip gradually decreases. In Ref. [8], for nickel-based single crystal superalloys, it has been noted that inelastic deformation corresponding to the [001] loading direction is dominated by octahedron slip systems and that inelastic deformation corresponding to the [111] loading direction is dominated by cubic slip systems. Secondary octahedral systems have been observed only after a long period of creep deformation. Creep fatigue interaction experiments have been conducted for the SRR99 nickel-based single crystal superalloy at 950 °C, and the phenomenological model of the nickel-based single crystal superalloy has been proposed based on the isotropous constitutive model. It can be seen that with an increase in temperature the primary octahedral slip system and the cubic slip system are both activated, and under the influence of creep, the secondary octahedral slip system is also activated. Therefore, at higher temperatures, there is a certain influence of the three slip systems on the fatigue life of nickel-based single crystals.

Some literature has determined the activated slip system by observing slip lines on the surface of specimens. In Ref. [9], monotonic tensile tests were carried out for PWA1480 notch specimens at room temperature. The primary octahedral slip system was considered active. The maximum resolved shear stress on the specimen surface near the notch was calculated. Slip lines based on maximum resolved shear stress were consistent with experimental observations of the surface. In Ref. [10], a cylindrical indentation experiment using PWA1480 at room temperature was carried out. It was also noted that the primary octahedral slip system was activated. The numerical calculation results of the surface slip lines of the specimens were consistent with experimental observations. For the PWA1480 notch specimens [11], a monotonic tensile experiment was carried out at room temperature, obtaining the same conclusions as those drawn in [9,10]. In Ref. [12], a monotonic tensile experiment of copper single crystal notch specimens at room temperature was conducted. It was noted that the primary octahedral slip system was activated, and it was postulated that single slips, double slips, and multiple slips might be generated with different loading directions, theoretically. However, there are manufacturing deviations and loading direction deviations due to the installation process of the specimens causing there to be only a single slip with a [001] loading direction. Thus, in the beginning, the largest resolved shear stress in primary octahedral slip systems plays a significant role. It can be seen that only the primary octahedral slip system is activated at room temperature and the corresponding maximum resolved shear stress plays a major role.

The theoretical calculation methods regarding resolved shear stress and shear strain of nickel-based single crystals are listed in the theoretical formulae section. Eight kinds of nickel-based single crystal fatigue life prediction models are listed in the next section, and the viewpoint that maximum resolved shear stress corresponding to the two slip systems in a nickel-based single crystal high temperature fatigue experiment work together is proposed. A nickel-based single crystal fatigue life prediction model based on modified resolved shear stress amplitude is proposed. In the third section, for the four groups of fatigue data, eight classical fatigue life prediction models are compared with the model proposed in this paper. The advantages and disadvantages of the current several nickel-based single crystal fatigue life prediction models and the model proposed in this paper are separately described and discussed.

2. Methods

2.1. Elastic Stress and Strain Calculation of Nickel-Based Single Crystal

The coordinate system consisting of the three principal axes is the material coordinate system $oxyz$. Correspondingly the calculation coordinate system is defined by $ox'y'z'$. The property of each material axis is described by three elastic parameters, respectively; these are the elastic modulus E, Poisson's ratio υ, and shear modulus G, where $G = \frac{E}{2(1+v)}$. According to the theory of elastic mechanics, in the material coordinate system, the stress-strain relation is $\varepsilon = C\sigma$ or $\sigma = D\varepsilon$, in which C is the flexibility matrix and D is the elasticity matrix. ε and σ represent the strain and stress vectors, respectively:

$$\varepsilon = \begin{bmatrix} \varepsilon_x \\ \varepsilon_y \\ \varepsilon_z \\ \gamma_{xy} \\ \gamma_{yz} \\ \gamma_{zx} \end{bmatrix}, \sigma = \begin{bmatrix} \sigma_x \\ \sigma_y \\ \sigma_z \\ \sigma_{xy} \\ \sigma_{yz} \\ \sigma_{zx} \end{bmatrix}, \tag{1}$$

Nickel-based single crystal material is one type of the commonly used materials utilized for turbine blades and is an orthotropic cubic symmetric material. In the three major axis directions of the material coordinate system, the elastic parameters of the material are equal, respectively. The elastic parameters in the three directions are: $E_x = E_y = E_z = E$, $\upsilon_{xy} = \upsilon_{yz} = \upsilon_{zx} = \upsilon$, and $G_{xy} = G_{yz} = G_{zx} = G$. l_i, m_i, and n_i are the cosines of the angles between the material coordinate system and the calculation coordinate system, respectively. The specific correspondence relationship is shown in Table 1.

Table 1. Direction cosines between coordinate axes in the different coordinate systems.

Direction Cosines	Material CSYS		
	x	y	z
x'	l_1	m_1	n_1
y'	l_2	m_2	n_2
z'	l_3	m_3	n_3

According to geometry relationships, the transformation matrix A and B of the material coordinate system and the calculation coordinate system can be expressed as

$$A = \begin{bmatrix} l_1{}^2 & m_1{}^2 & n_1{}^2 & 2l_1m_1 & 2m_1n_1 & 2l_1n_1 \\ l_2{}^2 & m_2{}^2 & n_2{}^2 & 2l_2m_2 & 2m_2n_2 & 2l_2n_2 \\ l_3{}^2 & m_3{}^2 & n_3{}^2 & 2l_3m_3 & 2m_3n_3 & 2l_3n_3 \\ l_1l_2 & m_1m_2 & n_1n_2 & l_1m_2+l_2m_1 & m_1n_2+m_2n_1 & l_1n_2+l_2n_1 \\ l_2l_3 & m_2m_3 & n_2n_3 & l_2m_3+l_3m_2 & m_2n_3+m_3n_2 & l_2n_3+l_3n_2 \\ l_1l_3 & m_1m_3 & n_1n_3 & l_1m_3+l_3m_1 & m_1n_3+m_3n_1 & l_1n_3+l_3n_1 \end{bmatrix} \tag{2}$$

Furthermore, $B = (A^{-1})^{\mathrm{T}}$. By using the transformation matrix, stress and strain in the calculation coordinate system can be obtained [13]:

$$\sigma' = A\sigma, \tag{3}$$

$$\varepsilon' = B\varepsilon, \tag{4}$$

From $\sigma = D\varepsilon$ and $\sigma' = D_{x'y'z'}\varepsilon'$, $D_{x'y'z'} = ADA^{\mathrm{T}}$ can be obtained; similarly, $C_{x'y'z'} = BCB^{\mathrm{T}}$ can be obtained. In actual engineering analysis, the material coordinate system and the calculation coordinate system are often not uniform and between them there is a certain angle. The coordinate transformation matrix A can be used to convert the elasticity matrix D in the material coordinate system to the elasticity matrix $D_{x'y'z'}$ in the calculation coordinate system. $D_{x'y'z'}$ contains 21 different elements:

$$D_{x'y'z'} = ADA^{\mathrm{T}} = \begin{bmatrix} C_{11} & C_{12} & C_{13} & C_{14} & C_{15} & C_{16} \\ & C_{22} & C_{23} & C_{24} & C_{25} & C_{26} \\ & & C_{33} & C_{34} & C_{35} & C_{36} \\ & & & C_{44} & C_{45} & C_{46} \\ & & & & C_{55} & C_{56} \\ & & & & & C_{66} \end{bmatrix} \tag{5}$$

When the elastic moduli $E_{[001]}$, $E_{[011]}$, and $E_{[111]}$ are known, the three independent material parameters E, υ, and G can be calculated using the coordinate transformation.

2.2. Nickel-Based Single Crystal Resolved Shear Stress and Resolved Shear Strain

In the material coordinate system, the relationship between resolved shear stress and the stress tensor is $\tau^{(a)} = \sigma : P^{(a)}$, in which $P^{(a)} = \frac{1}{2}(m^{(a)}n^{(a)\mathrm{T}} + n^{(a)}m^{(a)\mathrm{T}})$, $m^{(a)}$ is the sliding direction of the α slip system, and $n^{(\alpha)}$ is the normal direction of the plane of the α slip system. The solution to the resolved shear stresses of the 12 primary octahedral slip systems is showed in Equation (6). The solution to the resolved shear stresses of the 12 secondary octahedral slip systems is shown in Equation (7). The solution to the resolved shear stresses of the six cubic slip systems is shown in Equation (8). The corresponding resolved shear strains of the different slip systems were calculated by a similar formula to that used for the resolved shear stresses.

$$\begin{Bmatrix} \tau_1 \\ \tau_2 \\ \tau_3 \\ \tau_4 \\ \tau_5 \\ \tau_6 \\ \tau_7 \\ \tau_8 \\ \tau_9 \\ \tau_{10} \\ \tau_{11} \\ \tau_{12} \end{Bmatrix} = \frac{1}{\sqrt{6}} \begin{bmatrix} 1 & 0 & -1 & 1 & -1 & 0 \\ 0 & -1 & 1 & -1 & 0 & 1 \\ 1 & -1 & 0 & 0 & -1 & 1 \\ -1 & 0 & 1 & 1 & -1 & 0 \\ -1 & 1 & 0 & 0 & -1 & -1 \\ 0 & 1 & -1 & -1 & 0 & -1 \\ 1 & -1 & 0 & 0 & -1 & -1 \\ 0 & 1 & -1 & -1 & 0 & 1 \\ 1 & 0 & -1 & -1 & -1 & 0 \\ 0 & -1 & 1 & -1 & 0 & -1 \\ -1 & 0 & 1 & -1 & -1 & 0 \\ -1 & 1 & 0 & 0 & -1 & 1 \end{bmatrix} \begin{Bmatrix} \sigma_{xx} \\ \sigma_{yy} \\ \sigma_{zz} \\ \sigma_{xy} \\ \sigma_{yz} \\ \sigma_{zx} \end{Bmatrix} \tag{6}$$

$$\begin{Bmatrix} \tau_{13} \\ \tau_{14} \\ \tau_{15} \\ \tau_{16} \\ \tau_{17} \\ \tau_{18} \\ \tau_{19} \\ \tau_{20} \\ \tau_{21} \\ \tau_{22} \\ \tau_{23} \\ \tau_{24} \end{Bmatrix} = \frac{1}{3\sqrt{2}} \begin{bmatrix} -1 & 2 & -1 & 1 & 1 & -2 \\ 2 & -1 & -1 & 1 & -2 & 1 \\ -1 & -1 & 2 & -2 & 1 & 1 \\ -1 & 2 & -1 & -1 & -1 & -2 \\ -1 & -1 & 2 & 2 & -1 & 1 \\ 2 & -1 & -1 & -1 & 2 & 1 \\ -1 & -1 & 2 & 2 & 1 & -1 \\ 2 & -1 & -1 & -1 & -2 & -1 \\ -1 & 2 & -1 & -1 & 1 & 2 \\ 2 & -1 & -1 & 1 & 2 & -1 \\ -1 & 2 & -1 & 1 & -1 & 2 \\ -1 & -1 & 2 & -2 & -1 & -1 \end{bmatrix} \begin{Bmatrix} \sigma_{xx} \\ \sigma_{yy} \\ \sigma_{zz} \\ \sigma_{xy} \\ \sigma_{yz} \\ \sigma_{zx} \end{Bmatrix} \tag{7}$$

$$\begin{Bmatrix} \tau_{25} \\ \tau_{26} \\ \tau_{27} \\ \tau_{28} \\ \tau_{29} \\ \tau_{30} \end{Bmatrix} = \frac{1}{\sqrt{2}} \begin{bmatrix} 0 & 0 & 0 & 1 & 0 & 1 \\ 0 & 0 & 0 & 1 & 0 & -1 \\ 0 & 0 & 0 & 1 & 1 & 0 \\ 0 & 0 & 0 & 1 & -1 & 0 \\ 0 & 0 & 0 & 0 & 1 & 1 \\ 0 & 0 & 0 & 0 & 1 & -1 \end{bmatrix} \begin{Bmatrix} \sigma_{xx} \\ \sigma_{yy} \\ \sigma_{zz} \\ \sigma_{xy} \\ \sigma_{yz} \\ \sigma_{zx} \end{Bmatrix} \tag{8}$$

The formula $\sigma_{i'j'} = l_{i'i}l_{j'j}\sigma_{ij}$ can be used to convert the stress tensor in the $oxyz$ coordinate system to the $ox'y'z'$ coordinate system. The ox' axis in the new coordinate system $ox'y'z'$ corresponds to the normal direction of the slip plane and hence the normal stress of the slip plane can be calculated as

$$\sigma_{x'x'} = l_{x'i}l_{x'j}\sigma_{ij}, \text{ in which } l_{x'x} = \frac{\overrightarrow{ox'} \cdot \overrightarrow{ox}}{\left|\overrightarrow{ox'}\right| \times \left|\overrightarrow{ox}\right|}, l_{x'y} = \frac{\overrightarrow{ox'} \cdot \overrightarrow{oy}}{\left|\overrightarrow{ox'}\right| \times \left|\overrightarrow{oy}\right|}, \text{ and } l_{x'z} = \frac{\overrightarrow{ox'} \cdot \overrightarrow{oz}}{\left|\overrightarrow{ox'}\right| \times \left|\overrightarrow{oz}\right|}. \text{ The direction cosines}$$

between the old and new coordinate systems are shown in Table 2.

Table 2. Direction cosines between the two coordinate systems.

Direction Cosines	x	y	z
x'	l_{11}	l_{12}	l_{13}
y'	l_{21}	l_{22}	l_{23}
z'	l_{31}	l_{32}	l_{33}

2.3. Nickel-Based Single Crystal Fatigue Life Prediction Model

This article lists eight classic nickel-based single crystal fatigue life prediction models (see Table 3); parameters in these models may be calculated using Equations (6)–(8). The damage parameter of model (1) is the maximum resolved shear stress amplitude, which may be obtained by taking the maximum value of the 30 resolved shear stress amplitudes. The damage parameter of model (2) is the maximum value of the 12 resolved shear stress amplitudes corresponding to the primary octahedral slip system. The model (3) damage parameter is the maximum resolved shear strain amplitude, which may be obtained by taking the maximum value of the 30 resolved shear strain amplitudes. The damage parameter of model (4) is the maximum value of the 12 resolved shear strain amplitudes corresponding to the primary octahedral slip system. The model (5) damage parameter is the maximum shear stress range. Model (6) obtains the resolved shear strain range, the maximum resolved shear stress, the normal strain amplitude, and the maximum normal stress respectively corresponding to the 30 slip directions. The left side of model (6) is first obtained, the maximum value of which is taken as the damage parameter. For model (7), the resolved shear stress amplitudes and corresponding maximum normal stresses are obtained respectively considering the 30 slip directions. The left side of model (7) is first obtained, the maximum value of which is taken as the damage parameter. For model (8), the maximum resolved shear stress amplitude and the maximum normal stress in the 30 slip directions are calculated, the combination of which is used as the damage parameter. The difference between model (7) and model (8) is that for the former the resolved shear stress amplitude and the maximum normal stress correspond, while for the latter resolved shear stress amplitude does not necessarily correspond to the maximum normal stress.

Table 3. Nickel-based single crystal fatigue life prediction model [14,15].

(1) Maximum resolved shear stress amplitude	$\frac{\Delta\tau}{2} = a(N_f)^b$
(2) Maximum resolved shear stress amplitude (primary octahedral slip system)	$\frac{\Delta\tau_{oct\_prim}}{2} = a(N_f)^b$
(3) Maximum resolved shear strain amplitude	$\frac{\Delta\varepsilon_s}{2} = a(N_f)^b$
(4) Maximum resolved shear strain amplitude (primary octahedral slip system)	$\frac{\Delta\varepsilon_{s,oct\_prim}}{2} = a(N_f)^b$
(5) Maximum shear stress range	$\frac{\Delta\sigma}{2} = \Delta\tau_{max} = a(N_f)^b$
(6) Chu-Conle-Bonnen (CCB) model	$\left(2\Delta\varepsilon_s\tau_{max} + \frac{\Delta\varepsilon_n}{2}\sigma_{max}\right)_{max} = a(N_f)^b$
(7) Findley (Fin) model	$\left(\frac{\Delta\tau}{2} + k\sigma_{max}\right)_{max} = a(N_f)^b, k = 1$
(8) McDiarmid (McD) model	$\frac{\Delta\tau}{2} + k\sigma_{max} = a(N_f)^b, k = 0.1$

2.4. Modified Life Prediction Model

A single crystal fatigue life prediction model based on a modified resolved shear stress amplitude is proposed. M_1 and M_2 are the maximum and median values of the Schmid factor corresponding to the maximum resolved shear stress in the primary octahedral slip system, the secondary octahedral slip system, and the cubic slip system, respectively, which are obtained from the following two equations:

$$M_1 = \max\left\{\Delta\tau_{oct\_prim}, \Delta\tau_{oct\_sec}, \Delta\tau_{cub}\right\} / \frac{\Delta\sigma}{2} \tag{9}$$

$$M_2 = \text{median}\left\{\Delta\tau_{oct\_prim}, \Delta\tau_{oct\_sec}, \Delta\tau_{cub}\right\} / \frac{\Delta\sigma}{2} \tag{10}$$

The nickel-based single crystal fatigue life prediction model can be obtained as follows:

$$\frac{\Delta\sigma}{2} \times \frac{(M_1 + M_2)}{2} = a(N_f)^b \tag{11}$$

According to fatigue experiment data about different loading directions of nickel-based single crystal materials, compared to other classic nickel-based single crystal fatigue life models, the fatigue life prediction accuracy of the proposed model is higher. Compare the Schmid factor with the modified factor $(M_1 + M_2)/2$, as shown in Figure 1. Figure 1a corresponds to the maximum Schmid factor only considering the primary octahedral slip system. Figure 1b corresponds to the maximum Schmid factor only considering the secondary octahedral slip system. Figure 1c corresponds to the maximum Schmid factor considering the cubic slip system. Figure 1d corresponds to the maximum Schmid factor considering all three slip systems. In Figure 1e, compared to Figure 1d, only the primary octahedral slip system and the cubic slip system are activated. Figure 1f corresponds to the model proposed in this paper, namely, the modified factor $(M_1 + M_2)/2$. The plane formed by the x-axis and y-axis corresponds to the standard projection plane. Because of the spatial symmetry of the nickel-based single crystal material, all the loading directions can be represented by one point in the black line zone.

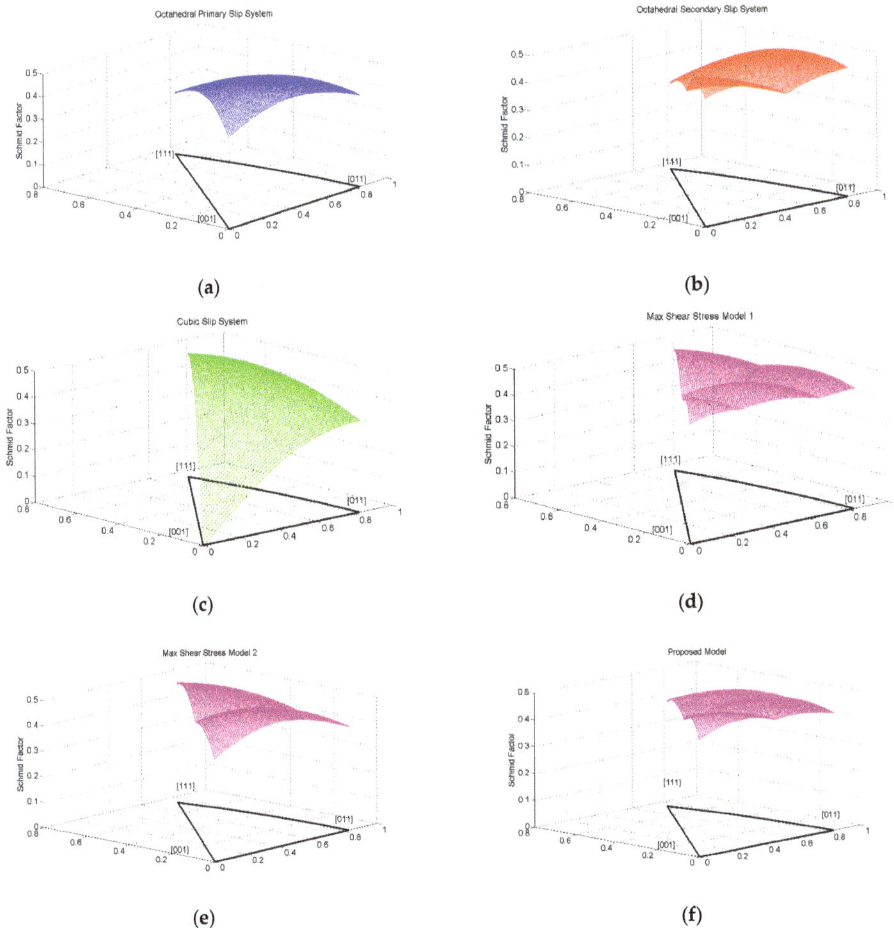

Figure 1. Schmid factors and modified factor: (**a**) primary octahedral slip system; (**b**) secondary octahedral slip system; (**c**) cubic slip system; (**d**) all three slip systems; (**e**) primary octahedral slip system and cubic system; and (**f**) modified factor.

When considering only the primary octahedral slip system, the Schmid factor corresponding to the [111] direction is less than that for the [001] and [011] directions. The experiment data shows that the stress level in the [111] direction is greater than that in the [001] and [011] directions in the uniaxial loading with the same fatigue life. However, when the fatigue life prediction method only considering the resolved shear stress of the primary octahedral slip system is adopted, the calculated resolved shear stress in the [111] direction tends to be smaller. According to [14,16], three slip systems for low cyclic fatigue at 648 °C can be activated for PWA1480 material and it has been concluded that the maximum resolved shear stress amplitude model has good prediction effect. For the primary octahedral slip system and the cubic slip system, the maximum Schmid factor for the [111] loading direction is greater than the maximum Schmid factor for the [001] and [011] loading directions. For the three slip systems, the maximum Schmid factor for the [111] loading direction is equal to the maximum Schmid factor for the [001] and [011] loading directions. However, for the monotonic tensile tests, the stress loading in the [111] direction is often greater than that in the [001] and [011] directions with the same fatigue life. Therefore, the nickel-based single crystal fatigue life prediction model which considers the three slip systems or the primary octahedral slip system and the cubic slip system is often inconsistent with its experimental results; this will be discussed further.

3. Results

For the above nine nickel-based single crystal fatigue life prediction models, the high cyclic fatigue data for DD6 material at 700 °C and 800 °C, the low cycle fatigue data for PWA1480 material at 648 °C, and the high cycle fatigue data for PWA1484 material at 593 °C were used to predict uniaxial fatigue life. The plasticity effect was not considered. The basic material properties of the three materials are shown in Table 4.

Table 4. Material properties.

Material Properties	DD6 (700 °C)	DD6 (800 °C)	PWA1480 (648 °C)	PWA1484 (593 °C)
E/GPa	107.0	102.2	106.2	108.2
G/GPa	100.2	85.3	108.3	109.8
ν	0.3740	0.3797	0.4009	0.3995

On the one hand, the fatigue life prediction model was selected by comparing the adjusted coefficient of determination (Adj. R_Square), where the coefficient of determination is $R^2 = 1 - \frac{\sum_{i=1}^{n}(y_i - \widehat{y}_i)^2}{\sum_{i=1}^{n}(y_i - \bar{y}_i)^2}$ and the adjusted coefficient of determination is $\text{Adj.R\_Square} = 1 - \frac{\sum_{i=1}^{n}(y_i - \widehat{y}_i)^2/(n-k-1)}{\sum_{i=1}^{n}(y_i - \bar{y}_i)^2(n-1)}$. n_i is the experimental sample freedom and k is the number of explanatory variables in the model (excluding the constants in the model). On the other hand, the fatigue life prediction model was selected by comparing the size of the fatigue life dispersion zone. In view of the fact that the calculation results are based on pure elastic static analysis, the prediction results of the single crystal life prediction models, including the elastic strain, tend to be poor, as shown by Model (3) and Model (4) in Table 3. Fatigue life prediction results of Model (3) and Model (4) are not listed.

3.1. High Cycle Fatigue Life Prediction for DD6 Material

(1) DD6 material at 700 °C

High cyclic fatigue data for DD6 material at 700 °C was selected for analysis [17]. The experiments were carried out according to the loading directions of [001], [011], and [111], with a stress ratio of −1. The results are shown in Figures 2 and 3.

For the high cyclic fatigue experiment using DD6 material at 700 °C with a stress ratio of −1, the fatigue life model with stress amplitude as the damage parameter, the fatigue life model with the maximum resolved shear stress amplitude considering 30 slip systems as the damage parameter, the Fin model, the McD model, and the model proposed in this paper produced relatively higher Adj. R_Square values and better life prediction results. The results of the fatigue life prediction model which only uses the resolved shear strain as the damage parameter were poor. The fatigue life prediction results obtained using the model which only considers the maximum resolved shear stress of the primary octahedral slip system as the damage parameter, and the Chu-Conle-Bonnen (CCB) model, were worse. The fatigue life prediction results of most data points using modified maximum resolved shear stress amplitude as the damage parameter proposed in this paper were within three times of the dispersion band. The fatigue life prediction model proposed in this paper can be used to predict fatigue life with high accuracy for high cyclic fatigue experimental data for DD6 material at 700 °C with a stress ratio of −1.

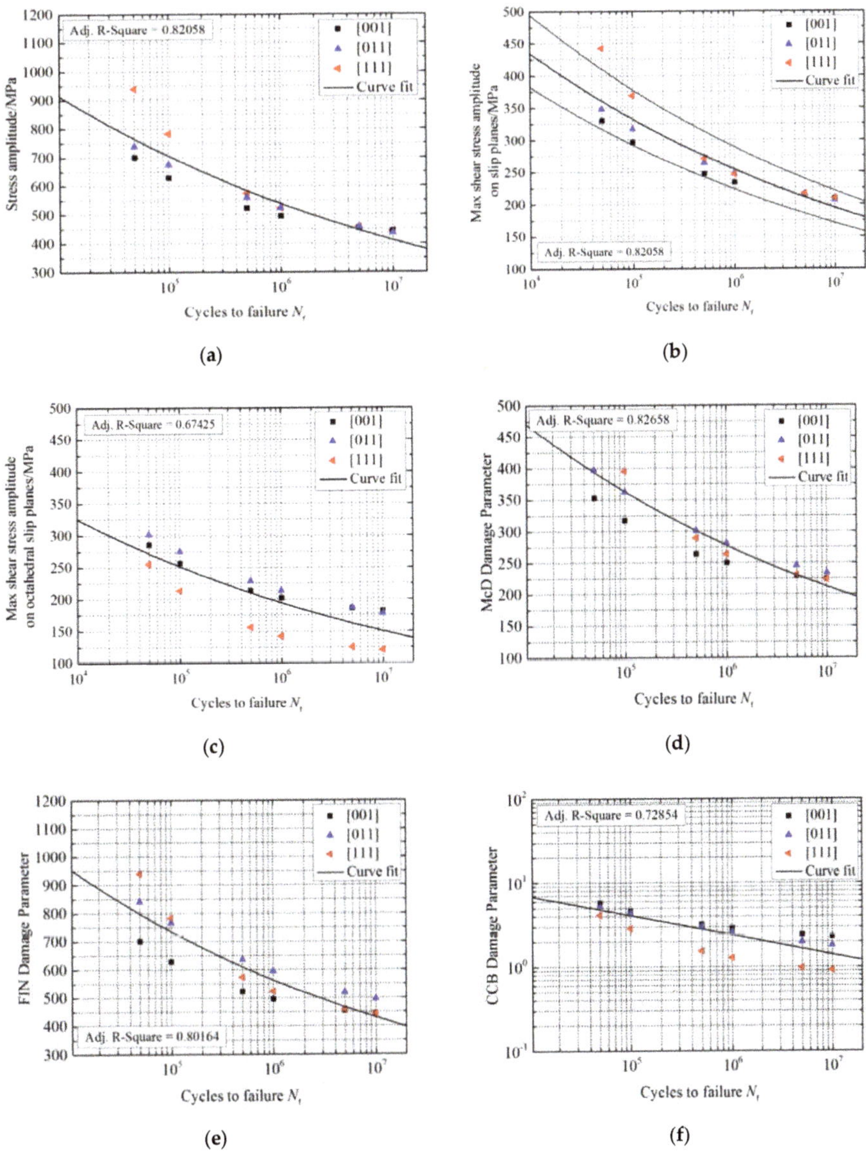

Figure 2. Fatigue life predictions for DD6 material at 700 °C: (**a**) stress amplitude; (**b**) maximum resolved shear stress amplitude; (**c**) maximum resolved shear stress amplitude (primary octahedral slip system); (**d**) McD model; (**e**) Fin model; (**f**) CCB model.

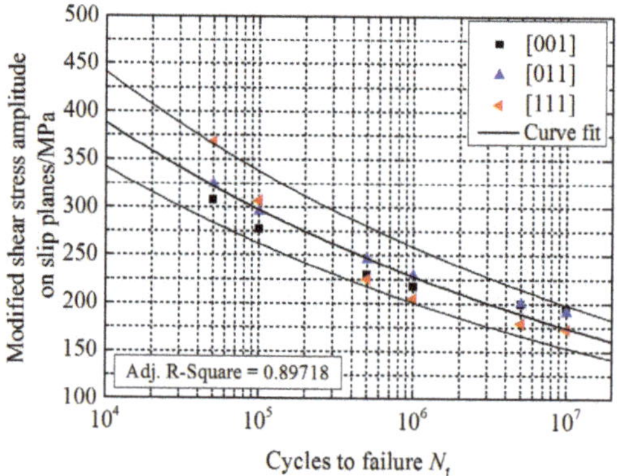

Figure 3. Fatigue life prediction for DD6 material at 700 °C using the proposed model.

(2) DD6 material at 800 °C

High cyclic fatigue data for DD6 material at 800 °C was selected for analysis [17]. Experiments were carried out according to the loading directions of [001], [011], and [111], and the stress ratio was −1. The fatigue life prediction results are shown in Figures 4 and 5.

For the high cyclic fatigue experiment for DD6 material at 800 °C with a stress ratio of −1, the fatigue life model with stress amplitude as the damage parameter, the fatigue life model with maximum resolved shear stress amplitude considering the 30 slip systems as the damage parameter, the Fin model, the McD model, and the model proposed model in this paper gave relatively higher Adj. R_Square values and better life prediction results. The comparison results are the same as for DD6 material at 700 °C. The fatigue life prediction results of all the data points using modified maximum resolved shear stress amplitude as the damage parameter proposed in this paper were within three times of the dispersion band. The fatigue life prediction model proposed in this paper can be used to predict fatigue life with high accuracy for high cyclic fatigue experimental data for DD6 material at 800 °C with a stress ratio of −1.

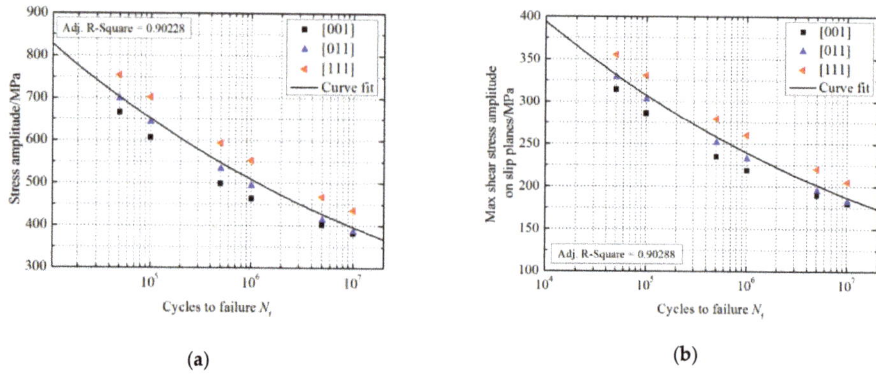

(a)

(b)

Figure 4. *Cont.*

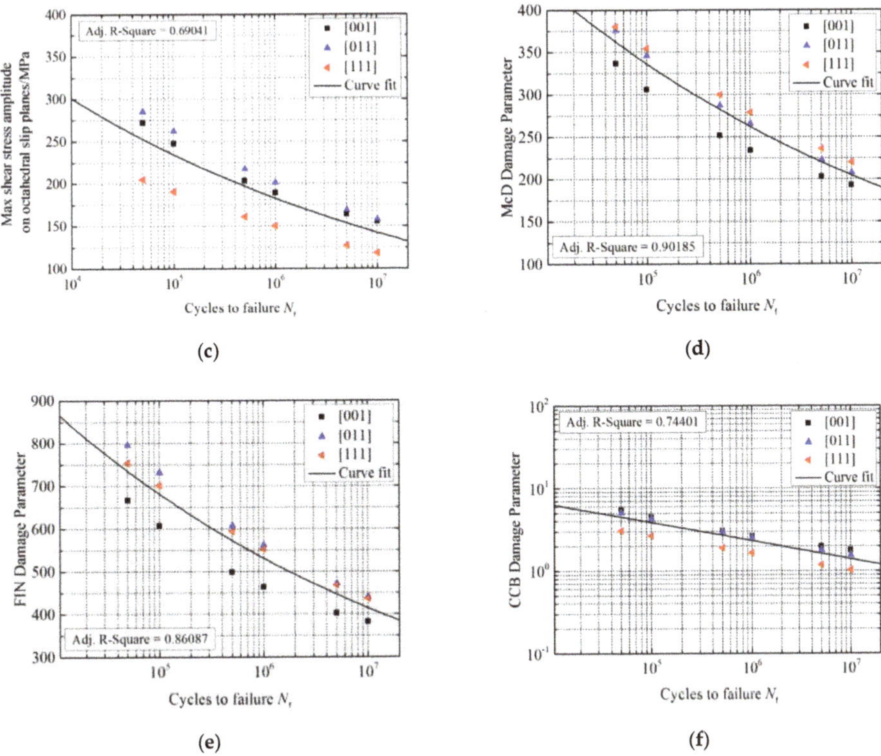

Figure 4. Fatigue life predictions for DD6 material at 800 °C: (**a**) stress amplitude; (**b**) maximum resolved shear stress amplitude; (**c**) maximum resolved shear stress amplitude (primary octahedral slip system); (**d**) McD model; (**e**) Fin model; (**f**) CCB model.

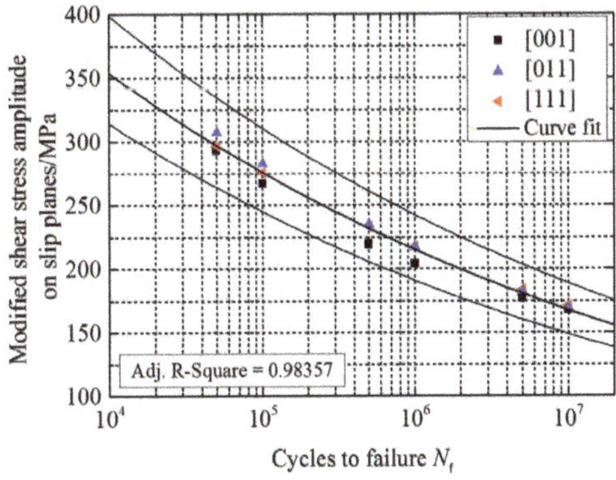

Figure 5. Fatigue life prediction for DD6 material at 800 °C by the proposed model.

3.2. Low Cycle Fatigue Life Prediction for PWA1480 Material

Analysis was performed for PWA1480 material with 648 °C low cycle fatigue data [14]. In order to facilitate comparison with the literature, experiments were conducted using the [001], [011], [111], and [213] loading directions, respectively, with different strain ratios controlled. The stress tensor and resolved shear stress were obtained by using the elasticity matrix [14]. The fatigue life prediction results are shown in Figures 6 and 7.

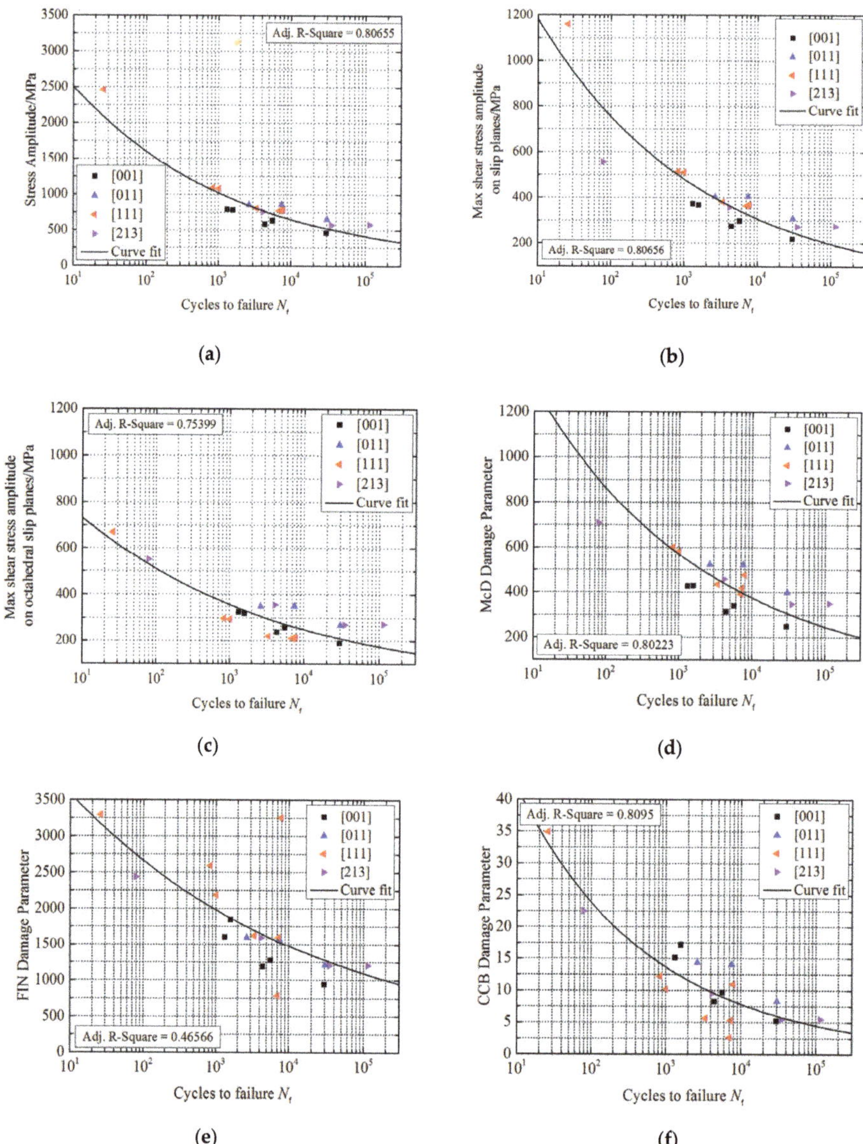

Figure 6. Fatigue life predictions for PWA1480 material at 648 °C: (**a**) stress amplitude; (**b**) maximum resolved shear stress amplitude; (**c**) maximum resolved shear stress amplitude (primary octahedral slip system); (**d**) McD model; (**e**) Fin model; (**f**) CCB model.

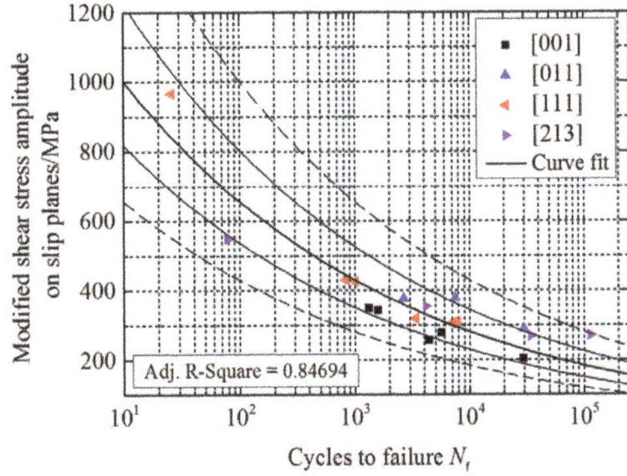

Figure 7. Fatigue life prediction for PWA1480 material at 648 °C using the proposed model.

For the low cyclic fatigue experiment using PWA1480 material at 648 °C with different strain ratios, the fatigue life model with stress amplitude as the damage parameter, the fatigue life model with maximum resolved shear stress amplitude considering the 30 slip systems as the damage parameter, the CCB model, the McD model, and the model proposed in this paper produced relatively higher Adj. R_Square values and better life prediction results. The fatigue life prediction results of most data points using modified maximum resolved shear stress amplitude as the damage parameter proposed in this paper were within three times of the dispersion band. The fatigue life prediction results of all the data points were within ten times of the dispersion band. The fatigue life prediction model proposed in this paper can be used to predict fatigue life with high accuracy for low cyclic fatigue experimental data for PWA1480 material at 648 °C with different strain ratios.

3.3. Fatigue Life Prediction of PWA1484 Material

Analysis was performed on the PWA1484 material at 593 °C with a stress ratio of 0.1 fatigue data [15]. Experiments were carried out according to the loading directions of [001], [011], and [111], respectively. The experiment loading frequency was 30 Hz. The fatigue life prediction results are shown in Figures 8 and 9.

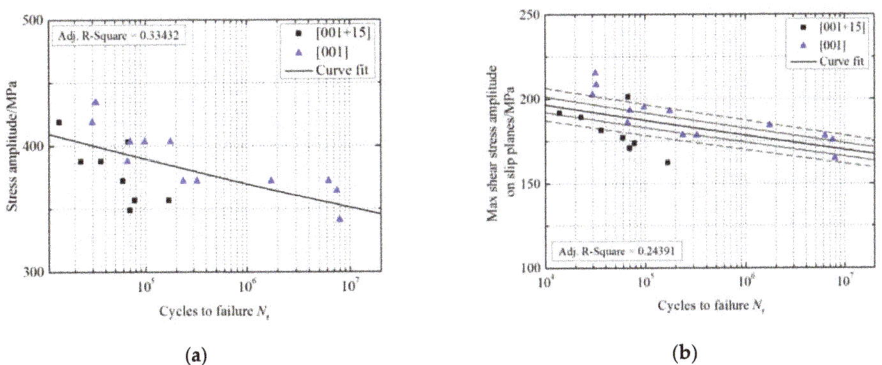

(a)

(b)

Figure 8. *Cont.*

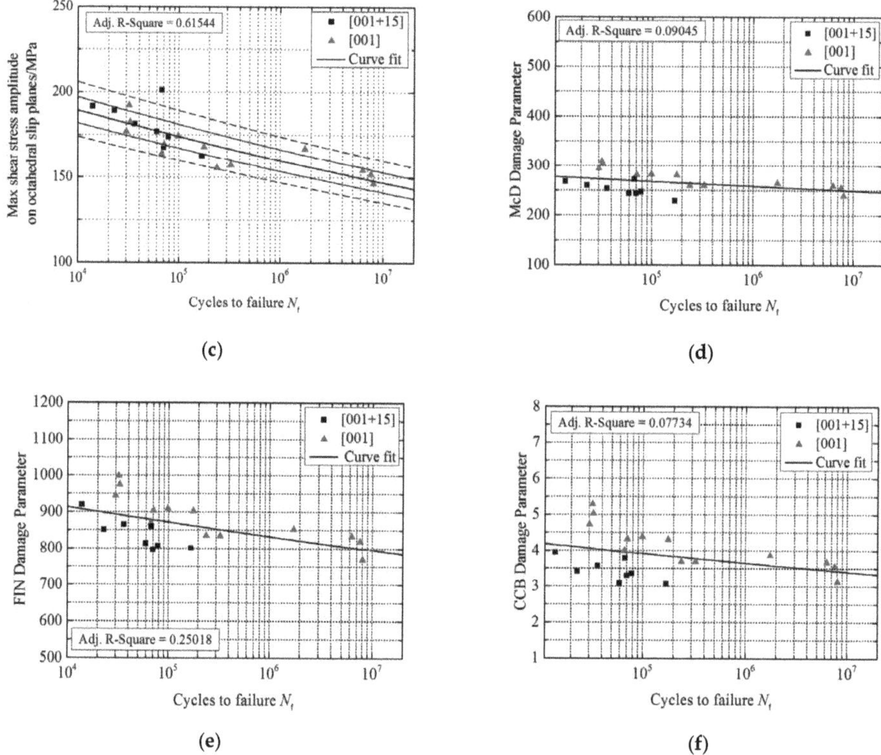

Figure 8. Fatigue life predictions for PWA1484 material at 593 °C: (**a**) stress amplitude; (**b**) maximum resolved shear stress amplitude; (**c**) maximum resolved shear stress amplitude (primary octahedral slip system); (**d**) McD model; (**e**) Fin model; (**f**) CCB model.

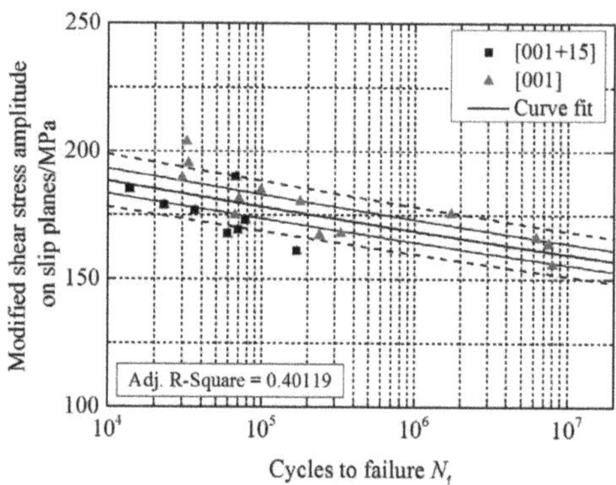

Figure 9. Fatigue life prediction for PWA1484 material at 593 °C using the proposed model.

For the fatigue experiment data for PWA1484 material at 593 °C with a stress ratio of 0.1, due to the large data dispersion, the fatigue life prediction results using the eight models mentioned above and the model proposed in this paper were worse. The life prediction model could not ensure that the prediction effect was located within ten times of dispersion. However, compared with the life prediction results of the other eight kinds of damage parameters, the Adj. R_Square value of the model proposed in this paper was higher and was only inferior to that value for the model with maximum resolved shear stress of the primary octahedral slip system. This was mainly due to the lower experimental temperature which resulted in the primary octahedral slip system playing a dominant role and the role of the secondary octahedral slip system and the cubic slip system not being obvious.

4. Discussion

Based on the prediction results of the four groups of nickel-based single crystal fatigue life data, all the fatigue life prediction models are summarized in Table 5.

Table 5. Nickel-based single crystal fatigue life prediction models.

(1) $\frac{\Delta\tau}{2} = a(N_f)^b$	With maximum resolved shear stress amplitude of the 30 slip systems as the damage parameter, all of the fatigue life prediction results are good.
(2) $\frac{\Delta\tau_{oct\_prim}}{2} = a(N_f)^b$	With maximum resolved shear stress amplitude of the primary octahedral slip system as the damage parameter, all of the fatigue life prediction results are worse.
(3) $\frac{\Delta\varepsilon_s}{2} = a(N_f)^b$, (4) $\frac{\Delta\varepsilon_{s,oct\_prim}}{2} = a(N_f)^b$	Models with resolved shear strain amplitude as the damage parameter generally have poor prediction results for nickel-based single crystal materials.
(5) $\frac{\Delta\sigma}{2} = \Delta\tau_{max} = a(N_f)^b$	The fatigue life prediction model with stress amplitude or maximum shear stress as the damage parameter has a higher life prediction accuracy.
(6) CCB $\left(2\Delta\varepsilon_s\tau_{max} + \frac{\Delta\varepsilon_n}{2}\sigma_{max}\right)_{max} = a(N_f)^b$	Life prediction results obtained by using the CCB model for PWA1480 low cycle fatigue data are better, while the life prediction results of the CCB model for DD6 high cycle fatigue data are worse.
(7) Fin $\left(\frac{\Delta\tau}{2} + k\sigma_{max}\right)_{max} = a(N_f)^b, k = 1$	Life prediction results for DD6 high cycle fatigue data obtained using the Fin model are better, while those for PWA1480 low cycle fatigue data obtained using the Fin model are worse.
(8) McD $\frac{\Delta\tau}{2} + k\sigma_{max} = a(N_f)^b, k = 0.1$	The McD model has a higher life prediction accuracy.
(9) $\frac{\Delta\sigma}{2}\frac{(M_1+M_2)}{2} = a(N_f)^b$ $M_1 = \max\{\Delta\tau_{oct\_prim}, \Delta\tau_{oct\_sec}, \Delta\tau_{cub}\}/\frac{\Delta\sigma}{2}$ $M_2 = \mathrm{median}\{\Delta\tau_{oct\_prim}, \Delta\tau_{oct\_sec}, \Delta\tau_{cub}\}/\frac{\Delta\sigma}{2}$	The nickel-based single crystal fatigue life prediction model based on modified resolved shear stress amplitude, which has been proposed in this paper, has higher life prediction accuracy.

Ref. [18] focuses on a fatigue experiment using thin plates with cooling holes at 900 °C for DD6 material, establishing the nickel-based single crystal plastic constitutive model and combining it with the critical distance method to predict its fatigue life. The fatigue life prediction method used in [18] which is based on a nickel-based single crystal fatigue constitutive model is complicated. It takes a long time to calculate the stress and strain of single crystal hollow blades with complex cooling structures. In addition, the program in [18] is unstable and is not easy to apply in engineering. Ref. [19] uses a modified Mücke's anisotropic model to predict fatigue life. Using a solution of a nonlinear equation to determine the model parameters in [19] is a tedious process. The multiaxial fatigue life prediction of a CMSX-2 nickel-based single crystal material at 900 °C with stress controlled for is performed in [20]. The result of the fatigue life prediction model with stress as the damage parameter is generally better than the fatigue life prediction model with strain as the damage parameter, which is consistent with the conclusion in this paper. Ref. [14,15,20] have adopted the same idea as this paper, selecting appropriate damage parameters and establishing nickel-based single crystal fatigue life prediction methods.

However, the model proposed in this paper does not consider the effect of plastic, which has no effect on predictions of high cycle fatigue life. The fatigue life prediction results are better than other models. For predictions of low cycle fatigue life, the material has often yielded, and the experiments are often strain controlled. The stress tensor and resolved shear stress are obtained by using the elasticity matrix; thus, the stress levels of some data points in Figure 6a will appear larger. In the absence of a stress gradient, the model proposed in this paper still has a good prediction accuracy of low cycle fatigue life. Although the corresponding damage parameters are fictitious, they still have a good one-to-one correspondence with life. Due to the deficiency of data, the effects of average stress and stress gradient are not considered in this paper. Ref. [21–23] have established life prediction models for the effects of stress gradient for uniaxial and multiaxial fatigue, respectively. As a next step, the right side of the model proposed in this paper can be modified to introduce the influence of average stress [24], and the stress gradient factor can be introduced into the model based on the nickel-based single crystal notch experiment [21].

In Ref. [25], monotonic tensile tests of a MD2 nickel-based single crystal superalloy at room temperature were conducted and it was observed that the cracks often initiated near the inclusions or holes. The initiation position was often located on the surface or near the surface of the specimen. In Ref. [5], monotonic tensile tests for SC16 at 950 °C were conducted, and it was concluded that micro-cracks on the surface and the casting holes of the specimens were often potential crack initiation sites. In Ref. [26], notch ultra-high cycle fatigue experiments for CMSX-4 and CM186LC at 850 °C were conducted, and crack initiation was often related to the interaction of the persistent slip band (PSB) with casting holes or carbides. At the same time, the life of crack initiation accounts for most of the total life. The actual life of the specimen often depends on the state of the material at the crack initiation position. Casting holes and inclusions often produce large stress concentrations, especially near the surface. Therefore, it is appropriate to predict the high cycle fatigue life of the material with crack initiation near the casting hole or the inclusion with stress as the damage parameter. To facilitate applications within engineering, it is still applicable to extend this method to low cycle fatigue life prediction.

5. Conclusions

- For the high cyclic fatigue experiments for DD6 material at 700 °C and 800 °C with a stress ratio of −1 and the low cyclic fatigue experiments for PWA1480 material at different strain ratios at 648 °C, several classical nickel-based single crystal fatigue life prediction models were compared. The fatigue life prediction model with stress amplitude as the damage parameter, the fatigue life prediction model considering maximum resolved shear stress of the 30 slip directions as the damage parameter, the McD model, and the modified maximum resolved shear stress amplitude fatigue life prediction proposed in this paper were observed to have higher accuracy.
- For fatigue data obtained for PWA1484 material at 593 °C with a stress ratio of −1, the fatigue life prediction model proposed in this paper and the model considering maximum resolved shear stress of the primary octahedral slip system as the damage parameter had better prediction results. At lower temperatures, the model considering maximum resolved shear stress of the primary octahedral slip system as the damage parameter has greater engineering application prospects.
- Fatigue life prediction models with strain as the damage parameter have poor prediction accuracy.
- By comparing the aforementioned nickel-based single crystal fatigue life prediction models, for nickel-based single crystals at higher temperature, the proposed method and the fatigue life model with stress amplitude directly as the damage parameter have higher prediction accuracy and hence greater engineering application value.

Author Contributions: Methodology, J.W., D.W. and Y.W.; Formal analysis, X.J.; Data curation, X.J.; Writing—original draft preparation, J.W.; Writing—review and editing, D.W. and Y.W.

Funding: This research was funded by the National Natural Science Foundation of China (grant No. 51475024).

Conflicts of Interest: The authors declare no conflict of interest. The funders had no role in the design of the study; in the collection, analyses, or interpretation of data; in the writing of the manuscript, or in the decision to publish the results.

Nomenclature

a	Fatigue strength coefficient
b	Fatigue strength exponent
A, B	Transformation matrix
C	Flexibility matrix
D	Elasticity matrix
E	Young's modulus
G	Shear modulus
l_i, m_i, n_i	Cosines of the angle between the material coordinate system and the calculation coordinate system
l_j	Firection cosines between the old and new coordinate systems
$m^{(a)}$	Sliding direction of the αth slip system
$n^{(a)}$	Normal direction of the plane of the αth slip system
M_1	Maximum value of the Schmid factor
M_2	Median value of the Schmid factor
N_f	Number of cycles to failure
ε	Strain tensor
$\Delta\varepsilon_{s,oct\_prim}$	Maximum resolved shear strain range (the primary octahedral slip system)
$\Delta\varepsilon_n$	Normal strain corresponding to the resolved shear strain
$\Delta\varepsilon_s$	Maximum resolved shear strain range
υ	Poisson's ratio
σ	Stress tensor
σ_{max}	Maximum normal stress
τ_{max}	Maximum shear stress
$\Delta\tau_{max}$	Maximum shear stress range
$\Delta\tau$	Maximum resolved shear stress range
$\tau^{(a)}$	Resolved shear stress of the αth slip system
$\Delta\tau_{s,oct\_prim}$	Maximum resolved shear stress range (the primary octahedral slip system)
$\Delta\tau_{oct\_sec}$	Maximum resolved shear stress range (the secondary octahedral slip system)
$\Delta\tau_{cub}$	Maximum resolved shear stress range (the cubic octahedral slip system)

References

1. Palmert, F.; Moverare, J.; Gustafsson, D.; Busse, C. Fatigue crack growth behaviour of an alternative single crystal nickel base superalloy. *Int. J. Fatigue* **2018**, *109*, 166–181. [CrossRef]
2. He, Z.W.; Zhang, Y.Y.; Qiu, W.H.; Shi, H.-J.; Gu, J. Temperature effect on the low cycle fatigue behavior of a directionally solidified nickel-base superalloy. *Mater. Sci. Eng. A* **2016**, *676*, 246–252. [CrossRef]
3. Ma, X.F.; Shi, H.J.; Gu, J.L.; Wang, Z.; Harders, H.; Malow, T. Temperature effect on low-cycle fatigue behavior of nickel-based single crystalline superalloy. *Acta Mech. Solida Sin.* **2008**, *21*, 289–297. [CrossRef]
4. Zhang, Y.Y.; Shi, H.J.; Gu, J.L.; Li, C.; Kadau, K.; Luesebrink, O. Crystallographic analysis for fatigue small crack growth behaviors of a nickel-based single crystal by in situ SEM observation. *Theor. Appl. Fract. Mech.* **2014**, *69*, 80–89. [CrossRef]
5. Wahi, R.P.; Auerswald, J.; Mukherji, D.; Dudka, A.; Fecht, H.-J.; Chen, W. Damage mechanisms of single and polycrystalline nickel base superalloys SC16 and IN738LC under high temperature LCF loading. *Int. J. Fatigue* **1997**, *19*, 89–94. [CrossRef]
6. Li, P.; Li, Q.Q.; Jin, T.; Zhou, Y.Z.; Li, J.G.; Sun, X.F.; Zhang, Z.F. Comparison of low-cycle fatigue behaviors between two nickel-based single-crystal superalloys. *Int. J. Fatigue* **2014**, *63*, 137–144. [CrossRef]
7. Wang, J.J.; Guo, W.G.; Su, Y.; Zhou, P.; Yuan, K. Anomalous Behaviors of a single-crystal nickel-base superalloy over a wide range of temperatures and strain rates. *Mech. Mater.* **2016**, *94*, 79–90. [CrossRef]
8. Li, S.X.; Smith, D.J. Development of an anisotropic constitutive model for single-crystal superalloy for combined fatigue and creep loading. *Int. J. Mech. Sci.* **1998**, *40*, 937–948. [CrossRef]
9. Sabnis, P.A.; Mazière, M.; Forest, S.; Arakere, N.K.; Ebrahimi, F. Effect of secondary orientation on notch-tip plasticity in superalloy single crystals. *Int. J. Plasticity* **2012**, *28*, 102–123. [CrossRef]

10. Sabnis, P.A.; Fores, S.; Arakere, N.K.; Yastrebova, V.A. Crystal plasticity analysis of cylindrical indentation on a Ni-base single crystal superalloy. *Int. J. Plasticity* **2013**, *51*, 200–217. [CrossRef]

11. Arakere, N.K.; Siddiqui, S.; Magnan, S.; Ferroro, L. Investigation of three-dimensional stress fields and slip systems for FCC single crystal superalloy notched specimens. *J. Eng. Gas Turbines Power* **2005**, *127*, 629–637. [CrossRef]

12. Gong, B.; Wang, Z.R.; Wang, Z.G. Cyclic Deformation behavior and dislocation structures of [001] copper single crystals—I Cyclic stress-strain response and surface feature. *Acta Mater.* **1997**, *45*, 1365–1377. [CrossRef]

13. Lekhnitskii, S.G. *Theory of Elasticity of an Anisotropic Elastic Body*; Society for Industrial and Applied Mathematics: Philadelphia, PA, USA, 1967; pp. 1–8.

14. Arakere, N.K.; Swanson, G. Effect of crystal orientation on fatigue failure of single crystal nickel base turbine blade superalloys. *J. Eng. Gas Turbines Power* **2002**, *124*, 161–176. [CrossRef]

15. Naik, R.A.; DeLuca, D.P.; Shah, D.M. Critical plane fatigue modeling and characterization of single crystal nickel superalloys. *J. Eng. Gas Turbines Power* **2004**, *126*, 391–400. [CrossRef]

16. Arakere, N.K. High-temperature fatigue properties of single crystal superalloys in air and hydrogen. *J. Eng. Gas Turbines Power* **2004**, *126*, 590–603. [CrossRef]

17. Editorial Board of Material Data Sheet of Aircraft Engine Design. *Material Manual for Aero Engine Design*, 4th ed.; Aviation Industry Press: Beijing, China, 2010; pp. 250–322.

18. Wen, Z.X.; Pei, H.Q.; Yang, H.; Wu, Y.; Yue, Z. A Combined CP theory and TCD for predicting fatigue lifetime in single-crystal superalloy plates with film cooling holes. *Int. J. Fatigue* **2018**, *111*, 243–255. [CrossRef]

19. Dong, C.L.; Yu, H.C.; Li, Y.; Yang, X.; Shi, D. Life modeling of anisotropic fatigue behavior for a single crystal nickel-base superalloy. *Int. J. Fatigue* **2014**, *61*, 21–27. [CrossRef]

20. Kanda, M.; Sakane, M.; Ohnami, M.; Hasebe, T. High temperature multiaxial low cycle fatigue of CMSX-2 Ni-base single crystal superalloy. *J. Eng. Mater. Technol.* **1997**, *119*, 153–160. [CrossRef]

21. Wei, D.S.; Wang, J.L.; Wang, Y.R.; Zhong, B. Experimental and numerical investigation of the creep behaviour of Ni-based superalloy GH4169 under varying loading. *Fatigue Fract. Eng. Mater. Struct.* **2018**, *41*, 1146–1158. [CrossRef]

22. Zhong, B.; Wang, Y.R.; Wei, D.S.; Wang, J. A New life prediction model for multiaxial fatigue under proportional and non-proportional loading paths based on the Pi-plane projection. *Int. J. Fatigue* **2017**, *102*, 241–251. [CrossRef]

23. Zhong, B.; Wang, Y.R.; Wei, D.S.; Zhang, K.; Wang, J. Multiaxial fatigue life prediction for powder metallurgy superalloy FGH96 based on stress gradient effect. *Int. J. Fatigue* **2018**, *109*, 26–36. [CrossRef]

24. Dowling, N.E.; Calhoun, C.A.; Arcari, A. Mean stress effects in stress-life fatigue and the walker equation. *Fatigue Fract. Eng. Mater. Struct.* **2009**, *32*, 163–179. [CrossRef]

25. Jiang, R.; Bull, D.; Evangelou, A.; Harte, A.; Pierron, F.; Sinclair, L.; Preuss, M.; Hu, X.T.; Reed, P.A.S. Strain accumulation and fatigue crack initiation at pores and carbides in a SX superalloy at room temperature: X-ray computed tomography, tabulated data and modelling datasets. *Int. J. Fatigue* **2018**, *114*, 22–33. [CrossRef]

26. Lukáš, P.; Kunz, L.; Svoboda, M. High-temperature ultra-high cycle fatigue damage of notched single crystal superalloys at high mean stresses. *Int. J. Fatigue* **2005**, *27*, 1535–1540. [CrossRef]

© 2019 by the authors. Licensee MDPI, Basel, Switzerland. This article is an open access article distributed under the terms and conditions of the Creative Commons Attribution (CC BY) license (http://creativecommons.org/licenses/by/4.0/).

Article

Precipitate Evolution and Fatigue Crack Growth in Creep and Artificially Aged Aluminum Alloy

Chi Liu [1], Yilun Liu [1], Liyong Ma [2,*], Songbai Li [1], Xianqiong Zhao [1] and Qing Wang [1]

[1] College of Mechanical and Electrical Engineering, Central South University, Changsha 410083, China; liuchi001@csu.edu.cn (C.L.); yilunliu@csu.edu.cn (Y.L.); csulsb@csu.edu.cn (S.L.); csuzxq@csu.edu.cn (X.Z.); ldlgwq@csu.edu.cn (Q.W.)

[2] School of Mechanical Engineering, Hebei University of Architecture, Zhangjiakou 075051, China

* Correspondence: maliyong@csu.edu.cn; Tel.: +86-189-3263-7968

Received: 8 November 2018; Accepted: 4 December 2018; Published: 7 December 2018

Abstract: The fatigue performance of high-strength Al-Cu-Mg alloys is generally influenced by the process of creep age formation when applied to acquire higher strength. The results show that creep aging accelerates the precipitation process, leading to a more uniform precipitation of strengthening phases in grains, as well as narrowed precipitation-free zones (PFZ). Compared with the artificially aged alloy, the yield strength and hardness of the creep aged alloy increased, but the fatigue resistance decreased. In the low stress intensity factor region ($\Delta K \leq 7$ MPa·m$^{1/2}$), the fatigue crack propagation (FCP) rate was mainly affected by the characteristics of precipitates, and the fatigue resistance noticeably decreased with the increased creep time. In a 4 h creep aged alloy, the microstructure was dominated by Cu-Mg clusters and Guinier-Preston (GP) zones, while S″ phases began to precipitate in the matrix, showing better fatigue resistance. After aging for 24 h, the needle shaped S′ phases were largely precipitated and coarsened, which changed the mode of dislocation slip, reduced the reversibility of slip, and accelerated the accumulation of fatigue damage. In stable and rapid crack propagation regions, the influence of precipitates on the FCP rate was negligible.

Keywords: precipitates; fatigue crack growth; creep aging; artificial aging

---

1. Introduction

A new aerospace high-strength Al-Cu-Mg alloy called AA2524, developed after the 2024 and 2124 aluminum alloys, exhibits excellent fatigue resistance [1–3]. Combined with creep age forming (CAF) technology, currently, AA2524 primarily finds its application in the manufacturing of wing coverings, fuselage panels, and other components [4,5], e.g., the upper wing skin of civil aircrafts such as the Airbus A330/340/380 [6], the wing panels of military aircrafts such as the B-1B bomber, C-17, and F-35 [7], as well as tank panels and melon petals in the American Saturn-5, Hercules-4, and Ariane-5 launch vehicles of the European Space Agency [4,8]. Creep age formation is a process that takes advantage of the creep property of metals to synchronize the forming and aging treatments. The formed products achieve excellent structural integrity and low residual stress [9–12]. Many scholars have conducted substantial research on the mechanical properties and springback prediction of creep aged materials. Zhan et al. [10] established a creep constitutive model and analyzed the relationship between the change of stress and precipitates and the aging strengthening in the creep aging process. Jeshvaghani et al. [13] asserted that for a 7075 aluminum alloy that has been creep aged at high and low temperatures in sequence, the springback rate decreased, and the exfoliation corrosion resistance was improved. Xu et al. [14] researched the creep aging behavior of AA2524 with the presence of pre-strain. The results showed that the increase of pre-strain can reduce the average size of S phases in the creep aged alloy and increase its density and uniformity, leading to a shortened time for peak aging and improved strength.

The current research on creep aged alloys focuses mainly on conventional mechanical properties and less on the fatigue properties. Influenced by frequent takeoff, landing, and airflow, creep aging-formed fuselage skin, wings, and other components are the most vulnerable to fatigue failure [15]. The fatigue crack propagation (FCP) resistance of aerospace structural components is an extremely important indicator. Yin [16] and Shou [17] et al. discussed the influence of grain size on the crack growth rate of 2524 aluminum alloy. Yin suggested that within the range of low stress intensity factor ΔK, the crack closure effect of the coarse grain samples was greater than that of the fine grain samples, and the grain refinement degraded the fatigue resistance. Shou demonstrated that when the grain size was between 50 and 100 μm, the crack growth rate was relatively low, and the crack growth path became more zigzagged. Srivatsan et al. [18] studied the effect of test temperature on the high-cycle fatigue and fracture properties of AA2524, indicating that the fatigue life decreased with the increasing test temperature. Baptista et al. [19] introduced an enhanced two-parameter exponential equation model to describe the subcritical FCP behavior of 2524-T3 aluminum alloy, which performed better than other test models. However, the aforementioned studies focused on the materials without thoroughly examining the impact of the creep age forming process.

Liu studied the fatigue behavior of creep aged AA2524 at 180 °C and suggested that the crack growth resistance of the alloy was reduced after treatment [20]. On the contrary, the research of Wenke Li [21] showed that the fatigue life of AA2524 was improved after creep aging. However, his research only considered the fatigue performance of the alloy under high stress and single stress, thus the results are not considered fully representative. Therefore, it is necessary to further study the effect of microstructure evolution on the fatigue performance of the creep formed alloy.

This current study achieved initial results. In this paper, the influence of creep aging and artificial aging on the microstructure, conventional mechanical properties, and FCP resistance of AA2524 is discussed. The goals of the present work are to characterize and correlate the evolution of precipitates and FCP resistance with creep age forming, and to provide a theoretical basis for AA2524 creep age forming technology.

2. Material and Experiments

The experimental material was a 5-mm-thick plate of AA2524-T3 alloy (Southwest Aluminum Group Co., Ltd., Chongqing, China), with chemical compositions (in wt.%) as follows: 4.26% Cu, 1.36% Mg, 0.57% Mn, 0.024% Zn, 0.01% Ti, 0.002% Cr, 0.089% Si, and a balance of Al. The CAF tests were completed in a vacuum autoclave. This process is shown in Figure 1. The creep temperature was set at 160 °C at a heating rate of 1.5 °C/min, and the creep aging times were 4 h, 9 h, and 24 h. The artificial aging test was performed using the same aging times and temperature, but without external stress applied, which is called stress-free aging (SFA).

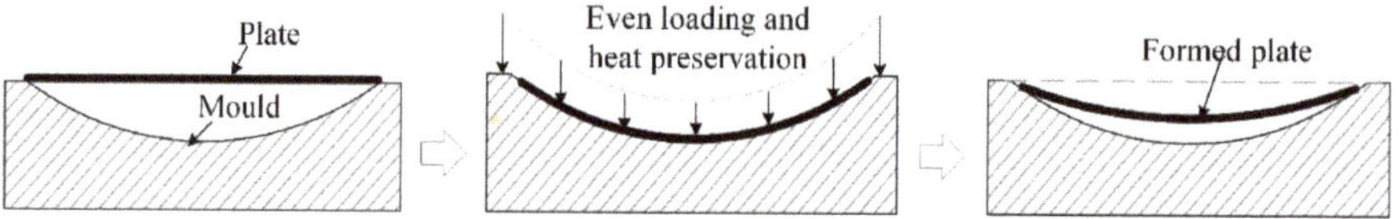

Figure 1. The process of creep age forming.

TEM observations were conducted on a Tecnai G220 (200 kV) transmission electron microscope (United States FEI limited liability company, Hillsboro, OR, USA). TEM samples were mechanically thinned to approximately 60~80 μm, then punched into 3-mm diameter discs and polished in an MTP-1 twin-jet electro-polisher in a 30% HNO_3 and 70% CH_3OH mixed solution at −30 °C~−25 °C with a voltage of 15 V.

Tensile tests and FCP tests were performed on a MTS810-50 KN (MTS Systems Corporation, Eden Prairie, MN, USA) electro-hydraulic servo fatigue machine. Tensile specimens were cut in the rolling direction of the plate and tested in accordance with ASTM-E8M-2004 at room temperature, with a strain rate of 2 mm/min, resulting in an average of three samples. The FCP rate tests were conducted in ambient air at a room temperature of 18 °C–25 °C and a relative humidity of 40–60% in accordance with ASTM-E647. The FCP specimens were of the compact tension (CT) geometry, as shown in Figure 2. FCP tests were characterized for constant amplitude loading at a frequency of 10 Hz and a stress ratio of 0.5. Crack length was measured by using a crack opening displacement (COD) gauge, as shown in Figure 3. The hardness tests were completed on a HVS-1000Z Vickers micro digital hardness machine (Huayin Testing Instrument Co., Ltd., Yantai, China) with a holding pressure of 3 KN for 15 s. The average of five test points per sample was taken as the hardness value.

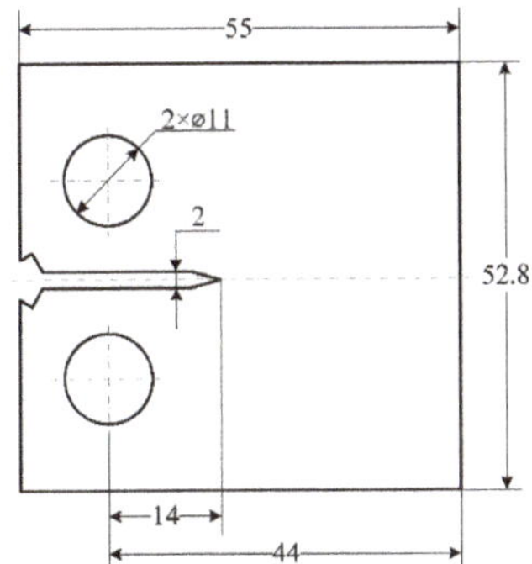

Figure 2. Compact tensile specimen (mm).

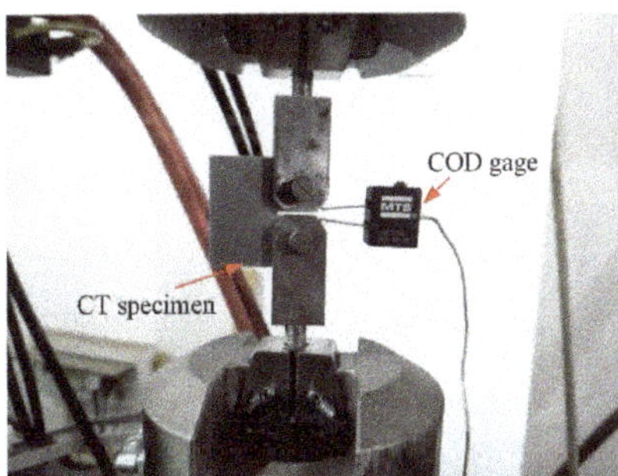

Figure 3. Fatigue crack propagation test with a crack opening displacement (COD) gauge.

3. Results and Discussion

3.1. Microstructure

The microstructures of AA2524 after creep aging and artificial aging are shown in Figure 4. The alloy contained rod-shaped Mn-rich phases (Figure 4b,d) and $Al_{20}Cu_2Mn_3$ (T phases) [22–24] with a size range of 0.2 µm~0.5 µm. These phases were formed during a homogenizing treatment and hot rolling, and were not re-dissolved in the subsequent heat treatment.

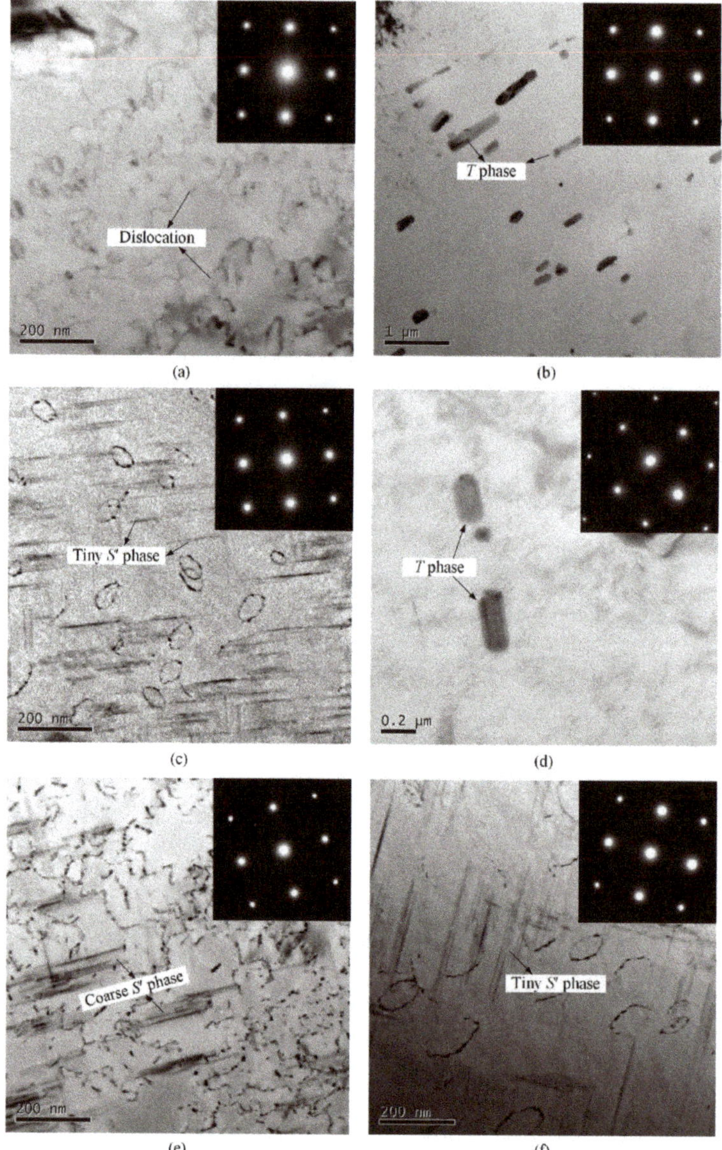

Figure 4. TEM bright field images and selected area electron diffraction (SAED) patterns along a <100>$_{Al}$ zone axis for AA2524 under different aging conditions: (**a**) creep aged (CAF)-4h; (**b**) stress-free aged (SFA)-4h; (**c**) CAF-9h; (**d**) SFA-9h; (**e**) CAF-24h; (**f**) SFA-24h.

Precipitates were not observed in the TEM bright field images of the alloy that was creep aged for 4 h (CAF-4h) (Figure 4a), but it was discovered that tiny S″ phases began to appear in the selected area electron diffraction (SAED) pattern in a <100>$_{Al}$ direction (Figure 5a). The S″ phase is a precursor of the S′ phase, and generally occurs in the early aging stage of the Al-Cu-Mg alloy. It is difficult to observe the S″ phases in the TEM bright field image [25] because they are only several nanometers in size and there is an ambiguous interface with the matrix. As these S″ phases appeared, a large number of dislocation loops and helical dislocation lines were observed in the grains (CAF-4h) (Figure 4a). However, no obvious precipitate was observed in the alloy that was stress-free aged for 4 h (SFA-4h) (Figure 4b), and there were no apparent diffraction spots or streaks in the SAED pattern [26]. Therefore, it can be determined that the microstructures in the SFA-4h alloy were mainly Cu-Mg clusters and Guinier-Preston (GP) regions.

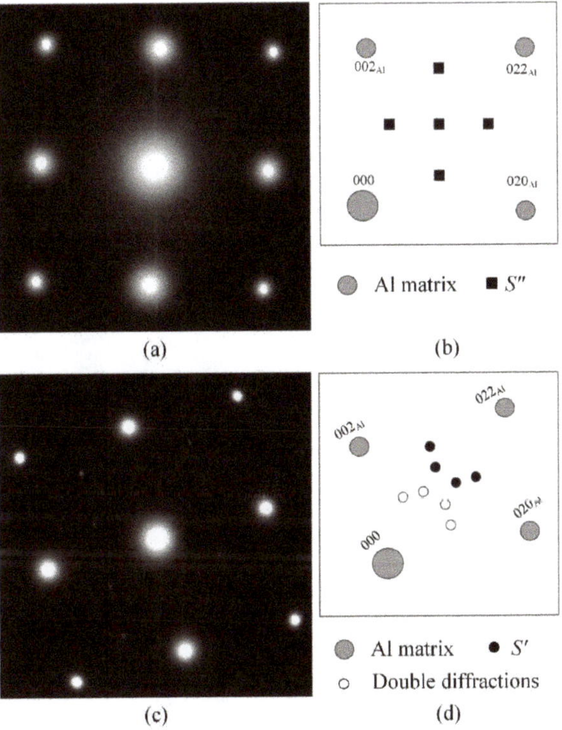

(a)

(b)

(c)

(d)

Figure 5. SAED patterns along a <100>_Al zone axis and corresponding schematic diagrams for AA2524: (**a,b**) CAF-4h; (**c,d**) CAF-24h.

Large quantities of needle-shaped, fine transition phases—the densely distributed S′ phases—were observed (Figure 4c) in the alloy that was creep aged for 9 h (CAF-9h), with sizes ranging from 150~250 nm. The precipitates in the alloy that was stress-free aged for 9 h (SFA-9h) were still not evident in the bright field images (Figure 4d). After creep aging for 24 h (CAF-24h), the precipitates continued to grow and thicken, reaching sizes of about 200~400 nm (Figure 4e), and the clearance between precipitates widened significantly. From the SAED pattern in Figure 5c, it is inferred that the main strengthening phase in grains occurs in the S′ phase. In the alloy that was stress-free aged for 24 h (SFA-24h), precipitates with sizes of about 200 nm~300 nm occurred (Figure 4f), which were smaller than those found in the CAF-24h alloy (Figure 4e). The precipitation behavior shows that the presence of stress during creep aging promoted the precipitation. Some scholars contend that a great deal of nucleation precipitates from the high-density dislocations that are caused by creep stress. Then, these dislocations act as fast diffusion channels to aggregate the solute atoms toward heterogeneous nucleation, thus promoting the growth of S′ phases [27].

Figure 6 shows the grain boundary feature of AA2524 samples that were creep aged and artificially aged for 9 h and 24 h. There was no apparent precipitation-free zone (PFZ) at the grain boundary (Figure 6a,c) in either the CAF-9h or SFA-9h alloys, and the precipitates at the grain boundary showed discontinuous distribution. After aging for 24 h, a distinct PFZ (Figure 6d) with a width of about

184 nm appeared at the grain boundary in the SFA-24h sample, while the PFZ (Figure 6b) in the CAF-24h sample was 140 nm wide—narrower than that of the stress-free aged sample. From this, it can be explained that creep stress generates a large number of dislocations in grains. These dislocations promote the surrounding preferential precipitation, resulting in more uniform precipitation kinetics of the strengthening phases in grains and grain boundaries, thus narrowing the PFZ.

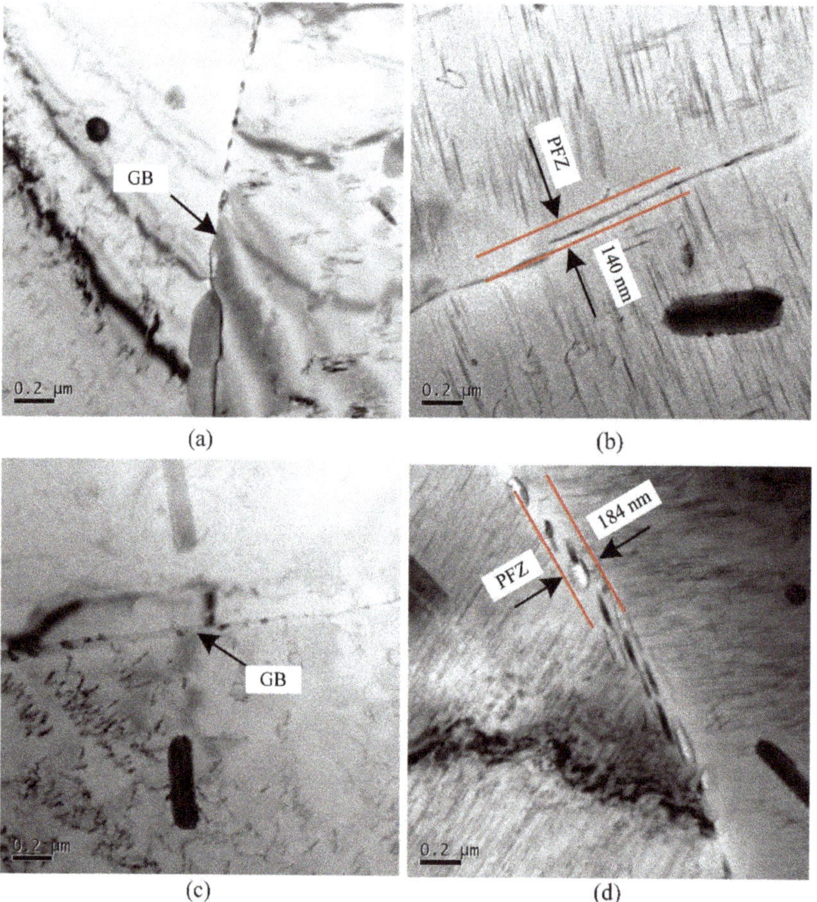

Figure 6. TEM images of AA2524 at the grain boundary under different treatment: (**a**) CAF-9h; (**b**) CAF-24h; (**c**) SFA-9h; (**d**) SFA-24h.

3.2. Conventional Mechanical Properties

Figures 7 and 8 show the conventional mechanical properties for AA2524 under creep aging and stress-free aging, respectively. As the aging time increased, the hardness of AA2524 first increased and then decreased (Figure 7), clearly reflecting characteristics that occur at three aging stages: under aging (4 h of aging), peak aging (9 h of aging), and over aging (24 h of aging). For the stress-free aged alloy, the hardness decreased within the aging time of 0 h~4 h (Figure 7) and increased from 9 h to 24 h. However, the hardness changed slightly during the 4 h~9 h range of time, indicating an apparent plateau region. The study done by Ringer et al. [28] on the low Cu/Mg ratio Al-Cu-Mg alloy confirmed that the hardening curves of the Al-Cu-Mg alloy, at an aging temperature ranging from 100 °C to 240 °C and within an α-S phase region, displayed an obvious plateau region. Experimental results in the present paper are consistent with this conclusion. In contrast, there was no hardness plateau region under creep aging. After aging for 9 h, the hardness of the CAF-9h alloy rapidly reached a peak value of 159 HV, 16.9% higher than that of the SFA-9h alloy (136 HV). This indicates that creep aging significantly accelerates the hardening rate and improves the hardness of the alloy.

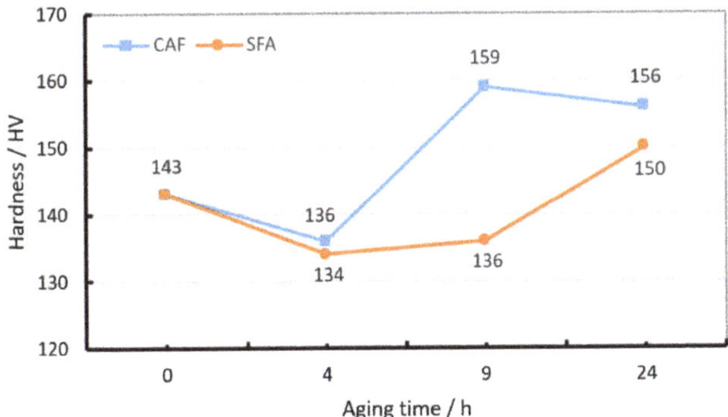

Figure 7. Hardness of AA2524 under creep age forming and artificial aging.

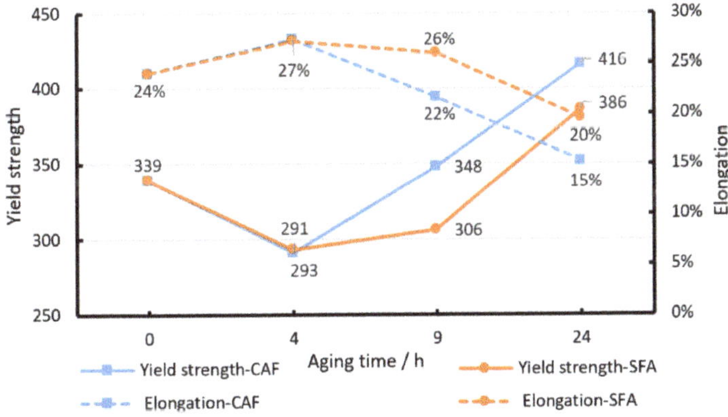

Figure 8. Yield strength and elongation of AA2524 under creep age forming and artificial aging.

After 4 h of aging, the yield strength ($\sigma_{0.2}$) of AA2524 had significantly degraded compared to AA2524-T3, the 0 h artificially aged alloy (Figure 8). With the increased aging time, the yield strength of the alloy was on the rise. With the same aging time, the yield strength of the creep aged alloys was higher than that of the artificially aged alloys. The yield strengths of the CAF-9h and CAF-24h alloys were 13.7% and 7% higher than those of the SFA-9h and SFA-24h alloys, respectively. All of the alloys were of superior plasticity, with an elongation (δ) above 15%. The variation trend of elongation was opposite that of the change in yield strength. Elongation of the 4 h aged alloys reached its highest point at 27%, followed by a downward trend. Compared with artificial aging, creep aging degraded the elongation of the alloy. After aging for 9 h, the elongation of the CAF-9h alloy was 18.2% lower than that of the artificially aged 9 h (AA-9h) alloy.

The changes in the mechanical properties of the alloy are closely related to its microstructure characteristics [29]. According to the aforementioned TEM observations, the main strengthening phases of AA2524 are the needle-shaped S (S′) phases, which can improve the strength and hardness but reduce the plasticity. The sizes of precipitates in the CAF-9h alloy (Figure 4c) were larger than those in the SFA-9h alloy (Figure 4d), and the effect of precipitation strengthening was more pronounced.

Therefore, creep aged alloys boast higher tensile strength and hardness than stress-free aged alloys, but lower elongation.

3.3. Fatigue Crack Growth Behavior

Figure 9 shows the FCP rate of AA2524 under different creep aging conditions. The fatigue crack growth behavior can be divided into three stages. In the low stress region of $\Delta K \leq 7$ MPa·m$^{1/2}$, the FCP rates of different creep aged specimens varied significantly, among which the FCP rate of the CAF-4h specimen was the lowest. With the increased aging time, the FCP rate was accelerated, and the FCP rate of the CAF-24h specimen was the highest. This difference was mainly related to the characteristics of precipitates and dislocation slips in the alloy matrix. However, with the increase of the stress intensity factor range ΔK, the FCP rates in the Paris region and the rapid fracture region tended to be consistent under different aging treatments, indicating that the effect of precipitate features in the alloy matrix on crack propagation resistance was no longer pronounced.

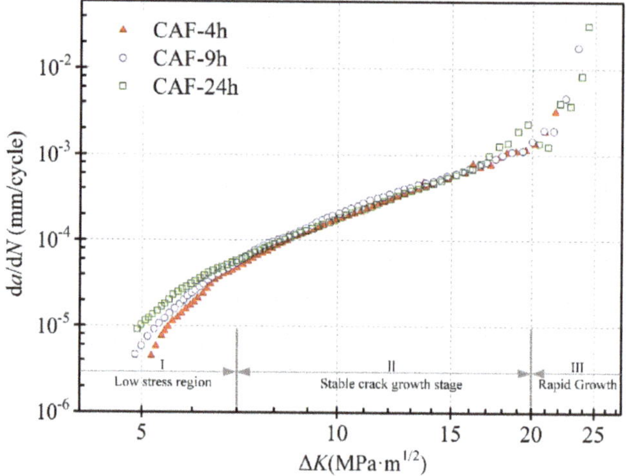

Figure 9. Fatigue crack growth rates of AA2524 under different creep age forming conditions.

The CAF-4h alloy mainly contained fine S″ phases (Figure 4a), Cu-Mg clusters, and GP zones. These coherent clusters promoted the planar slip of dislocations under cyclic loading, which greatly increased the reversibility of dislocation slip, reduced the accumulation of fatigue damage, and improved the fatigue resistance of the alloy. The precipitates in both the CAF-9h and CAF-24h alloys were mainly needle-shaped S′ phases (Figure 4c,e), with larger sized S′ phases in the CAF-24 h alloy. These needle-shaped coarse phases changed the dislocation slip mode from single slip to cross slip, which degraded the reversibility of dislocation slip and accelerated the FCP rate. A PFZ appeared in the grain boundary of the CAF-24h alloy (Figure 6b), leading to a decrease in the grain boundary strength. These softer PFZs caused stress concentration around the grain boundary and accelerated the FCP rate [30]. To summarize, PFZs and larger needle-shaped precipitates lead to the highest FCP rate of the CAF-24h alloy.

In the stable crack growth stage—namely when $\Delta K = 7\text{~}16$ MPa·m$^{1/2}$—the Paris region precipitates, with a higher ΔK level, showed a weakened effect on the FCP rate. Figure 9 displays that the crack growth rate of the CAF-9h sample was slightly higher, but the difference in FCP rates under these three conditions was much smaller than that of the low ΔK level. When $\Delta K \geq 16$ MPa·m$^{1/2}$, $da/dN \geq 10^{-3}$ mm/cycle, and the da/dN-ΔK curves showed an apparent knee, this indicated that the crack propagation had entered a rapid growth stage.

Figure 10 shows a comparison between the FCP rates of creep aged and stress-free aged AA2524 for aging times of 4 h, 9 h, and 24 h. In Figure 10a, the FCP rates of 4 h aged alloys under two different aging conditions were very close because both alloys contained Cu-Mg clusters and GP zones (Figure 4a,b). Although fine S″ phases could be found in the SAED pattern of the CAF-4h alloy, these phases had little influence on the FCP rate of the alloy due to the S″ phases remaining coherent with the matrix. As seen in Figure 10b,c, after aging for 9 h and 24 h, the FCP rates of the creep aged alloys were significantly higher in the low stress intensity factor region than those of the stress-free aged alloys.

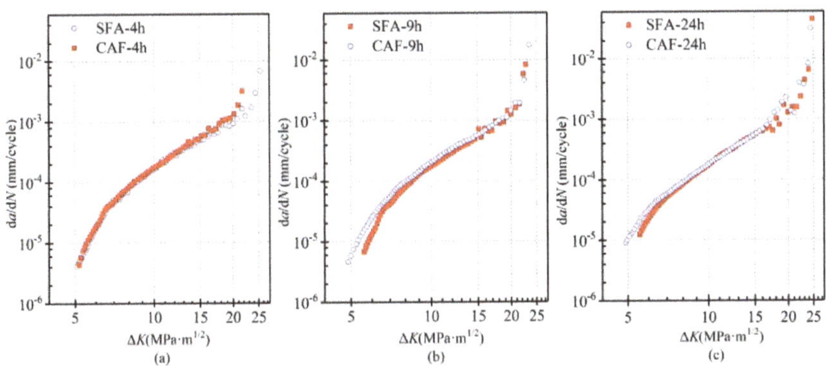

Figure 10. Comparison of the fatigue crack propagation (FCP) rates of creep aged and stress-free aged AA2524 for various aging times: (**a**) 4 h; (**b**) 9 h; (**c**) 24 h.

After 9 h of aging, needle-shaped S′ phases that were semi-coherent with the matrix (Figure 4c,d) mainly precipitated. However, the presence of stress in the creep aging process promoted the precipitation, so the sizes of the S′ phases in the CAF-9h alloy were larger. Under the cyclic loading, dislocations could not cut through part of the coarse S′ phases and instead bypassed them, thus reducing the reversibility of cyclic slip, accumulating a large plastic deformation at the crack tip, and promoting crack propagation. In the region of $\Delta K \leq 7.5$ MPa·m$^{1/2}$, the fatigue crack propagation resistance of the CAF-9h alloy was lower than that of the SFA-9h alloy.

After 24 h of aging, the sizes of precipitates showed a similar rule (Figure 4e–f). In the low stress region, the FCP rate of the SFA-24h alloy was lower than that of the CAF-24h alloy. However, the PFZ in the grain boundaries of the SFA-24h alloy (Figure 6d) was wider than it was (Figure 6b) in the CAF-24h alloy, which accelerated the FCP rate to some extent. Under the influence of both precipitates and PFZs, the effect of microstructures on the FCP resistance disappeared in advance. Therefore, the da/dN-ΔK curves of the two alloys (SFA-24h and CAF-24h) were basically consistent when $\Delta K \geq 7$ MPa·m$^{1/2}$.

In the lower stress region, we chose to analyze the corresponding fatigue fracture at $\Delta K = 6$ MPa·m$^{1/2}$, in which the crack was in the stable growth stage. Under the SEM, a large area of regular and parallel fatigue striations could be observed, as shown in Figure 11. When encountering the particles in the matrix, the striations would bypass the particles and continue to expand (Figure 11b,c), as the crack propagation direction conveys with the dotted arrow in Figure 11. The width of fatigue striations marked in Figure 11 shows the general rule that the width of fatigue striations increases with the increasing creep time. The fatigue striation width of the creep aged specimen was larger than that of the stress-free aged one when tested with the same aging time. The fatigue striation was caused by the repeated sharpening of the crack tip during the cyclic loading. The fatigue striation width corresponds to the length of a certain cyclic fatigue crack propagation, which can represent the fatigue crack growth rate to some extent. That is, the larger the fatigue striation width, the higher the fatigue crack growth rate. This conclusion is also consistent with the previous da/dN-ΔK curve analysis.

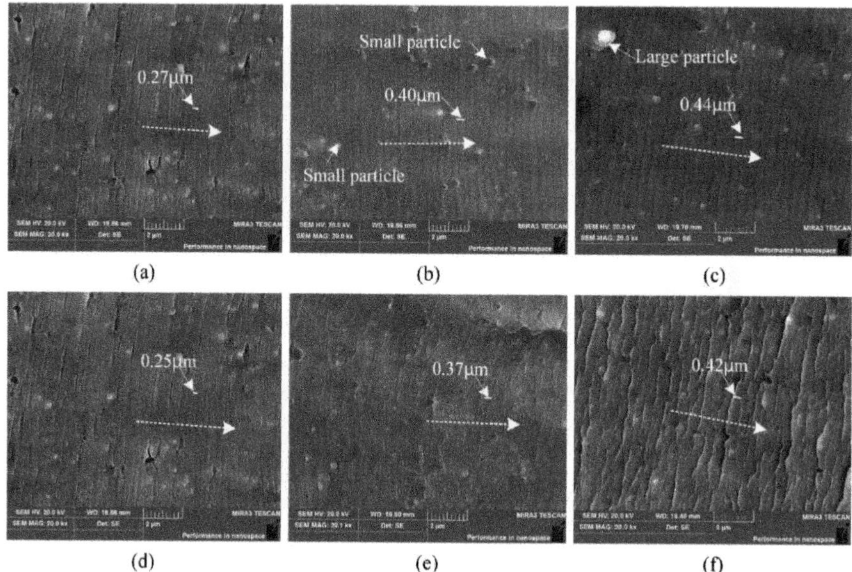

(a)   (b)   (c)

(d)   (e)   (f)

Figure 11. Fatigue striation morphology at fractures of different specimens when $\Delta K = 6$ MPa·m$^{1/2}$: (**a**) CAF-4h; (**b**) CAF-9h; (**c**) CAF-24h; (**d**) SFA-4h; (**e**) SFA-9h; (**f**) SFA-24h.

It can be concluded that the FCP rate of AA2524 in the low stress region is mainly affected by the precipitate features. With the increase of ΔK, precipitates gradually show a weakened effect on the FCP rate, while PFZs in the grain boundaries accelerate the FCP of the alloy to some extent.

4. Conclusions

(1) Creep age forming generated a large quantity of dislocations in the grains of AA2524, which provided nucleation locations for the heterogeneous nucleation of the second phase. The resulting precipitates in the creep aged AA2524 were larger in number and size than those in the stress-free aged AA2524 after the same aging time.

(2) The yield strength of AA2524 increased with increasing aging time. After the same aging time, the yield strength of the creep aged alloy was higher, while the hardening curve of the stress-free aged alloy had a plateau region for approximately 5 h. This indicates that creep aging can significantly accelerate the age hardening process, with the alloy quickly reaching the peak aging state.

(3) In the low stress intensity factor region, the crack growth behavior of AA2524 was mainly affected by precipitates. The FCP rate was accelerated with an increased creep aging time. Creep aging promoted the precipitation and growth of S″ and S′ phases in the alloy. The needle-shaped coarse S″(S′) phases altered dislocation slip mode in the alloy, reduced the reversibility of dislocation slips, and degraded the crack propagation resistance. The FCP rate of the creep aged alloy was higher than that of the stress-free aged alloy after the same aging time.

(4) In the stable crack propagation region, the FCP rates were generally consistent under different aging treatments, and the effect of precipitate features in the alloy matrix on crack propagation resistance was negligible.

Author Contributions: C.L. and Y.L. conceived and designed the experiment; C.L., Q.W. performed the experiments; C.L., L.M., and X.Z. analyzed the data; S.L. and X.Z. contributed reagents, materials, and analysis tools; C.L. wrote the paper; M.L. was responsible for the revision.

Funding: This work was supported by the National Natural Science Foundation of China (No. 51375500), the Fundamental Research Funds for the Central Universities of Central South University (No. 2015zzts038), Hebei Provincial Science and Technology Plan Self-Financing Project (No. 17211828), Hebei Province Higher Education Science and Technology Research Youth Fund Project (QN2018013), and the Hunan Science and Technology Plan Project (No. 2016GK2005).

Conflicts of Interest: The authors declare no conflicts of interest.

References

1. Nakai, M.; Eto, T. New aspect of development of high strength aluminum alloys for aerospace applications. *Mater. Sci. Eng. A* **2000**, *285*, 62–68. [CrossRef]
2. Starke, E.A., Jr.; Staley, J.T. Application of modern aluminum alloys to aircraft. *Prog. Aerosp. Sci.* **1996**, *32*, 131–172. [CrossRef]
3. Warner, T. Recently-developed aluminium solutions for aerospace applications. *Mater. Sci. Forum* **2006**, *519*, 1271–1278. [CrossRef]
4. Zeng, Y.; Huang, X. Forming technologies of large integral panel. *Acta Aeronaut. Et Aeronaut. Sin.* **2008**, *29*, 721–727.
5. Zhan, L.; Lin, J.; Dean, T.A. A review of the development of creep age forming: Experimentation, modelling and applications. *Int. J. Mach. Tools Manuf.* **2011**, *51*, 1–17. [CrossRef]
6. Watcham, K.; Watcham, K.; Watcham, K. Airbus A380 takes creep age-forming to new heights. *Mater. World* **2004**, *12*, 10–11.
7. Holman, M.C. Autoclave age forming large aluminum aircraft panels. *J. Mech. Work. Technol.* **1989**, *20*, 477–488. [CrossRef]
8. Liu, X. Forecasts on crucial manufacturing technology develeopment of heavy lift launch vehicle. *Aerosp. Manuf. Technol.* **2013**, *1*, 002.
9. Jeunechamps, P.P.; Ho, K.C.; Lin, J.; Ponthot, J.P.; Dean, T.A. A closed form technique to predict springback in creep age-forming. *Int. J. Mech. Sci.* **2006**, *48*, 621–629. [CrossRef]
10. Zhan, L.; Lin, J.; Dean, T.A.; Huang, M. Experimental studies and constitutive modelling of the hardening of aluminium alloy 7055 under creep age forming conditions. *Int. J. Mech. Sci.* **2011**, *53*, 595–605. [CrossRef]
11. Zhang, J.; Zhang, S.N.; Liang, E.W. Blazar anti-sequence of spectral variation within individual blazars: Cases for Mrk 501 and 3C 279. *Astrophys. J.* **2013**, *767*, 8. [CrossRef]
12. Liu, L.; Zhan, L.; Li, W. Creep aging behavior characterization of 2219 aluminum alloy. *Metals* **2016**, *6*, 146. [CrossRef]
13. Jeshvaghani, R.A.; Zohdi, H.; Shahverdi, H.R.; Bozorg, M.; Hadavi, S.M.M. Influence of multi-step heat treatments in creep age forming of 7075 aluminum alloy: Optimization for springback, strength and exfoliation corrosion. *Mater. Charact.* **2012**, *73*, 8–15. [CrossRef]
14. Xu, Y.; Zhan, L.; Li, W. Effect of pre-strain on creep aging behavior of 2524 aluminum alloy. *J. Alloys Compd.* **2017**, *691*, 564–571. [CrossRef]
15. Schijve, J. Fatigue damage in aircraft structures, not wanted, but tolerated? *Int. J. Fatigue* **2009**, *31*, 998–1011. [CrossRef]
16. Yin, D.; Liu, H.; Chen, Y.; Yi, D.; Wang, B.; Wang, B.; Shen, F.; Fu, S.; Tang, C.; Pan, S. Effect of grain size on fatigue-crack growth in 2524 aluminium alloy. *Int. J. Fatigue* **2016**, *84*, 9–16. [CrossRef]
17. Shou, W.B.; Yi, D.Q.; Liu, H.Q.; Tang, C.; Shen, F.H.; Wang, B. Effect of grain size on the fatigue crack growth behavior of 2524-T3 aluminum alloy. *Arch. Civil Mech. Eng.* **2016**, *16*, 304–312. [CrossRef]
18. Srivatsan, T.S.; Kolar, D.; Magnusen, P. The cyclic fatigue and final fracture behavior of aluminum alloy 2524. *Mater. Des.* **2002**, *23*, 129–139. [CrossRef]
19. Baptista, C.A.R.P.; Adib, A.M.L.; Torres, M.A.S.; Pastoukhov, V.A. Describing fatigue crack growth and load ratio effects in Al 2524 T3 alloy with an enhanced exponential model. *Mech. Mater.* **2012**, *51*, 66–73. [CrossRef]
20. Liu, C.; Liu, Y.; Li, S.; Ma, L.; Zhao, X.; Wang, Q. Effect of creep aging forming on the fatigue crack growth of an AA2524 alloy. *Mater. Sci. Eng. A* **2018**. [CrossRef]
21. Li, W.; Zhan, L.; Liu, L.; Xu, Y. The effect of creep aging on the fatigue fracture behavior of 2524 aluminum alloy. *Metals* **2016**, *6*, 215. [CrossRef]
22. Chen, Y.Q.; Yi, D.Q.; Jiang, Y.; Wang, B.; Xu, D.Z.; Li, S.C. Twinning and orientation relationships of T-phase precipitates in an Al matrix. *J. Mater. Sci.* **2013**, *48*, 3225–3231. [CrossRef]

23. Chen, Z.; Chen, P.; Li, S. Effect of Ce addition on microstructure of $Al_{20}Cu_2Mn_3$ twin phase in an Al-Cu-Mn casting alloy. *Mater. Sci. Eng. A* **2012**, *532*, 606–609. [CrossRef]

24. Liu, C.; Liu, Y.; Ma, L.; Yi, J. Effects of solution treatment on microstructure and high-cycle fatigue properties of 7075 aluminum alloy. *Metals* **2017**, *7*, 193. [CrossRef]

25. Chen, Y.Q.; Pan, S.P.; Liu, W.H.; Cai, Z.H.; Tang, S.W.; Tang, C.P. Effect of precipitates on creep behaviors of Al-Cu-Mg alloy. *Chin. J. Nonferrous Metals* **2015**, *25*, 900–909.

26. Ringer, S.P.; Polmear, I.J.; Sakurai, T. Effect of additions of Si and Ag to ternary Al-Cu-Mg alloys in the α + s phase field. *Mater. Sci. Eng. A* **1996**, *217*, 273–276. [CrossRef]

27. Shen, F.; Wang, B.; Yi, D.; Liu, H.; Tang, C.; Shou, W. Effects of heating rate during solid-solution treatment on microstructure and fatigue properties of AA2524 T3 Al–Cu–Mg sheet. *Mater. Des.* **2016**, *104*, 116–125. [CrossRef]

28. Ringer, S.P.; Sakurai, T.; Polmear, I.J. Origins of hardening in aged Al-Cu-Mg-(Ag) alloys. *Acta Mater.* **1997**, *45*, 3731–3744. [CrossRef]

29. Zhan, L.H.; Yan-Guang, L.I.; Huang, M.H. Microstructures and properties of 2124 alloy creep ageing under stress. *J. Cent. S. Univ.* **2012**, *43*, 926–931.

30. Krol, T.; Baither, D.; Nembach, E. The formation of precipitate free zones along grain boundaries in a superalloy and the ensuing effects on its plastic deformation. *Acta Mater.* **2004**, *52*, 2095–2108. [CrossRef]

© 2018 by the authors. Licensee MDPI, Basel, Switzerland. This article is an open access article distributed under the terms and conditions of the Creative Commons Attribution (CC BY) license (http://creativecommons.org/licenses/by/4.0/).

Article

Fatigue Damage Accumulation Modeling of Metals Alloys under High Amplitude Loading at Elevated Temperatures

Jarosław Szusta * and Andrzej Seweryn

Faculty of Mechanical Engineering, Bialystok University of Technology, 45C Wiejska Str., 15-351 Bialystok, Poland; a.seweryn@pb.edu.pl

* Correspondence: j.szusta@pb.edu.pl; Tel.: +48-085-746-9300; Fax: +48-085-746-9210

Received: 16 November 2018; Accepted: 3 December 2018; Published: 6 December 2018

Abstract: This article presents an approach related to the modeling of the fatigue life of constructional metal alloys working under elevated temperature conditions and in the high-amplitude load range. The article reviews the fatigue damage accumulation criteria that makes it possible to determine the number of loading cycles until damage occurs. Results of experimental tests conducted on various technical metal alloys made it possible to develop a fatigue damage accumulation model for the LCF (Low Cycle Fatigue) range. In modeling, the material's damage state variable was defined, and the damage accumulation law was formulated incrementally so as to enable the analysis of the influence of loading history on the material's fatigue life. In the proposed model, the increment of the damage state variable was made dependent on the increment of plastic strain, on the tensile stress value in the sample, and also on the actual value of the damage state variable. The model was verified on the basis of data obtained from experiments in the field of uniaxial and multiaxial loads. Samples made of EN AW 2024T3 aluminum alloy were used for this purpose.

Keywords: elevated temperature; low cycle fatigue; damage accumulation; uniaxial and multiaxial loading

---

1. Introduction

During operation, machine parts are frequently subjected to fatigue loading at elevated temperatures. Such operating conditions have an adverse effect on the materials' behavior. An elevated temperature accelerates wear processes and contributes to increasing the rate of fatigue crack propagation [1,2]. Aluminum alloys, e.g., EN AW 2024T3, are among the materials used to make constructions capable of working at an elevated temperature. It is widely used in the motorization and aviation industries, which place strong emphasis on ensuring high strength and light weight at the same time [3]. These features make it suitable for the manufacture of parts working in close proximity to combustion and jet engines where elevated temperature zones are present. In the work of Szusta and Seweryn [4,5], the influence of elevated temperature on the fatigue life of this material was tested. The authors demonstrated that strength parameters decreased as the test temperature increased. Based on the tests conducted, the authors determined the parameters of the Manson-Coffin fatigue life curve as a function of temperature:

$$\frac{\Delta\varepsilon_{eq}}{2} = \frac{\sigma'_f(T)}{E(T)}(2N_f)^{b(T)} + \varepsilon'_f(T)(2N_f)^{c(T)}, \tag{1}$$

where N_f is the number of cycles until failure; $E(T)$ is Young's modulus determined at temperature T; $\sigma_f{}'(T)$ and $\varepsilon_f{}'(T)$ are, respectively, coefficients of the elastic fatigue life curve and plastic fatigue life curve for the analyzed temperature T; and $b(T)$ and $c(T)$ are, respectively, exponents of the elastic and plastic fatigue life curve for temperature T. Furthermore, based on observations of the material's cracking mechanisms at an elevated temperature, the authors proposed a semi-empirical model for estimating the material's fatigue life. In this model, it was accepted that the material's fatigue life drops as the test temperature (isothermal) increases. The lower the material's melting temperature, the more substantial this drop is. Moreover, it was assumed that there exists a limit temperature T_m at which the given material no longer has any immediate strength. Therefore, knowing the value of this temperature and the function defining the reduction in fatigue life as temperature increases, calculations of the number of cycles until crack initiation can be performed with engineering accuracy for a given temperature level and strain amplitude. For uniaxial tensile-compressive loads, the function of the temperature's influence on the material's fatigue life has been defined as follows:

$$N_f(T) = AT + B, \tag{2}$$

where A and B are material constants determined in periodical tensile-compression tests at temperature T.

In this model, the material constant B is a function of the material's recrystallization temperature T_m ($B = -AT_m$). The value of parameter A is calculated on the basis of the experimentally determined number of cycles to failure (obtained for a given load course at room temperature) related to the value of temperature difference: the material recrystallization and room temperature. Parameter A is determined from the following equation:

$$A\left(\frac{\Delta\varepsilon_{eq}}{2}\right) = -\frac{N_f\left(RT, \frac{\Delta\varepsilon_{eq}}{2}\right)}{(T_m - RT)}, \tag{3}$$

where $N_f\left(RT, \frac{\Delta\varepsilon_{eq}}{2}\right)$ is determined from Manson–Coffin–Basquiun equation at room temperature RT for the set total strain amplitude $\frac{\Delta\varepsilon_{eq}}{2}$.

Modified cast irons and cast steels, which are used in the manufacture of brake disks, are another group of materials that can be exploited at elevated temperatures. In the works of Samec et al. [6], Peve et al. [7], and Li et al. [8], the authors analyzed the fatigue life of the following materials: EN-GJS-500-7, EN-GJL-250, and Cr-Mo-V. Material cracking mechanisms were studied within the LCF range at temperature up to 700 °C, and the influence of temperature on the material's physical properties was defined.

Alloy steels are another group of materials used to make constructions which are exposed to the action of elevated temperatures. Here, some examples include 8Cr-2WvTa, (RAFM) JLF-1, or 617M steels, which are used to build thermoreactors and machine parts exposed to the action of high temperatures. In the works of Ishii et al. [9], Mariappan et al. [10], and Shankar et al. [11], the characteristics of these steels' cyclic properties were determined as a function of temperature. Degradation mechanisms of materials subjected to the action of fatigue loads and temperatures were analyzed. In the article by Ishii et al. [9], an approach to modeling fatigue life of the material was also presented, accounting for the influence of creep on the material's fatigue life:

$$\varepsilon_{eq}(N_f, T) = \varepsilon_e(N_f, T) + \varepsilon_{pl}(N_f, T) + \varepsilon_{cr}(N_f, T), \tag{4}$$

where ε_{eq} is the equivalent strain, ε_e is the elastic strain, ε_{pl} is the plastic strain, and ε_{cr} is the creep-induced strain.

Works by Nagode and Zingsheim [12] and Nagode and Hack [13] present the results of experimental tests performed on steels for work at elevated temperatures: 10 CrMo 9 10 and X22CrMoV121. Fatigue damage accumulation models of the material under non-isothermal conditions at elevated temperatures were developed on the basis of these results. These models were developed for estimation of the fatigue life of materials which undergo changes of the load amplitude and temperature transient over the course of its exploitation. These models do not account for the influence of creep or hardening of the material on its fatigue life. The authors presented a stress and strain approach in which it is assumed that the instant working temperature will be accounted for. In models, stabilized hysteresis loops were described by means of the Ramberg–Osgood equation. The damage parameter P_{SWT} was identified similarly as in the SWT (Smith–Watson–Topper) model:

$$P_{SWT} = \sqrt{\left(\sigma_f'^2(T_e)(2N_f)^{2b(T_e)} + \sigma_f'(T_e)E(T_e)\right)\varepsilon_f'(T_e)(2N_f)^{b(T_e)+c(T_e)}}, \text{ where}$$

$$\varepsilon_f'(T_e) = \left(\frac{\sigma_f'(T_e)}{K'(T_e)}\right)^{\frac{1}{n'(T_e)}},$$

$$c(T_e) = \frac{b(T_e)}{n'(T_e)},$$

(5)

where T_e is the actual temperature; $\sigma'_f(T_e)$ and $b(T_e)$ are, respectively, the coefficient and exponent of the fatigue life curve at the tested temperature; $K'(T_e)$ and $n'(T_e)$ are, respectively, the coefficient and exponent of the cyclic hardening curve at the tested temperature; and $\varepsilon'_f(T_e)$ and $c(T_e)$ are, respectively, the coefficient and exponent of the plastic fatigue life curve.

Elements of gas turbines, e.g., rotor blades working under extremely difficult conditions, are also made from Inconel 718 alloy. Test results for this material was given in the work of Schlesinger et al. [14]. Here, a dispersion-hardening nickel-chromium alloy was analyzed with respect to fatigue loading at temperatures up to 650 °C. Similarly, in a different study [15], the same authors studied the Inconel 718 alloy. They proposed a model for estimation of the fatigue life of material working at elevated temperatures. The Manson–Coffin equation was used during modeling, in which parameters of the fatigue life curve were made functionally dependent on the structure and grain size in the given material G, test temperature T, and strength parameter S:

$$\frac{\Delta\varepsilon_{eq}(G,T,S)}{2} = \frac{\sigma_f'(G,T,S)}{E(T)}(2N_f)^{b(G,T,S)} + \varepsilon_f'(G,T)(2N_f)^{c(G,T)}$$

(6)

Materials adapted for manufacturing tools like casting molds and forging dies play an important role among materials intended for work at elevated temperature. In the study by Tunthawiroon et al. [16], the single-phase Co-29Cr-6Mo steel alloy, intended for aluminum casting molds, was tested. The study included analysis of the alloy's cracking mechanisms at temperatures up to 700 °C. A simple fatigue damage accumulation model at elevated temperatures was also proposed based on the Arrhenius equation:

$$D = N_f^{-1} = A\exp(Q/RT)$$

(7)

where N_f is the number of cycles until crack initiation, T is the test temperature, R and A are material constants, and Q is the activation energy.

Similarly, the works of Gopinath et al. [17] and He et al. [18] present the results of experimental tests conducted on 720Li and HAYNES HR-120 superalloys within the low-cycle fatigue range at temperatures up to 980 °C. The material was tested under uniaxial tensile-compressive loads. The mechanisms accompanying the cracking of these materials were defined.

ACI HH50 austenitic stainless steels can also work at elevated temperatures. The results given in the work by Kim and Jang [19] can serve as evidence of this. Due to its high resistance to elevated temperatures, high fatigue strength, and resistance to pitting and corrosion, the material in question finds applications in parts of combustion engines and power unit structures of nuclear power plants. The paper investigates the material's degradation mechanisms and presents a model estimating fatigue life in the LCF range. The model assumes that fatigue crack initiation will occur when damage reaches a limit value given by the following function:

$$\left(\frac{N_f}{N_0}\right)\left(\frac{\Delta\varepsilon_p}{\Delta\varepsilon_{p0}}\right)\sinh\left(\frac{\Delta\varepsilon_p}{\Delta\varepsilon_{p0}}\right) = 1, \quad \text{where}$$
$$N_0 = \frac{\alpha \nu C}{\dot{\varepsilon}_0 \exp\left(\frac{Q}{RT}\right)},$$
$$\Delta\varepsilon_{p0} = \frac{\dot{\varepsilon}_0 \exp\left(\frac{-Q}{RT}\right)}{\nu}, \tag{8}$$

where α is the material constant associated with dislocation movement over the course of the material loading process, ν is the frequency of load change during the test, $\dot{\varepsilon}_0$ and C are material constants independent of test temperature, Q is activation energy, R is the gas constant, and T is the test temperature.

The non-linear creep damage accumulation model under uniaxial loads at elevated temperatures (700 °C) that was also developed on the basis of tests performed on stainless steel X-8-CrNiMoNb-16-16 was presented by Pavlou 2001 [20]. The model was developed and verified by other researchers. In the work of Grell et al. [21], attempts have been made to estimate durability under the conditions of uniaxial isothermal loads (constant and incremental) at elevated temperatures using the dependencies proposed by Pavlou 2001 [20]. The work demonstrates greater accuracy in predicting durability with the use of non-linear models of the accumulation of defects in relation to linear models.

Titanium alloys are the next group of materials used to manufacture components working at elevated temperatures. One study [22] analyzes the influence of different methods of the surface hardening of the Ti-6Al-4V titanium alloy on the fatigue life of this material under uniaxial tensile-compressive loading conditions at temperatures up to 555 °C. The influence of temperature on the evolution of the cyclic properties of the TNB-V2 titanium alloy is also presented in the work [23]. Specimens of the material were subjected to constant-amplitude strains within the low-cycle loading range at temperatures within the range of 550–850 °C. Due to its good mechanical properties at elevated temperatures, this alloy finds applications in motorization and aviation and in engine parts. The authors tested the material within the entire safe temperature range, and determined that, in the case of this alloy, the Manson–Coffin fatigue life model does not allow for the prediction of the material's fatigue life at elevated temperatures due to the length of the phase between fatigue crack initiation and its propagation.

The fatigue damage accumulation model proposed in this paper was designed to predict the fatigue life of the material operated at elevated temperatures. The concept of elevated temperature can be defined here, according to the works by Chen et al. [24], as the temperature corresponding to two-thirds of the melting point of the bulk material T_c. In most cases, as temperature increases, a material's strength and fatigue properties are reduced in a predictable manner until a certain threshold temperature value is reached [25]. Under these conditions, the operation of technical machinery may be safe when the material's response to loading conditions and the working environment are known. However, if the conventional upper threshold of elevated temperature is crossed even slightly, the material's properties cannot be predicted directly. The material's strength suddenly drops (in a non-linear manner), and use of machinery under such conditions may pose a safety risk.

Uncontrolled loss of durability may lead to costly failures, standstills on process lines, and sometimes even to the death of those nearby when a structural component cracks. This is why information about how long and under what loading and temperature conditions a structural component can work safely is important. The fatigue damage accumulation models of materials working under low-cycle loading conditions at elevated temperatures presented in the literature are modified functions formulated to estimate the material's fatigue life at room temperature. The equations mentioned earlier can serve as an example: Manson–Coffin–Basquin, Ramberg–Osgood, and Smith–Watson–Topper. In these models, material parameters are determined independently for each of the analyzed temperatures. The process of calculating these parameters is not complicated in itself; however, preparing the data for determining them is very expensive and time-consuming and requires specialized testing apparatus. These models are dedicated for specific loading cases and metal alloys. Due to the low number of functioning fatigue criteria in this scope, there is a justified need to create new models that will provide a more accurate description of the material cracking process, thus enabling calculation of fatigue life with greater accuracy. That is also why it this paper set out to create a new fatigue damage accumulation model that will make it possible to determine the number of cycles until failure of a material working under conditions of fatigue loads of constant amplitude at elevated temperatures with engineering accuracy.

2. Materials and Methods

2.1. Fatigue Life of EN AW-2024T3 Aluminum Alloy at Elevated Temperatures

To develop a fatigue damage accumulation model, it is indispensable to perform experimental tests. They constitute the physical basis of formulated numerical dependencies. The fatigue damage accumulation model proposed in this paper was created based on an experiment in which EN AW-2024T3 aluminum alloy samples were subjected to the action of high-amplitude fatigue loading in the form of uniaxial tension-compression and complex loading (tension-compression with simultaneous torsion) at elevated temperatures. More information about the experiment can be found in other works [4,5].

Experimental tests of fatigue life were carried out in accordance with the guidelines included in ISO 6892-1, ISO 12106: 2017, ASTM E8-04, ASTM E 606, GB/T 15248-2008, GB/T4338-2006, and (J IS) Z2201. Cylindrical and tubular test specimens (Figure 1) were made using machining. The working surfaces of the specimens were polished until satisfactory smoothness was obtained.

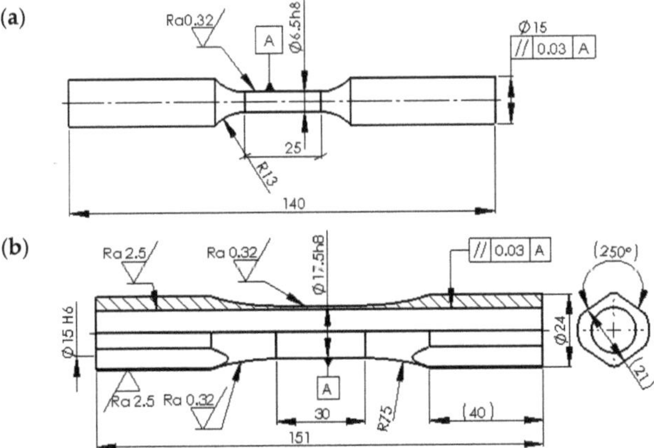

Figure 1. Diagram of test specimens for determination of the fatigue life of the material at elevated temperatures: (**a**) cylindrical specimen used in uniaxial loading tests (unit: mm) and (**b**) tubular specimen used in multiaxial loading tests (unit: mm).

The scope of performed tests covered, first of all, monotonic tests determining the influence of elevated temperatures on the mechanical properties of the EN AW-2024 T3 aerospace aluminum alloy. Table 1 presents the obtained results.

Table 1. Material parameters of the EN AW 2024 T3 alloy obtained for different temperature values [4].

T (°C)	E (GPa)	v	σ_{yp} (MPa)	σ_{uts} (MPa)	A_5 (%)
20	72	0.33	370	536	16.7
100	64	0.33	368	516	18.4
200	53	0.33	367	500	18.8
300	52	0.33	308	418	21.6

Next, fatigue tests were carried out. Samples were subjected to uniaxial loading, periodically varying (tension-compression), until their failure at different set temperatures. Tests were performed at temperatures of 20, 100, 200, and 300 °C and at a constant value of the strain change range ε, frequency $f = 1$ Hz, and $R_\varepsilon = -1$. The following values of control variable ε were applied: 0.015, 0.01, 0.0095, 0.008, and 0.006. The tests were performed on three samples for every value of the control variable's range and every test temperature.

Based on the performed tests, parameters describing the process of cyclic deformation of the material (as expressed by the Ramberg–Osgood equation) at elevated temperatures were determined by the following equation:

$$\varepsilon_a = \varepsilon_a^e(T) + \varepsilon_a^p(T) = \frac{\sigma_a}{E(T)} + \left(\frac{\sigma_a}{K(T)}\right)^{1/n(T)} \tag{9}$$

where σ_a is the amplitude of normal stress induced by the action of periodic axial force, $K(T)$ and $n(T)$ are the coefficient and exponent of Ramberg-Osgood cyclic strain curve, which are dependent on temperature.

Table 2 presents the calculated parameters of the cyclic strain curve.

Table 2. Ramberg–Osgood equation parameters for the case of cyclic tension of the EN AW 2024T3 alloy at different temperature values [4].

T	20 °C	100 °C	200 °C	300 °C
$K(T)$	605.23	652.57	996.42	492.98
$n(T)$	0.0651	0.0893	0.1881	0.0602

In the next stage, parameters of the Manson–Coffin–Basquin strain-based fatigue life curve were determined. Its parameters, calculated for various temperatures, are given in Table 3.

Table 3. Coefficients and factors of the fatigue curve for the EN AW-2024T3 alloy at different temperature values [4].

$\frac{\Delta\varepsilon_{eq}}{2} = \frac{\sigma_f'(T)}{E(T)}(2N_f)^{b(T)} + \varepsilon_f'(T)(2N_f)^{c(T)}$			
T	20 °C	100 °C	200 °C	300 °C
$\sigma_f'(T)$	850 MPa	592 MPa	700 MPa	591 MPa
$b(T)$	−0.086	−0.0498	−0.083	−0.065
$\varepsilon_f'(T)$	0.22	0.43	0.168	0.25
$c(T)$	−0.462	−0.562	−0.446	−0.544

Next, samples were subjected to the action of cyclic loading with torque moment and tensile or compressive force (during loading of samples, a constant quotient of maximum longitudinal and shear strain was assumed over the course of a loading cycle: $\varepsilon_a/\gamma_a = \sqrt{3}$). In all cases, loading histories formed oriented, multi-segment loops. Their configuration and parameters are given in Figure 2. Just as in the case of uniaxial tests, the loading process was controlled by means of increments of strain tensor components.

Mark	Load History	Load Paths

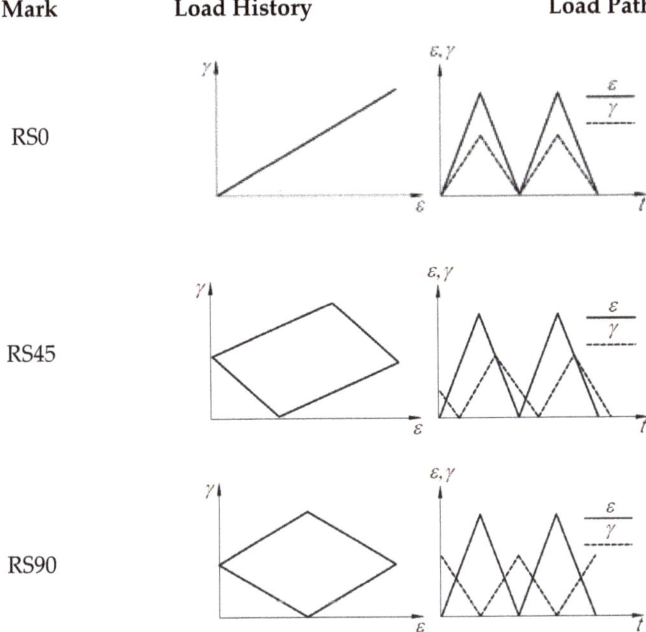

Figure 2. Evolutions of multiaxial loadings and the values of the control variable corresponding to them under the assumption of a constant quotient of maximum strains $\varepsilon_{max} = 0.003$ and $\gamma_{max} = 0.00173$ [5].

Tubular specimens (Figure 1b) were used in the experiment, and they were subjected to proportional and non-proportional tension-torsion cyclic loading with cycle asymmetry factor $R = 0$ ($\varepsilon_{min}/\varepsilon_{max}$; $\gamma_{min}/\gamma_{max}$). Fatigue tests were conducted for different combinations of axial and shear strains as well as for four temperature values: 20, 100, 200, and 300 °C. The specimen's kinematic input was controlled by means of the MTS 632.68F-08 (Eden Prairie, MN, USA) biaxial extensometer, and increments of strain components (linear and shear), averaged over a 25 mm-long segment of the measuring base, were used for this purpose. Three tests were conducted for every load with frequency of load changes $f = 1$ Hz.

The crack initiation moment was determined on the basis of the analysis of the maximum force value in the loading cycle. A 10% reduction of the force value relative to its maximum value at a given strain level was accepted as the criterion for crack initiation.

A compilation of the results of experimental tests (number of loading cycles until crack initiation) is given in Table 4.

The registered hysteresis loops for selected multiaxial loading configurations (individual loading components) have been presented in Figure 3.

Table 4. Results of experimental fatigue life tests for multiaxial loading configurations (tension/compression with torsion).

Load Path	T20		T100		T200		T300	
	N_f	$\overline{N_f}$	N_f	$\overline{N_f}$	N_f	$\overline{N_f}$	N_f	$\overline{N_f}$
RS0	6663 6638 5826	6376	3765 3456 3020	3414	1276 1316 1705	1432	770 992 920	894
RS45	6740 7638 10482	8287	3522 3911 3891	3775	2361 2002 1683	2015	846 1120 1379	1115
RS90	10290 8026 10,023	9446	4822 3556 4332	4237	2100 2295 2075	2157	1500 1390 987	1292

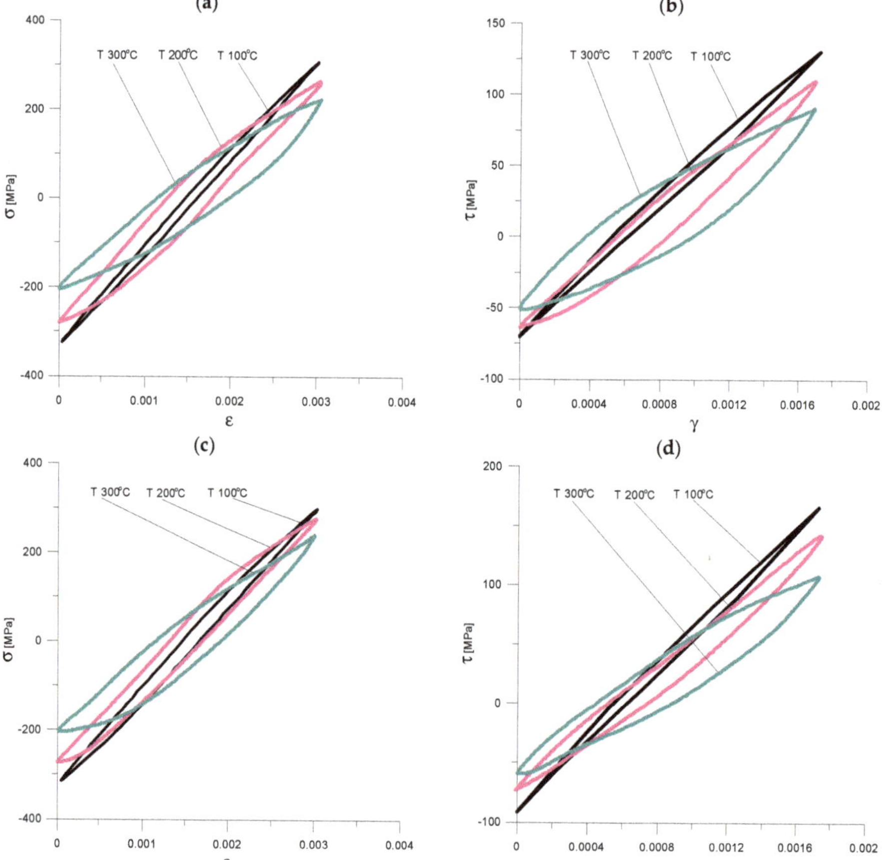

Figure 3. Hysteresis loops obtained at half of the fatigue life of the loading progression: (**a**) axial component for RS0; (**b**) non-dilatational component for various elevated temperature values for RS0; (**c**) axial component for RS45;(**d**) non-dilatational component for various elevated temperature values for RS45.

2.2. Modeling of Damage Accumulation and Fracture of Material at Elevated Temperatures

Preliminary tests based on the cyclic loading of a flat sample and observation of the amount of energy dissipating from the material due to the action of plastic strains were performed in order to investigate the fatigue damage accumulation process. The energy released as a result of fatigue loading was in the form of heat and was observed through a thermal imaging camera. Tests were performed at room temperature on the stand shown in Figure 4.

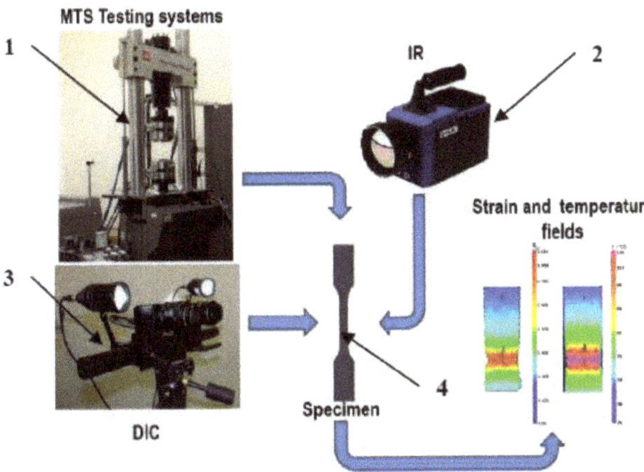

Figure 4. Test stand: 1—Fatigue testing machine, 2—thermal imaging camera IR, 3—*Aramis* video system DIC (Digital Image Correlation), 4—test sample.

During the tests, changes in the deformation fields (Aramis 4M-DIC system) and temperature fields (Cedip Titanium-IRT system) were recorded simultaneously. On their basis, areas of plastic deformation were identified and the distribution of plastic strain and energy converted into heat was determined. As can be seen in Figure 5, the temperature distribution corresponds to the strain distribution. In connection with the above, it can be assumed that the change in the temperature of the object caused by the action of external load is an indicator of damage generated during the loading process.

An example of temperature and strain fields is presented in Figure 5. Those are the fields corresponding to the localization of the plastic strain. This localization was manifested by the non-uniform temperature field and by the heterogeneity of the strain field on the specimen surface. The evolution of the temperature field due to the plastic deformation was used to determine the evolution of the field of heat dissipated by the specimen. The distribution of the heat and the plastic work on the surface of the deformed specimen was used to determine the failure mechanism.

The amount of heat emitted over the course of the tests can be related to the number of faults (damage) appearing in the material as a result of the action of external load. As the number of damages in the material increases, the object's temperature increases. Figure 6 presents the function of the maximum sample temperature at the crack initiation site as a function of test time (number of loading cycles).

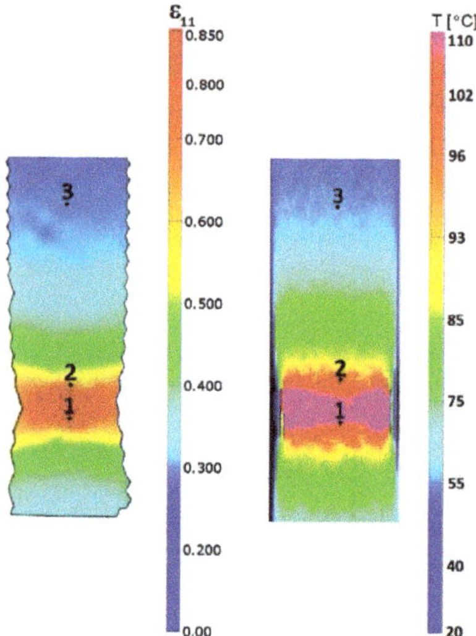

Figure 5. Fields of strain and temperature at the selected time during cyclic loading (five cycles until failure).

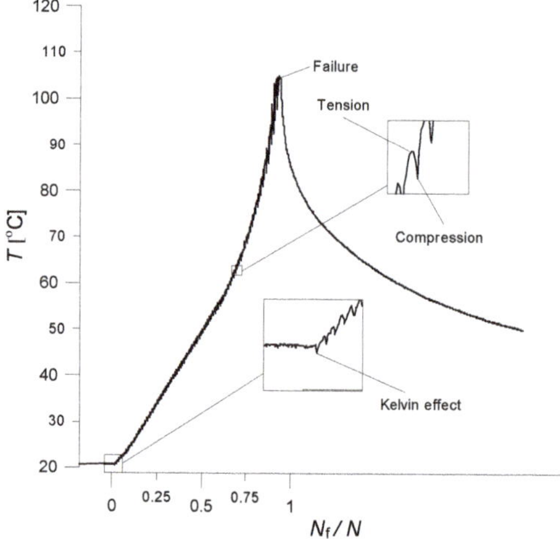

Figure 6. Evolution of temperature over the course of the process of cyclic deformation of a flat sample ENAW 2024T3.

As can be seen in the initial phase of the loading process, damage accumulated according to the linear hypothesis of damage summation. At a certain time in the loading process, there was a sharp spike in the amount of damage, observed as an increase in released thermal energy. Damage then increased exponentially. In the tensile half-cycle of loading, the object's temperature increased as a result of growing damage. Meanwhile, the compressive half-cycle slowed down the damage accumulation process, resulting in a drop of the sample's temperature.

Based on observations of fatigue cracking planes obtained over the course of experimental tests, it was noted that slip lines form first on the samples' polished surfaces under the influence of cyclic loadings at elevated temperature, even with values less than that of fatigue strength, and they transform into fatigue slip bands (numerous parallel faults) as the number of loading cycles grew (Figure 7).

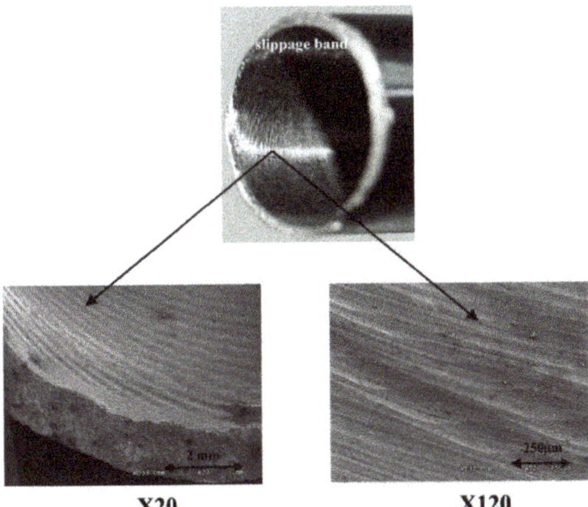

Figure 7. Slip bands (parallel faults) observed on the inner surface of a damaged tube sample (RS0—100 °C).

Similarly, as in the case of using room temperature [26], a fatigue damage accumulation criterion can be proposed based on the assumption that the material's damage state mainly depends on variable slips on physical planes and the normal stresses on these planes.

Considering the above, this paper proposes a damage accumulation model intended for the analysis of the fatigue life of constructional elements working under uniaxial and multiaxial loading conditions at elevated temperatures. This model assumes that permanent slip bands are the site of crack nucleation and defect formation [26], where crack initiation is the result of the accumulation of this damage. In this case, damage is induced by the action of the stress vector's normal component. Here, it is accepted that tensile stress in the loading cycle is responsible for generating new faults and accelerating the development of existing faults (temperature increase in loading cycle; Figure 6), while compressive stress retards the damage accumulation process (temperature drop in loading cycle; Figure 6). This fact was accounted for in the definition of the damage state variable $d\omega_n$.

The presented law of damage accumulation induced by plastic strains under elevated temperature conditions was linked to the stress-based damage accumulation function Ψ_p, the value of which is dependent on the value of normal stresses σ_n, damage state variable ω_n on the physical plane, and temperature T. The increment of the damage state variable on the physical plane $d\omega_n$, as caused by the development of plastic strains at elevated temperature T, was made dependent on the increment of plastic shear strains $d\gamma_n{}^p$ on the same plane [27], namely:

$$\begin{cases} d\omega_n(\sigma, d\varepsilon, T) = A_p(T)\, \Psi_p(\sigma_n, \omega_n, T) \left| d\gamma_n^p \right| & \text{for } \sigma > 0 \text{ i } d\varepsilon > 0 \\ d\omega_n(\sigma, d\varepsilon, T) = 0 & \text{for } \sigma \leq 0 \text{ i } d\varepsilon \leq 0 \end{cases} \tag{10}$$

where $A_p(T)$ is the material variable describing the evolution of the material's plastic properties depending on the actual stress state and temperature. In numerical simulations, the following values of parameter A_p were accepted: $A_p(20\ °C) = 0.55$; $A_p(100\ °C) = 1.4$; $A_p(200\ °C) = 1.8$; and $A_p(300\ °C) = 1.95$.

The damage accumulation mechanism described by this law is presented in Figure 8.

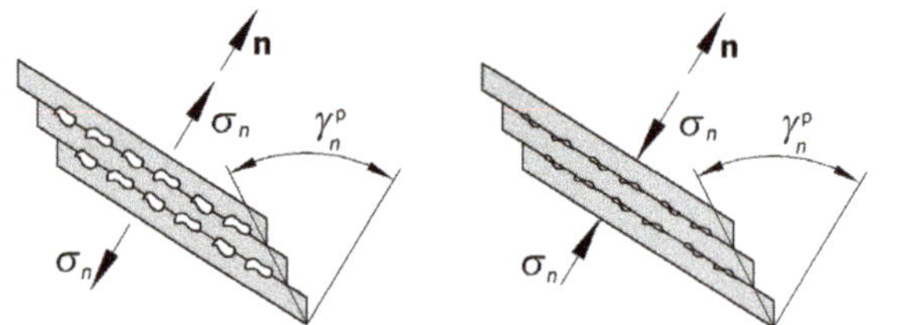

Figure 8. Generation of defects on the physical plane by slip band [22].

Similar to the dependencies presented in the work [26], the damage accumulation function Ψ_p can be proposed in the following form:

$$\Psi_p(\sigma_n, \omega_n, T) = \left(1 - \frac{1}{3}R_\sigma(\sigma_n, \omega_n, T)\right)^{1/c(T)} \tag{11}$$

where R_σ is the stress function of cracking given by the condition of maximum normal stresses, which is dependent on the test temperature:

$$R_\sigma(\sigma_n, \omega_n, T) = \frac{\sigma_n}{\sigma_c(\omega_n, T)} \tag{12}$$

$c(T)$ is the exponent present in the Manson–Coffin–Basquin equation, which is also dependent on temperature (values of parameter $c(T)$ used in numerical simulations are given in Table 3).

The crack initiation criterion can be written in the form of the condition of maximum normal stresses (linked to the physical plane), namely:

$$R_{f\sigma} = \max_{(n)} R_\sigma(\sigma_n, \omega_n, T) = \max_{(n)}\left(\frac{\sigma_n}{\sigma_c(\omega_n, T)}\right) = 1 \tag{13}$$

where $R_{f\sigma}$ is the stress cracking coefficient dependent on the state of stress and damage in the material as well as on the test temperature T, and σ_c is the actual value of normal failure stresses for the material at the given temperature, which is dependent on the damage state variable and temperature, namely:

$$\sigma_c(\omega_n, T) = \sigma_{c0}(T)(1 - \omega_n), \tag{14}$$

where $\sigma_{c0}(T)$ is the critical stress for undamaged material, which is dependent on temperature. The value of parameter $\sigma_{c0}(T)$ can be equated with the value of the coefficient of the Manson–Coffin fatigue life curve $\sigma_f'(T)$. Critical stress values according to Table 3 were used to calculate the accumulation of fatigue damage by means of the proposed model.

An additional cracking condition should also be introduced, in which it is accepted that crack initiation will occur when the damage state variable induced by plastic strains on any physical plane reaches critical value, that is:

$$\max_{(n)} \omega_n = 1. \tag{15}$$

At that moment, the material has no strength ($\sigma_c = 0$).

In the presented model, stress and strain values were determined independently for each test temperature by using the generalized Hooke's law, the Huber-von Mises yield criterion, the gradient flow law, and Mroz–Garud multiple-surface material hardening model associated with the yield criterion. They are an integral part of the numerical model of damage accumulation, making it possible to determine stress and strain state in any loading state [28].

When estimating the material's fatigue life, the principle of additivity of components of elastic and plastic strain increment tensors was adopted. Increments of elastic strain were determined by means of the generalized Hooke's law, and increments of plastic strain were calculated by means of the gradient flow law, associated with the yield surface, namely (T = const):

$$\mathrm{d}\varepsilon_{ij} = \frac{1 + \nu(T)}{E(T)}\mathrm{d}\sigma_{ij} - \frac{\nu(T)}{E(T)}\mathrm{d}\sigma_{kk}\delta_{ij} + \mathrm{d}\lambda(T)\frac{\partial f(T)}{\partial\sigma_{ij}},\tag{16}$$

where both the proportionality coefficient $\mathrm{d}\lambda$ and the elasticity constants E and ν are dependent on temperature. In numerical simulations, the values of parameters determined in monotonic tensile tests were used according to Table 1. The material's hardening curve was plotted according to the parameters given in Table 2.

Yield surface $f = 0$ was determined using the Huber-von Mises criterion:

$$f = \frac{3}{2}\left(s_{ij} - \alpha_{ij}\right)\left(s_{ij} - \alpha_{ij}\right) - R(T)^2 = 0,\tag{17}$$

where $R(T)$ defines the temperature-dependent size of the yield surface.

3. Results

Simulations of the fatigue life of the aerospace aluminum alloy within the range of uniaxial loads (tension-compression) at elevated temperatures were conducted for the modified Manson-Coffin-Basquin model (Equation (1)), semi-empirical model S-S_1 (Equation (2)), and the proposed approach S-S_2 (Equation (10)). Damage accumulation simulations were carried out for four different temperatures and five strain levels. The results of the simulations are provided in Table 5.

Table 5. Experimental data and results of numerical simulations of fatigue damage accumulation for uniaxial loads at elevated temperatures.

Strain Level	Calculations N_f (cycle)			Exp. N_f
	M-C-B	**S-S_1**	**S-S_2**	**(cycle)**
20 °C				
0.015	575	734	534	572
0.01	2510	3358	2181	2704
0.0095	3205	4151	3529	4156
0.008	6724	6717	5231	6458
0.006	12,838	12,316	14,164	14,597
100 °C				
0.015	695	452	424	464
0.01	3100	2135	2252	2164
0.0095	2200	3282	3381	3209
0.008	6250	5098	5308	5168
0.006	12,000	11,524	10,630	10,755
200 °C				
0.015	550	301	368	355
0.01	1955	1424	1589	1679
0.0095	3700	2188	2315	2178
0.008	4960	3399	4527	4435
0.006	9520	7683	9119	8328
300 °C				
0.015	250	151	218	225
0.01	850	712	509	464
0.0095	1056	1094	865	870
0.008	1623	1700	1867	1882
0.006	4450	3842	3870	3775

Figure 9a presents the evolution of the damage state variable over the course of the fatigue loading process, plotted based on Equation (10) for the analyzed levels of temperature and strain $\varepsilon_a = 0.015$. As can be seen in its evolution, the fatigue damage accumulation process progresses in a near-linear manner in the first loading stage until it reaches a certain limit state, and after this limit state is crossed, there is a sudden spike in damage. This was confirmed by thermograms (Figure 9b), which show a sudden temperature increase in the final phase of periodic loading. The characters of the fatigue damage accumulation process and of the evolution of the sample's temperature were similar over the course of cyclic loading. It can be surmised that the object's temperature may be an indicator of the number of damage accumulated over the course of the material loading process.

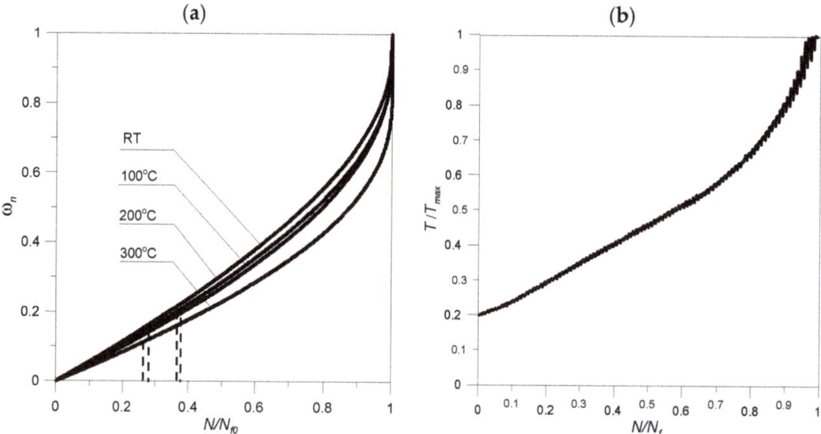

Figure 9. (a) Evolution of the damage state variable over the course of the action of fatigue load for different temperatures and strain levels $\varepsilon_a = 0.015$. (b) Evolution of sample temperature over the course of cyclic loading at RT–Room Temperature.

Temperature influences the rate at which the material subjected to the action of cyclic loads reaches the state of "damage saturation". Samples loaded at 300 °C reached the limit state after working 0.26 of the number of cycles until failure, while at room temperature, this state was reached after 0.38 of the number of cycles until failure. This state can be explained by the fact that movement of dislocations is facilitated as temperature increases, since they do not encounter obstacles that could slow them down on their path due to the elevated mobility of atoms. After crossing the "limit" state, further damage accumulation progresses exponentially. Microcracks appear in the material, leading to the material's decohesion as they are joined.

Figure 10 presents a comparison of numerical calculations of fatigue life (for the modified Manson-Coffin-Basquin model (Equation (1)), semi-empirical model S-S_1 (Equation (2)), and the proposed approach S-S_2 (Equation (10)) with experimental data for uniaxial tension-compression of ENAW2024T3 aluminum alloy samples at elevated temperature: (a) 20 °C; (b) 100 °C; (c) 200 °C, (d) 300 °C.

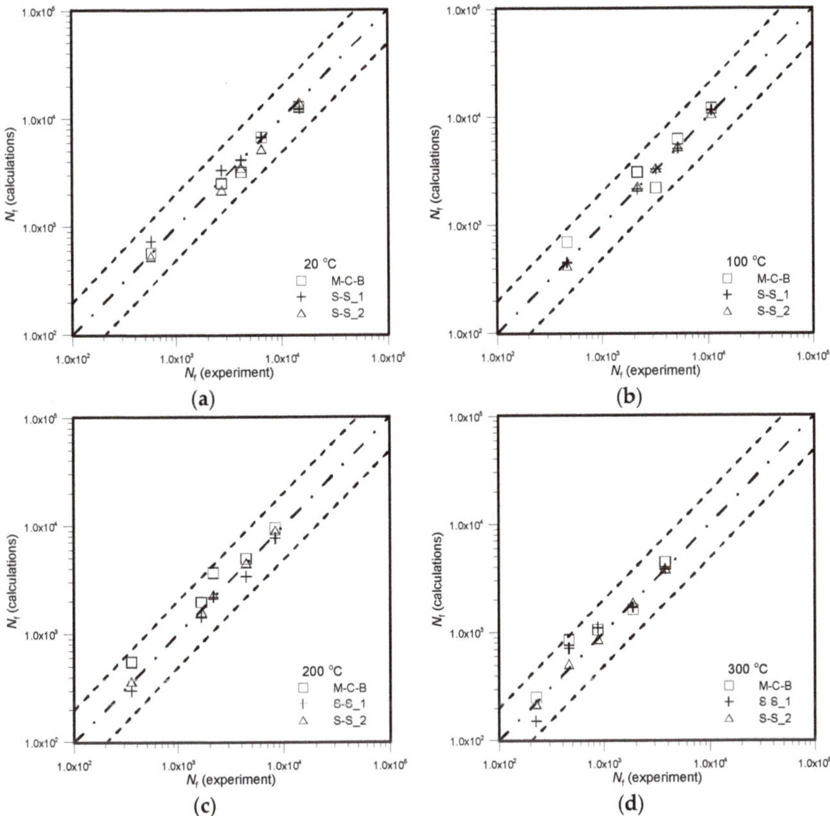

Figure 10. Comparison of numerical simulations of fatigue life with experimental data for uniaxial tension-compression of ENAW2024T3 aluminum alloy samples at elevated temperature: (**a**) 20 °C; (**b**) 100 °C; (**c**) 200 °C, (**d**) 300 °C.

Next, the proposed fatigue damage accumulation model was also verified for the case of multiaxial loads. The results of original experimental tests performed on the EN AW-2024T3 aerospace aluminum alloy were used for this purpose [5]. Figure 11 presents normalized evolutions of the damage state variable, plotted for load paths RS0 and RS90 at the analyzed temperature. A comparison between the results of experimental tests and of numerical simulations using the proposed SS2 approach (Equation (10)) and the SS1 semi-empirical model (Equation (2)) described in Reference [5] is given in Figure 12 and in Table 6.

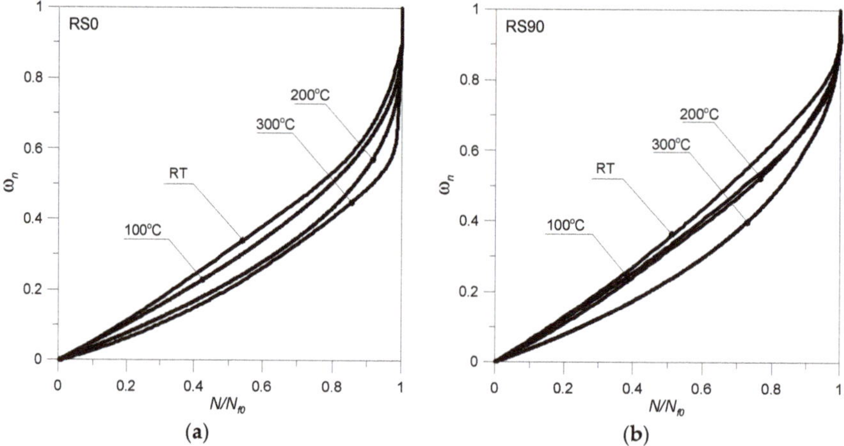

Figure 11. Evolution of damage state variable normalized with respect to cycles until crack initiation at the analyzed temperature for load path: (**a**) RSO; (**b**) RS90.

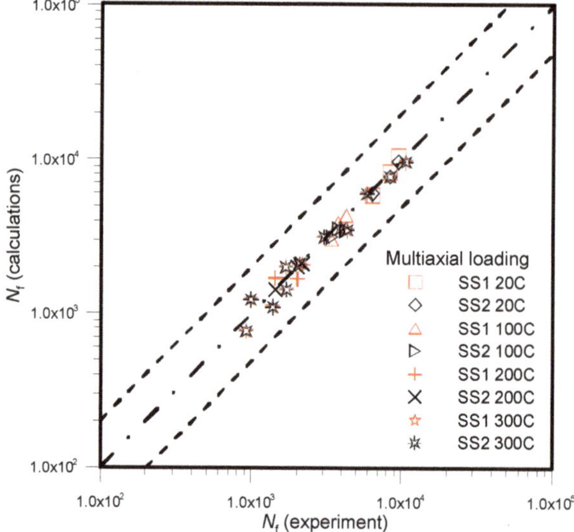

Figure 12. Comparison of the results of numerical simulations (strain-based approach) with experimental data for the EN AW-2007 aluminum alloy for complex loading histories.

Table 6. Comparison of the results of numerical simulations using the SS1 model (Szusta Seweryn 2015) and SS2 model (Equation (10)) and experimental tests of the number of cycles N_f for different loading paths (EN AW-2024 T3).

| Load Path | Calculations N_f (cycle) | | Exp. N_f |
	SS1	SS2	(cycle)
20 °C			
RS0	5876	5983	6376
RS45	8347	7739	8287
RS90	10,583	9655	9446
100 °C			
RS0	3020	3122	3414
RS45	3891	3538	3775
RS90	4332	3459	4237
200 °C			
RS0	1705	1426	1432
RS45	1683	1995	2015
RS90	2075	2103	2157
300 °C			
RS0	920	764	894
RS45	1379	1099	1115
RS90	987	1230	1292

The conducted simulations indicate good consistency between results obtained using the proposed strain-based model and the authors' original experimental results, within the scope of both uniaxial and complex loading.

4. Conclusions

This paper presents an approach that makes it possible to estimate the fatigue life of materials working under conditions of cyclic loading at elevated temperatures. An undoubted advantage of the model is its capability of accounting for the loading history in the fatigue damage accumulation process. This is done when determining stress or strain tensor components depending on the method of effecting force (force/displacement). The tests verifying the developed model show that, in the initial loading phase, as a result of the action of cyclic loads, damage accumulates gradually with every cycle of high-amplitude loading until the material reaches a certain limit state. After this state is exceeded, the "damage density" is high enough that the material is no longer able to safely carry the load applied to it. The process related to the appearance of microcracks in the material then occurs, and this leads to a sharp rise in damage during further loading. After that, faults increase exponentially until macrocracks form, leading to decohesion of the material. Elevated temperatures foster damage accumulation by facilitating the yielding of the material; dislocations can move more easily within the material. Moreover, the material's temperature changes the character of cracking and the time until failure.

Author Contributions: J.S. was the author of the concept of the article, made calculations, and analyzed the results. A.S. was a scientific consultant.

Funding: The investigation described in this paper is part of the research project no. G/WM/5/2017 Sponsored by the Polish National Science Centre and realized in Bialystok University of Technology.

Conflicts of Interest: The authors declare no conflicts of interest.

References

1. Tomczyk, A.; Seweryn, A.; Gradzka-Dhalke, M. The effect of dynamic recrystallization on monotonic and cyclic behaviour of Al-Cu-Mg alloy. *Materials* **2018**, *11*, e874. [CrossRef] [PubMed]
2. Juijerm, P.; Altenberger, I. Effect of temperature on cyclic deformation behavior and residual stress relaxation of deep rolled under-aged aluminium alloy AA6110. *Mater. Sci. Eng.* **2007**, *452–453*, 475–482. [CrossRef]
3. Karakas, O.; Szusta, J. Monotonic and Low Cycle Fatigue Behaviour of 2024-T3 Aluminium Alloy between Room Temperature and 300 °C for Designing VAWT Components, Fatigue. *Fract. Eng. Mater. Struct.* **2016**, *39*, 95–109.
4. Szusta, J.; Seweryn, A. Damage accumulation modeling under uniaxial low cycle fatigue at elevated temperatures. *Eng. Fail. Anal.* **2015**, *56*, 474–483. [CrossRef]
5. Szusta, J.; Seweryn, A. Experimental study of the low-cycle fatigue life under multiaxial loading of aluminum alloy EN AW-2024-T3 at elevated temperatures. *Int. J. Fatigue* **2017**, *96*, 28–42. [CrossRef]
6. Samec, A.; Potrc, I.; Sraml, M. Low cycle fatigue of nodular cast iron used for railway brake discs. *Eng. Fail. Anal.* **2011**, *18*, 1424–1434. [CrossRef]
7. Pevec, M.; Oder, G.; Potrc, I.; Sraml, M. Elevated temperature low cycle fatigue of grey cast iron used for automotive brake discs. *Eng. Fail. Anal.* **2014**, *42*, 221–230. [CrossRef]
8. Li, Z.; Han, J.; Li, W.; Pan, L. Low cycle fatigue behavior of Cr–Mo–V low alloy steel used for railway brake discs. *Mater. Des.* **2014**, *56*, 146–157. [CrossRef]
9. Ishii, T.; Fukaya, K.; Nishiyama, Y.; Suzuki, M.; Eto, M. Low cycle fatigue properties of 8Cr-2WVTa ferritic steel at elevated temperatures. *J. Nucl. Mater.* **1998**, *258–263*, 1183–1186. [CrossRef]
10. Mariappan, K.; Shankar, V.; Sandhya, R.; Laha, K. Low cycle fatigue design data for India-specific reduced activation ferritic-martensitic (IN-RAFM) steel. *Fusion Eng. Des.* **2016**, *104*, 76–83. [CrossRef]
11. Shankar, V.; Kumar, A.; Mariappan, K.; Sandhya, R.; Laha, K.; Bhaduri, A.K.; Narasaiah, N. Occurrence of dynamic strain aging in Alloy 617M under low cycle fatigue loading. *Int. J. Fatigue* **2017**, *100*, 12–20. [CrossRef]
12. Nagode, M.; Zingsheim, F. An online algorithm for temperature influenced fatigue–life estimation: Strain–life approach. *Int. J. Fatigue* **2004**, *26*, 155–161. [CrossRef]
13. Nagode, M.; Hack, M. An online algorithm for temperature influenced fatigue life estimation: Stress–life approach. *Int. J. Fatigue* **2004**, *26*, 163–171. [CrossRef]
14. Schlesinger, M.; Seifert, T.; Preussner, J. Experimental investigation of the time and temperature dependent growth of fatigue cracks in Inconel 718 and mechanism based lifetime prediction. *Int. J. Fatigue* **2017**, *99*, 242–249. [CrossRef]
15. Maderbacher, H.; Oberwinkler, B.; Gänser, H.-P.; Tan, W.; Rollett, M.; Stoschka, M. The influence of microstructure and operating temperature on the fatigue endurance of hot forged Inconels 718 components. *Mater. Sci. Eng. A* **2013**, *585*, 123–131. [CrossRef]
16. Tunthawiroona, P.; Lib, Y.; Koizumia, Y.; Chibaa, A. Strain-controlled iso-thermal fatigue behavior of Co–29Cr–6Mo used for tooling materials in Al die casting. *Mater. Sci. Eng. A* **2017**, *703*, 27–36. [CrossRef]
17. Gopinath, K.; Gogia, A.K.; Kamat, S.V.; Balamuralikrishnan, R.; Ramamurty, U. Low cycle fatigue behaviour of a low interstitial Ni-base superalloy. *Acta Mater.* **2009**, *57*, 3450–3459. [CrossRef]
18. He, Y.H.; Chen, L.J.; Liaw, P.K.; McDaniels, R.L.; Brooks, C.R.; Seeley, R.R.; Klarstrom, D.L. Low-cycle fatigue behavior of HAYNES HR-120 alloy. *Int. J. Fatigue* **2002**, *24*, 931–942. [CrossRef]
19. Kim, Y.-J.; Ho, J. High temperature fatigue resistance of an ACI HH50-type cast austenitic stainless steel. *Mater. Sci. Eng.* **2010**, *527*, 5415–5420. [CrossRef]
20. Pavlou, D.G. Creep life prediction under stepwise constant uniaxial stress and temperature conditions. *Eng. Struct.* **2001**, *23*, 656–662. [CrossRef]
21. Grell, W.A.; Niggeler, G.H.; Groskreutz, M.E.; Laz, P.J. Evaluation of creep damage accumulation models: Considerations of stepped testing and highly stressed volume. *Fatigue Fract. Eng. Mater. Struct.* **2007**, *30*, 689–697. [CrossRef]
22. Altenberger, I.; Nalla, R.K.; Sano, Y.; Wagner, L.; Ritchie, R.O. On the effect of deep-rolling and laser-peening on the stress-controlled low- and high-cycle fatigue behavior of Ti–6Al–4V at elevated temperatures up to 550 °C. *Int. J. Fatigue* **2012**, *44*, 292–302. [CrossRef]

23. Heckel, T.K.; Christ, H.J. Thermomechanical Fatigue of the TiAl Intermetallic Alloy TNB-V2. *Exp. Mech.* **2009**, *50*, 717–724. [CrossRef]
24. Chen, L.J.; Eang, Z.G.; Yao, G.; Tian, J.F. 1The influence of temperature on low cycle fatigue behavior of nickel base superalloy GH4049. *Int. J. Fatigue* **1999**, *21*, 791–797. [CrossRef]
25. Bar-Cohen, Y. *High Temperature Materials and Mechanisms*; CRC Press: Boca Raton, FL, USA, 2017.
26. Szusta, J.; Seweryn, A. Low-cycle fatigue model of damage accumulation—The strain approach. *Eng. Fract. Mech.* **2010**, *77*, 1604–1616. [CrossRef]
27. Szusta, J.; Seweryn, A. Fatigue damage accumulation modelling in the range of complex low-cycle loadings—The strain approach and its experimental verification on the basis of EN AW-2007 aluminum alloy. *Int. J. Fatigue* **2011**, *33*, 255–264. [CrossRef]
28. Seweryn, A.; Buczynski, A.; Szusta, J. Damage accumulation model for low cycle fatigue. *Int. J. Fatigue* **2008**, *30*, 756–765. [CrossRef]

 © 2018 by the authors. Licensee MDPI, Basel, Switzerland. This article is an open access article distributed under the terms and conditions of the Creative Commons Attribution (CC BY) license (http://creativecommons.org/licenses/by/4.0/).

MDPI

St. Alban-Anlage 66

4052 Basel

Switzerland

Tel. +41 61 683 77 34

Fax +41 61 302 89 18

www.mdpi.com

Metals Editorial Office

E-mail: metals@mdpi.com

www.mdpi.com/journal/metals

www.ingramcontent.com/pod-product-compliance
Lightning Source LLC
LaVergne TN
LVHW071245170726
843515LV00006BA/1335